IP TELEPHONY

IP Telephony

Walter J. Goralski
and
Matthew C. Kolon

McGraw-Hill
New York San Francisco Washington, D.C.
Auckland Bogotá Caracas Lisbon London
Madrid Mexico City Milan Montreal New Delhi
San Juan Singapore Sydney Tokyo Toronto

McGraw-Hill

*A Division of The **McGraw-Hill** Companies*

1 2 3 4 5 6 7 8 9 0 AGM/AGM 0 6 5 4 3 2 1 0

ISBN 0-07-213082-2

The sponsoring editor for this book was Steven Elliot, the managing editor was Jennifer Perillo, the editing supervisor was Ruth W. Mannino, and the production supervisor was Claire Stanley. It was set in New Century Schoolbook by Victoria Khavkina of McGraw-Hill's desktop composition unit in cooperation with Spring Point Publishing Services.

Printed and bound by Quebecor/Martinsburg.

This book is printed on recycled, acid-free paper containing a minimum of 50% recycled de-inked fiber.

To my family, with thanks for their continued support as the scramble to find time to create books becomes increasingly hectic. So this one is for Jodi, Christopher, Alexander, and Arianna. Thanks for being there and realizing that there just might be something to this writing thing.

—Walter Goralski

To my wife Patty, without whose faith, patience, and love my half of this book would never have been possible.

—Matt Kolon

CONTENTS

PREFACE

Many people are familiar with Cisco® Systems (more properly, cisco Systems), the company that recently claimed that most Internet traffic passes through Cisco routers. If anything, the advertisements are overly modest, but Cisco has devoted a lot of effort recently to voice networking as well as data networking.

Cisco Systems, located at two major sites in San Jose, California, is in the process of replacing the traditional telephone systems used by businesses everywhere with IP telephony systems, even to the desktop. In the end, the voice over IP (VoIP) system will connect the 40 buildings on the two campus sites and have some 30,000 telephone lines. Like most major corporations, Cisco depends on private branch exchange (PBX) equipment, close to 100 pieces of them, and voice mail for telephone needs. This dedicated voice-only equipment will be phased out as part of the VoIP project.

New buildings on the site are constructed with VoIP needs in mind, and in the near future, new building construction may never need special telephone wire for the exclusive use of voice (and other analog) services again. Eventually, VoIP will be extended to Cisco's worldwide multiprotocol wide-area network (WAN), which connects 225 sites in 77 countries, probably over the company's existing Asynchronous Transfer Mode (ATM) network backbone for now.

Why would a company such as Cisco go through the time, effort, and expense to migrate voice services to a data network? This is what this book is all about. Right now, there is the global Internet, used primarily for data communications, and the global telephone network, used primarily for voice communication. But this will not be true in the future. Voice over networks traditionally used for data purposes are picking up momentum. This book will explain why.

People pick up a telephone attached to the global public switched telephone network (PSTN) when they want to talk to someone else, and sit down at a computer attached to the global public Internet when they want to send e-mail to someone else. Few people would imagine using the Internet to make a telephone call. After all, these are two different networks, aren't they?

Not only are they apparently physically different, they are used for voice and data respectively. The PSTN links telephones all over the world for what is known as circuit-switched, physically channelized (time division multiplexed) voice service. The Internet links computers all over the world for packet-switched, logical circuit (unchannelized) data services. Readers are not expected to be equally familiar with any of these terms, if at all. All will be made clear, especially the reasons people may some day call you on the Internet as easily as they call you on the telephone network today.

There are in fact many similarities and overlaps between the PSTN voice service infrastructure and the Internet Protocol (IP)-based Internet service infrastructure. Many people, especially at home, will use a dial-up voice circuit to access the nearest node attached to the Internet backbone, known as the internet service provider (ISP) point of presence (POP). Even those totally unfamiliar with networking concepts know that this is what their modem is for: to allow the digital 0s and 1s generated by a computer to travel over the analog local access line (local loop) to the Internet. And the Internet backbones themselves are typically made up of high-speed digital links leased from the telephone companies (which are all also ISPs today, of course). Once this local and backbone use of circuits is acknowledged, then the split between PSTN and Internet is not as absolute as it once seemed to be.

So why not use IP for telephone calls over the Internet or some other IP-based network? If the Internet can be used to transfer video and audio between a server and client computer, then why not voice? Voice is just considered to be interactive audio in this case. And modern voice digitization techniques actually make it difficult to distinguish packet streams with voice bits from packet streams with data bits. "Bits is bits," as the saying goes, and it has never been truer.

This is what this book is about: voice over IP. The IP-based network could be the public Internet, or it could be some kind of "managed" IP-based network service. A managed IP service is typically one in which the service provider not only gives the customer links to the global Internet, but also has circuits (virtual or real) that have the link's bandwidth dedicated exclusively for the use of that particular customer. The IP network could even be a local area network (LAN) within an organization's sole site or office campus complex. No matter: It is the presence of IP that counts. Voice over IP is still VoIP whether the IP runs on a LAN, across LANs linked by point-to-point private lines, a managed Internet service, or the global public Internet. There are important differences in the voice quality of service (QoS) that may be expected, the savings involved, and the equipment needed, but all these considerations will be discussed here as well.

Even a few short paragraphs into this work, many key terms and concepts have been tossed about as if everyone that ever logged onto the Internet or made a telephone call is familiar with them. These few paragraphs used *channelized* and *unchannelized, circuits, time division multiplexing, ISP, POP,* and *QoS,* just to sample a few of the most important.

This familiarity may not be the case, especially since VoIP combines terms and concepts from the voice world of the PSTN and the data world of the Internet, worlds that traditionally have spoken their own languages. Thus every effort will be made in this book to clarify and illuminate the key terms and concepts used in VoIP, starting from a very basic foundation.

This is the reason that this particular book on VoIP has been written. There are many VoIP books for IP people and fewer VoIP books for PSTN people, but there are no VoIP books for just people. Anyone who has made a telephone call and has logged onto the Internet can read this book. The more technical the background, the better, of course. But no assumptions as to background or training have been made, other than basic user familiarity with telephones and PCs.

This book has a structure that builds from simple topics to more complex, and intersperses chapters geared toward those readers who are more familiar with voice networks with those designed for readers who are more familiar with data networks. The chapters of the book are as follows.

Chapter 1: The Promise of IP Telephony. This chapter examines the allure of VoIP from a number of different perspectives. The chapter begins by investigating the costs of telephone service from an international perspective, with regard to telemarketing (perhaps even more annoying than the "save big money on long distance" ads), with respect to call centers, and finally, with regard to the use of facsimile machine (fax) services.

The chapter then moves on to explore the allure of VoIP and Internet telephony. Four aspects of VoIP are discussed in detail. The fact that the world today forms one interconnected, global economy is explored, along with the fact that the global Internet has grown up right along with this phenomenon. The rise of the PC contributed to this growth, which then exploded with the advent of the World Wide Web. Both PC and Web are examined in full, and the section ends with a look at recent, but early, efforts to combine voice and data (and even some video) networks in one unified whole.

The chapter then offers a model for VoIP, placing business and residential, both incoming and outgoing, aspects of VoIP into a unified framework. Different aspects of this framework will be detailed in other chapters. The chapter ends with a look at the related technical, economic, and social factors that have made VoIP such a hot topic today. Regulation and deregulation plays a key role in this discussion.

Chapter 2: Digital Voice Fundamentals. This chapter investigates how analog voice came to be digitized in the first place. The chapter begins with a comparison between analog voice networks and digital voice networks, all with the idea of allowing the reader to grasp the essential differences between network devices such as modems and DSU/CSUs. The chapter outlines the reasons for voice digitization and introduces the digital signal processor (DSP) chipset as the workhorse of VoIP.

The chapter then explores how DSP chipsets actually digitize voice. The three key steps of sample, quantize, and code are given full treatment, and the end result is a full appreciation of 64 kb/s PCM voice, still the most common digital voice standard.

The chapter closes with a look at newer DSP techniques designed to digitize voice at speeds below 64 kb/s. Voice quality is considered, and a survey of current techniques closes the chapter.

Chapter 3: Telephony for IP People. There are lots of introductory expositions of the functions and history of telephone networks. This chapter tries to be a little different by assuming a basic knowledge of routers and the Internet, and attempts to emphasize similarities and differences between the PSTN and the Internet. The chapter starts off with a long look at the structure of the PSTN, including details on loops, trunks, and switches.

All the differences between analog voice trunking and digital voice trunking are explored, and SONET/SDH is introduced as well. The form and function of the basic telephone number in North America is investigated next, which leads to an examination of just what happens when a telephone call is made (and contrasted with sending Internet packets). This leads into a discussion of signaling networks in the PSTN.

A brief history of the Bell System is given to facilitate an understanding of the current deregulated PSTN structure in the United States (perhaps to be emulated around the world). Some of the mysteries of LECs, CLECs, and ISPs are explored, along with LATAs, access charges, and other concepts that complicate the world of the PSTN today.

Chapter 4: Packet Voice Fundamentals. Digitizing voice is much different from *packetizing* voice. This chapter explains why packet voice networks are much more challenging to build and operate than networks that support digital voice circuits. The chapter begins by exploring the differences between circuit switching and packet switching and introduces X.25 as the international standard for public packet-switching networks.

The chapter then introduces two connectionless packet-switching networks: the SS7 signaling network for the PSTN and the Internet. SS7, used primarily to establish and maintain voice connections on the PSTN, is investigated first, and then a very brief history of the Internet is given. The whole point is that these two seemingly vastly different networks have various features and characteristics in common. This is important, because VoIP calls often will involve both the SS7 signaling network and the Internet.

The heart of the chapter examines the issues that must be resolved to effectively packetize voice. Seven aspects of digital voice are discussed, and each one is given a solution that allows voice to be carried on packet networks. The chapter closes with a look at G.729B, the international standard way of providing several of the features needed in a packet voice network.

Chapter 5: Packets for Telephony People. This is the companion chapter to Chapter 3, which introduced the PSTN to those familiar with the Internet and its protocols. But where Chapter 3 explained the mysteries of signaling to those familiar with routing, this chapter explains connectionless networks to those familiar with telephone calls.

This chapter begins with a more detailed look at circuits and connections, all with the idea of establishing the use of virtual circuits for packet networks. This section introduces the concept of network nodes for connectionless networking, and the features of switches and routers are discussed. The important ideas of packets and datagrams are investigated. This section also includes a comparison of networks with switches as the network nodes (e.g., X.25) and networks with routers as the network nodes (e.g., the Internet).

The chapter includes a look at the Internet Protocol suite (TCP/IP). The identification and operation of common TCP/IP protocols is included. The chapter closes with a look at the World Wide Web in more detail, including Web graphics, animation, and (most important) audio.

Chapter 6: Voice-Enabled Web Sites. This chapter explores the use of VoIP to add voice capability to a Web site. This is almost the same as adding the Web to a traditional inbound call center for sales and tech support or an outbound call center used for telemarketing, so both are discussed in this chapter as well.

After examining the reasons for voice enabling the Web, three ways of doing so are investigated, ending with full "Web VoIP call" capability. Two case studies applying VoIP to the call center follow, one exploring the tech support center and the other exploring interactive voice sales. This leads to a fuller discussion of the integration of VoIP and the Web, as well as the operation of inbound and outbound call centers and how VoIP integration plays a role in the modern voice/Web integrated call center/Web site.

The chapter closes with a look at distributed call centers using VoIP and another short case study on how the Web site and call center can be brought closer than ever before with VoIP. Finally, some of the issues involved in voice-enabled Web sites and call centers are detailed.

Chapter 7: Standards for IP Telephony. This chapter details the two leading families of standards for VoIP and IP telephony: H.323 and the Session Initialization Protocol (SIP). The chapter begins by showing just what happens when an IP telephone call is made, and then introduces the H.323 family of standards from the ITU-T. H.323 is the most popular platform for the implementation of IP telephone systems, and all aspects of H.323 are investigated, from terminal types, to gateways, to control signaling, to set up connections.

The chapter continues with a look at SIP as an alternative to H.323. All the details of the SIP architecture are discussed, including addressing and naming, directory issues, and network requirements. More details about the Real-Time Transport Protocol (RTP) and the Real-Time Control Protocol (RTCP), both also used with H.323 implementations of IP telephony, are examined in this section as well.

The chapter closes with a brief look at some of the issues regarding quality of service (QoS) in IP networks. This includes looks at differentiated services (DiffServ) and QoS support in IPv4 and IPv6.

Chapter 8: IP Telephony Gateways. This chapter examines the functional components, internal design, and operation of the most important component of most VoIP networks. It goes on to describe the signaling that gateways use to communicate with the PSTN and SS7 network, and describes the operation of MGCP as a solution for making VoIP gateways appear like standard telephony switches to the PSTN.

Different models for gateway design are then examined, from the simplest software-only solution to VoIP cards for integration of packet voice into existing PBX switches. The chapter closes with a brief discussion of the role of gatekeepers and their relationship to the gateway devices.

Chapter 9: LAN Telephony. This chapter discusses the use of IP telephony as an alternative to the traditional private branch exchange (PBX) for office or campus telephony. The chapter begins by examining the particular qualities of the PBX environment, including a discussion of the roles that PBX traditionally has played within an organization. A discussion of the typical LAN environment follows, with a discussion of the suitability—or lack thereof—of the LAN environment for voice. The chapter discusses the particular challenges that implementers of LAN telephony face, such as lack of device power, as well as solutions to these problems.

Finally, this chapter concludes with a survey of the benefits that the early adopters of LAN telephony can expect, and examines the future of what promises to be a revolution in the way premises phone systems are deployed.

Chapter 10: Internet Fax. This chapter begins with a surprising history of facsimile communications, and goes on to discuss the reasons for the phenomenal acceptance of the fax machine into today's business environment. After detailing the benefits and liabilities of fax services, the chapter continues by describing the basic operation of the T.30 and T.4 protocols—the building blocks of fax technology.

The chapter continues with a discussion of the efforts by the ITU-T and the IETF to produce workable standards for the operation of fax over IP, noting the reasons why fax traffic cannot be treated as voice traffic by VoIP systems. The T.38 and IETF IFax efforts are outlined in this section. The chapter closes with an examination of the types of fax connectivity products that are available on today's market, as well as a look at the role of IP-based fax service providers.

Chapter 11: International Internet Bypass. This chapter starts out with a summary of the international telephony environment using the global PSTN and the attractions of using VoIP on the global Internet to bypass some of the charges that accrue with international calling. The general issue of "free" Internet calling is examined, which leads to a discussion of how Internet telephony should be billed to and paid for by customers.

Differences between the structure of the global PSTN and the Internet ISPs is explored next, in order to investigate the price structures of PSTN and Internet voice. This topic leads to a detailed discussion of separations, access charges, and international settlements. The LATA structure in the United States is examined and contrasted with those of other countries. The issue of global deregulation and privatization is also examined. Various ways of providing the equivalent of international settlements for Internet telephony are then detailed. The chapter closes with a look at international Internet bypass as a kind of IP tele-"phony" and examines Internet bypass as a tool to lower international settlement rates.

Chapter 12: Quality of Service in Voice over IP. This chapter begins with a detailed but necessary look at just what is meant by the term *quality of service* (QoS) when QoS is applied to a network. The provision of QoS is defined as the ability of users to specify the values of certain parameters, especially bandwidth, on an application-by-application basis. Key related concepts such as granularity, performance guarantees, and service-level agreements are also discussed.

Six QoS parameters are introduced and explained—bandwidth, delay, jitter, information loss, availability, and security. The role of bandwidth as the principal QoS parameter is explored and different philosophies regarding network QoS are investigated. Many ways of adding QoS to an IP network are surveyed, and the challenges that the Internet router environment poses for IP telephony are fully examined.

Chapter 13: IP Telephony Case Studies. This chapter presents three simple case studies, with network diagrams and product specifications, of IP telephony installations. The three studies are of an engineering firm seeking to reduce long distance charges, a textiles manufacturer in need of a new PBX, and an ISP wishing to add voice and fax capability to its managed network offering.

Chapter 14: The Future of Voice over IP and Internet Telephony. This chapter examines the future of voice over IP and the integration of voice and data traffic from several points of view: that of the users, who are notoriously fickle regarding the quality of telephone service; the service provider, who perhaps stands to gain the most from the convergence trend; and the technologists, who fight to develop the next new technology that will revolutionize voice systems.

The chapter concludes with speculation regarding the future of the PSTN itself, and whether or not it eventually will become a packet-switched, multiservice network providing many more services than are available today.

Walter J. Goralski
Matthew C. Kolon

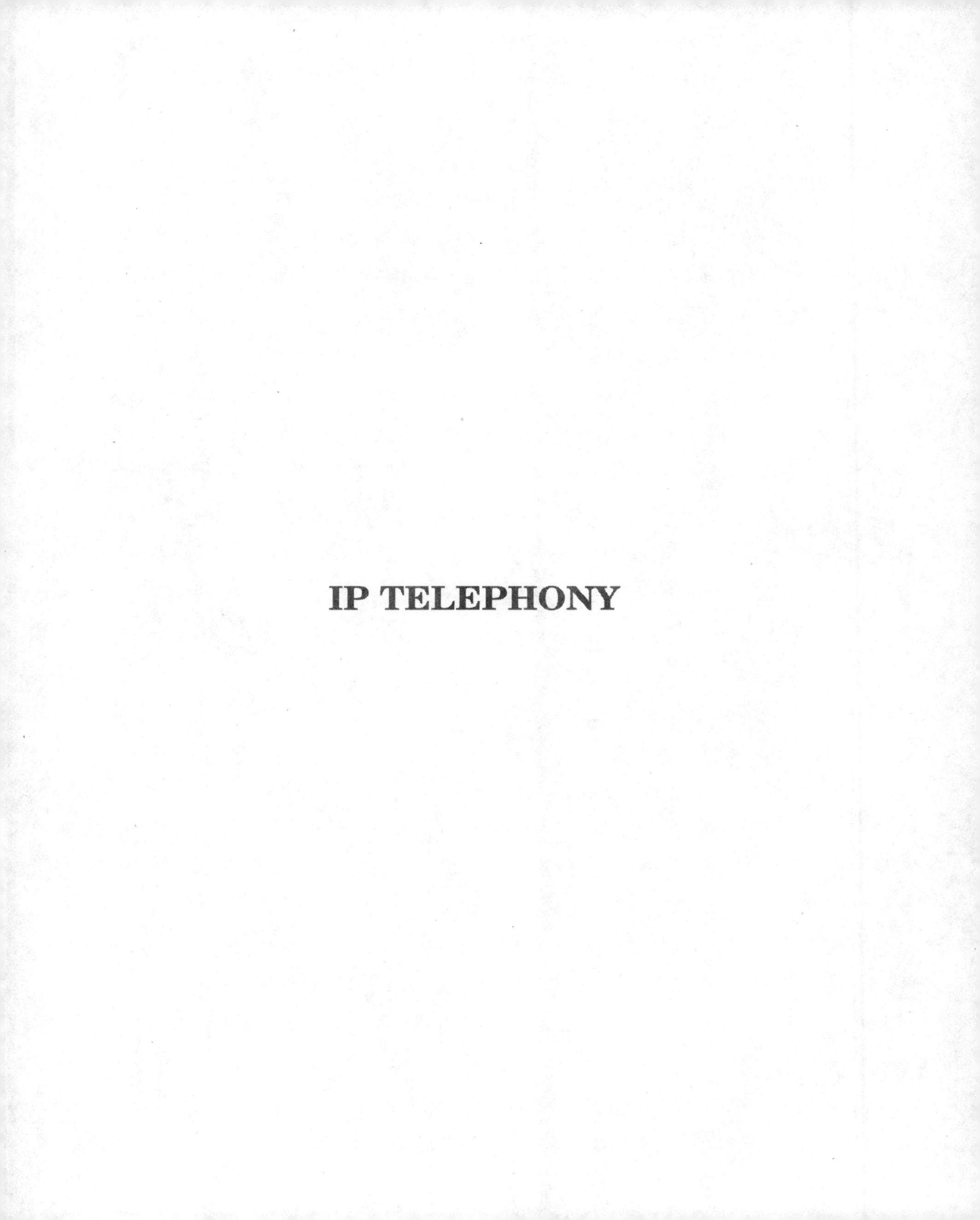

IP TELEPHONY

CHAPTER 1

The Promise of IP Telephony

Who has not seen those often annoying television advertisements featuring people whose lives have been made so much better because they use 10-10-555 or some other dialing prefix when they make telephone calls? In the United States, where telephones and telephone service are abundant and inexpensive, many viewers tend not to get excited about such services. Some are convinced that there is a hidden gimmick somewhere, while others cannot see the difference between 9 cents a minute and 8 cents a minute as being all that important in their lives. Yet many of the services are popular enough to at least stay in business, if not prosper (the cost of the ads alone, which run thousands of times, can be quite a drain on a company's resources).

Outside the United States, cost differentials matter more. Even within the United States in the not-so-distant past, saving money on telephone calls was something worth the effort to accomplish. And given enough volume, even pennies do add up. Businesses tend to look at telephony costs differently than residential users, and lower costs for higher-volume billing packages were offered much more commonly to businesses before telephone companies offered them to residential customers. Businesses are more cost conscious than even the most attentive families. Businesses always have an eye on the cash flow of the operation, but few families bother to compute their monthly rate of return or worry about the bottom line unless there is a real crisis when the time comes to pay the bills.

This book is about voice over IP (VoIP). This chapter is about Internet Protocol (IP) telephony. It mentions voice over packet networks. What's the difference? In these pages, VoIP just means that the IP protocol is used to deliver digital voice over a network. The network could be a local-area network (LAN), wide-area network (WAN), or public data network like the Internet. The term *IP telephony* is used here to mean that the IP protocol is used not only to deliver voice but also to perform some of the related functions that the voice network must include to make the whole network into a full system. Some of these functions include special features (e.g., call forwarding, call waiting, and so on), collect calling, gateways into the public voice network, and associated actions. As time goes on, and IP becomes more used for everything, VoIP and IP telephony will come to mean one and the same thing. *Voice over packet* (VOP) just means that the digital voice is delivered over a packet network of some kind, perhaps the Internet, but maybe not. Not all packets are IP packets. But unless otherwise noted, both VoIP and IP telephony are used here to indicate IP voice carried over the Internet, with gateways and special devices used to add functionality to the basic IP transport of digital voice.

But VoIP has something for everybody. While businesses look to save and make things more convenient for customers, residential users typically look for convenience first and cost savings later. There are exceptions, of course. Residential customers still used alternative telephony services in the early days when such use involved two separate dial tones and as many as 20 extra dialed digits instead of only 7. What the businesses and people had in common was a desire to use more cost-effective measures to make telephone calls. Perhaps the best place to begin, therefore, is with a look at the costs of telephony as a whole.

The Costs of Telephony

Telephony was not always the inexpensive commodity item associated with normal household and business costs such as coffee and electricity. Once telephony was a luxury, and an expensive luxury at that. Alexander Graham Bell invented the telephone as a business tool for manufacturers to call suppliers, hospitals to call doctors, and so on. Bell was astonished that ordinary people would want telephones in their homes. Ordinary people wrote letters to communicate (the mail then came twice a day except on Sunday, as in the *Saturday Evening Post*). In fact, to the day he died, Bell had no telephone in his office, mainly because he hated to be interrupted by the ringing.

However, being able to converse with someone who was not right there next to you in real time, like a kind of "interactive letter," fascinated almost everyone who experienced it. Mark Twain was among the first to have a telephone installed in his home, in Hartford, Connecticut, where it can still be seen today. Twain also was among the first to complain about the quality of service on the noisy line. Businesses did use the telephone as well, but usually only for the boss or high-level managers who had private offices in which to converse.

In the 1930s in the United States, local telephone service cost between $18 and $24 per year, depending on the region. The more urbanized Northeast had lower rates than the more isolated Midwest and Far West, at least outside the cities. This cost does not sound like much, but the amount must be put in perspective. The average farm worker in 1930 made less than $200 per month, for an annual salary of under $2400. A half-gallon of milk cost 28 cents, and a pound of bacon went for 40 cents. In 1929, three hot dogs could be bought for a dime, and a soft drink to go with it cost 3 cents. Looked at this way, $20 a year for a telephone could buy at least 160 meals, or enough food to live on for about 3 months. This is one reason that right before the United States entered World War II in 1941, only some 40 percent of Americans saw fit to have a telephone in their homes. And that was only because of extraordinary efforts by the telephone companies to keep subscribers during the dark days of the Depression in the early 1930s. These efforts included even more widespread use of party lines (access lines that supported more than one subscriber) and flat-rate local service to make monthly charges more predictable in the tight monthly budget.

As expensive as local service was, long-distance telephone calls were even more expensive, especially international calls. For example, in 1927, a 5-minute telephone call from New York to London cost $75. This is almost $700 in year 2000 dollars, so it is no wonder that international calling was mostly a business expense. Even domestic long-distance was awkward and costly to use until the mid-1960s. A long-distance call usually had to be set up by hand from operator to operator (the caller was told, "I'll call you back when your party has been reached") and could take anywhere from 7 to 12 minutes to route across the United States. By 1961, the cost for an AT&T coast-to-coast call was down to $12 per minute, at a time when minimum wage was about a dollar an hour. There were even cartoons showing people rushing from the shower, dripping wet, and wrapped in a towel to take a long-distance call. This was only etiquette: The caller was paying that $12 a minute, except for collect calls, and no one took a collect call unless it was from a close relative anyway.

It was only after World War II that even the business use of the telephone took off. The economy in the United States throughout the 1950s was booming, and the major business problem was how to find enough qualified people to open a new sales office, not downsize people to close one. No wonder older people look at the 1950s as some long-lost Golden Age. The gross domestic product doubled from 1945 to 1960, and this vast influx of wealth allowed businesses and people to view technology in a whole new light. The manager has a telephone and works more effectively because of it? Give all the workers a telephone!

The interplay between technology, economics, and social factors quickly became apparent in those days as well. A photograph of an office in the 1930s and 1940s shows pretty much the same thing: row upon row of clerks at desks with typewriters and the manager standing at his (hardly ever a *her*, except in the telephone company) office door in the rear. Only the manager had a telephone because only the managers had an office to talk in without disturbing those around him (not to mention the privacy issue). Now put telephones on all the workers' desks, as happened in the 1950s and 1960s. Suddenly, the noise level increased dramatically. The solution, of course, was to give everyone a small measure of privacy with the cubicle, which is the most widely lampooned aspect of office life today.

When it comes to telephony, in many cases trends start out in the business world and make their way to the residential marketplace. For instance, competitive long-distance services in the 1970s and 1980s targeted businesses and only slowly began to be offered to residential users. Thus many of the perceived advantages for VoIP apply to businesses first and foremost but should make their way down the chain to ordinary residential users quite quickly.

When it comes to the business advantages of VoIP, all of them boil down to cost factors in one way or another. Four VoIP benefits are usually cited with regard to business (and might apply to residential services soon enough). These are

1. Reduced long-distance costs
2. More calls with less bandwidth
3. More and better enhanced services
4. Most efficient use of IP

Consider each benefit in turn.

Reduced Long-Distance Costs

Businesses usually make a lot more long-distance calls than residential users do, although the mobility of American families today requires the use of long-distance service to keep in touch much more than even 30 years ago. Therefore, reducing long-distance costs is always a worthwhile goal. VoIP costs are usually cited to be about 30 percent of calls using the Public Switched Telephone Network (PSTN), although this figure has been challenged by the telephone companies (not surprisingly). International calling, a related topic, is important enough to mention separately a little later.

It is no surprise, however, to see this as benefit number one for VoIP business use. Businesses generate a lot of long-distance calls. In fact, given the dispersed nature of businesses today due to mergers and acquisitions, *long-distance calling* (defined here as any telephone call that requires a different area code to be used) is more common and essential than ever before.

Mention has been made of the initial coast-to-coast *direct-distance dialing* (DDD, the marketing name given to using area codes to complete calls without operator intervention) cost of $12 per minute. The rate soon dropped to about $6 per minute, only because not enough people were using the DDD service. As a result, the revenues anticipated by AT&T failed to materialize. At lower rates, long-distance calling grew in popularity, although relatively slowly. It was not until the wildly successful "reach out and touch someone" campaigns of the late 1970s that people did not worry about the cost of a long-distance telephone call.

This brings up a related point that will be examined more fully when the regulation of telephony and Internet telephony is considered later in this chapter. Local telephone companies are typically regulated businesses, usually called a *monopoly*, although the term has very negative connotations and probably should be avoided. Regulated businesses make money by maximizing revenues. In a competitive business environment, such as the current long-distance telephony environment in the United States, businesses make money by minimizing expenses.

Telephony rates can reflect either philosophy depending on who is offering the service. Local telephone companies, being mostly regulated (despite the widely touted Telecommunications Act of 1996, or TA96), tend to try to preserve rate levels to prevent revenue erosion. This is a polite way of saying that regulated businesses like to keep rates high so that it is easier to maximize allowed statutory revenues. Long-distance telephone companies, given the cut-throat competitive environment in the United States today, try to make money by minimizing expenses in order to keep their rates competitive.

In any case, VoIP is more or less by definition competitive with both local and long-distance telephone companies. Does this automatically make a company that offers VoIP a telephone company? This issue has been debated hotly and also will be explored in full in Chapter 11. For now, the point is that VoIP competes with other, more traditional telephony services at all levels and makes every market from local to long-distance to international calling concerned with minimizing expenses.

More Calls with Less Bandwidth

As far as handling more calls with less bandwidth, this might require a few words of explanation. More details on this key VoIP and voice in general concept will be examined later. For now, it is enough to point out that traditional voice digitization techniques require 64 kb/s (thousands of bits per second) to function properly. Business access lines are often digital in nature as soon as they leave the customer premises, but residential access lines most often carry analog voice conversation. Voice digitization is the topic of the next chapter, so only the briefest mention of this analog/digital voice dichotomy is made here. However, even in the case where the voice line in question leads to a residential user, the analog voice is digitized as soon as it reaches the major PSTN component, the voice switch, and sometimes even before.

Voice digitized at 64 kb/s is most often called *64-kb/s PCM voice* or just *PCM voice*. PCM stands for *pulse code modulation*. The point is that PCM voice occupies the entire bandwidth on the voice circuit for the entire duration of the call (conversation). This is the essence of the time division multiplexed, circuit-switched approach to networking. This "all the bandwidth, all the time" happens even if one person is totally silent while listening (ambient, or background, noise is typically still present, digitized, and sent) or even if both people put the telephone on hold.

One of the reasons that voice was and still is digitized at 64 kb/s is that people do not want to pay for voice calls beneath a certain quality (see Mark Twain above). Thus 64 kb/s gave the best quality (going above 64 kb/s improved quality only marginally under the noisiest of conditions). Also, when the first practical voice digitization techniques arrived in the early 1960s, 64 kb/s was considered state-of-the-art operation.

None of this is true anymore today. Very good voice quality can be achieved far below 64 kb/s using modern voice digitization methods. And the state-of-the-art no longer remains so for years, as before. State-of-the-art today can sometimes be measured in months. Today, state-of-the-art toll quality (the quality that generates few complaints about voice service) voice can be achieved at as low as 2 kb/s, but is usually provisioned at 8 kb/s for a variety of reasons (discussed later). Methods of generating digitized voice below 64-kb/s PCM voice is often referred to as *voice compression*, but not all digitized voice below 64 kb/s is produced by compressing 64-kb/s PCM voice. And the 8-kb/s digitized voice is typically only produced when someone is actually talking. *Silence suppression* is also a key component of many of these techniques.

Why should anyone care? Simply because the telephone companies always chopped up their network bandwidth into *voice channels* or circuits running at 64 kb/s. These voice channels are technically called *digital signal level 0* (DS-0) circuits and remain the most common type of line leased by the month or sold to organizations as private lines. This is true even when the DS-0 leased line is used for data purposes, such as to connect routers at two separate locations. Of course, many organizations used a series of DS-0s to connect their private voice switches, or private branch exchanges (PBXs). Common groupings of DS-0s used for this purpose gathered 6, 12, or 24 voice channels into a single circuit.

Consider a DS-0 voice circuit used to carry voice conversations between PBXs. A single DS-0 can carry one 64-kb/s PCM voice call. That's all. However, if the end equipment on the customer premises (called *customer premises equipment*, or CPE) is capable of generating 8-kb/s digitized voice, then the same bandwidth could carry eight voice calls between the sites for the *same monthly price* for the DS-0. The customer increases its voice-carrying bandwidth capacity by eight times. If the price of the end equipment is less than the cost of seven additional private lines (which must be paid for by the month), then the organization has a real incentive to use the newer voice technique.

This bandwidth optimization is shown in Figure 1-1, using a single DS-0 as an example. This is not too realistic, but this is only a simple example used here to explore the concept.

No mention has been made yet regarding IP itself. IP is the Internet Protocol, part of the Internet protocol suite. The Internet protocol suite is usually called *Transmission Control Protocol/Internet Protocol* (TCP/IP), but TCP and IP are only a few of the protocols that comprise the entire Internet protocol suite.

So far, 8-kb/s digitized voice still uses all the bandwidth all the time, just less of it. But silence suppression makes the digitized voice into a *bursty* stream of bits. That is, the digitized voice bits are generated by a *voice burst* (also called a *voice spurt*), and even when the person is talking continuously, several voice bursts are generated because of the way the newer voice compression techniques function. Similar bursts of bits also characterize data applications, and IP is the most popular protocol used today for routing bursts of bits from one location to another.

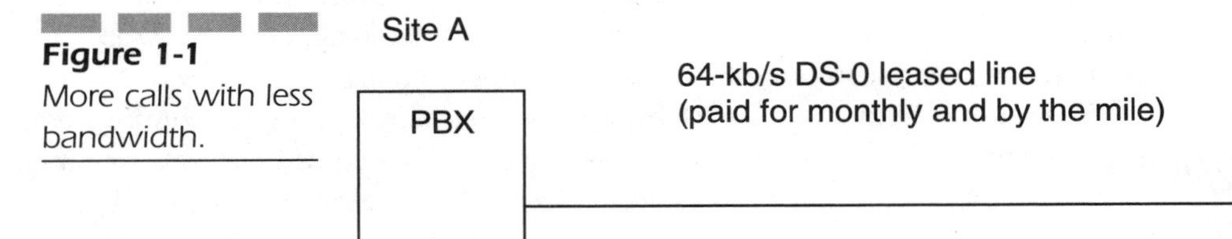

Figure 1-1 More calls with less bandwidth.

So why not just wrap the voice burst bits in an IP packet, a packet indistinguishable in terms of header and content from any other packet on the network, and send it out? This is what VoIP is: compressed and silence suppressed voice burst bits wrapped in an IP packet and sent out over an IP-based network. There are also data packets as well, but as long as there are not too many simultaneous voice and data bursts, which would add delay due to buffering or information loss because of discarding packets, the same circuit could handle both voice and data. A DS-0 will do just fine for handling this combined voice and data traffic.

Figure 1-2 shows the same DS-0 now being used to carry not only a voice conversation between the organization's sites but also regular IP router to router packets as well. Something seems strange about the network shown in the figure, however. The PBX must somehow share the link with the router. Networking is easy when diagrams consist of lines and boxes. Just draw a line, and things are connected! In the real world, it is a little more complicated. PBXs have interfaces and use procedures appropriate for circuit-switching networks, and IP routers have interfaces and use procedures appropriate for packet-switched networks.

Maybe the voice PBX should blend in better with the data environment. This brings us to the third benefit: more and better enhanced services, mostly from the data perspective for now, but potentially involving traditional enhanced voice services as well.

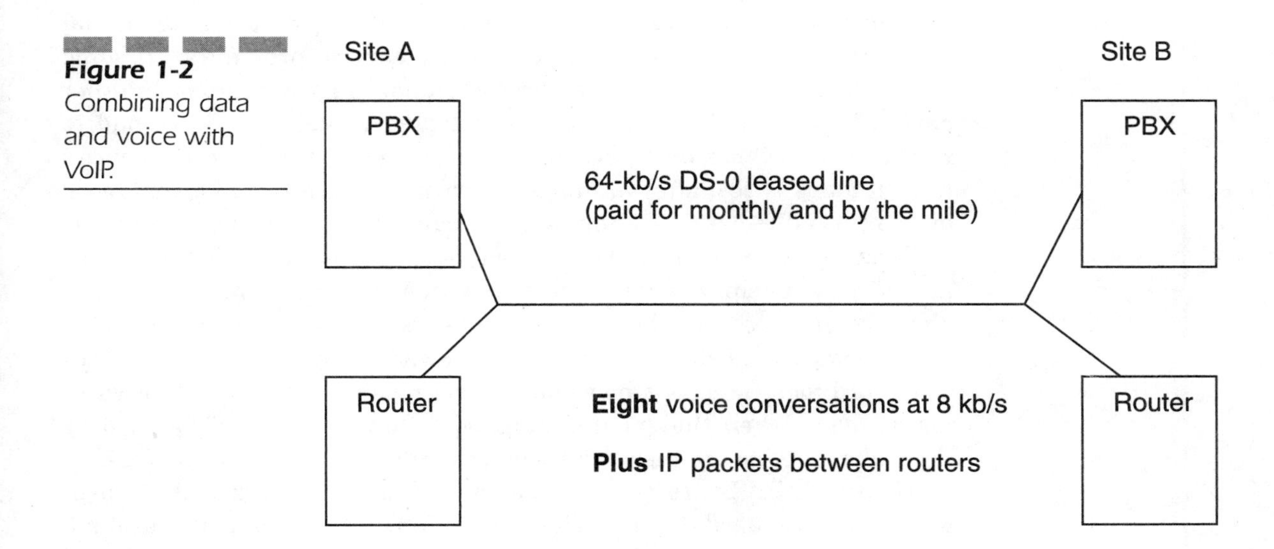

Figure 1-2 Combining data and voice with VoIP.

More and Better Enhanced Services

Talking on the telephone is the *basic service* of the PSTN. But the PSTN supports many *enhanced services* as well, although as time goes on people tend to take many of these enhanced services for granted. Enhanced services are those which add value to the basic voice conversations supported by simple telephony. Since enhanced services add value to the conversation, there is usually an extra charge involved. There are numerous examples, and enhanced services cover a wide range in terms of features and capabilities.

It sounds hard to believe today, but even using touch-tone dialing instead of rotary (pulse) dialing was considered to be an enhanced service (and might still be in some parts of the United States, depending on local regulation). More useful enhanced services include things like call forwarding, call waiting, third-party calling (charging the call to a telephone other than the caller or the recipient), collect calling (charge reversal), caller identification, and so on. These features tend to be heavily dependent on telephone company switch and signaling software in use and formerly were available only by buying the enhanced service or services in question for a whole month. These features usually were bundled together and offered as a package for a small monthly fee. Today, enhanced services generally are unbundled and available on a per-use basis, typically by dialing a special code before the telephone number. This special code is most often * for touch-tone operation and the prefix 11 for rotary telephones to initiate the enhanced service.

Enhanced services have had a spotty record of generating additional residential service revenue for telephony service providers. In some areas, the services have been embraced and are highly popular. In other parts of the United States, the local telephone companies have had to run advertisements on television to encourage the enhanced service's use. In many cases, after an initial spurt in use, the revenue generated by the special enhanced service such as call waiting falls off rapidly.

Enhanced services have been much more successfully tried and marketed in the business environment, but ironically not through the telephone companies. Most business uses of enhanced services are available on the company's PBX, the private voice switch that even smaller offices can afford today. People who would not often think of using call forwarding at home when they visited neighbors think nothing of forwarding their telephone to another office when there is a meeting in progress.

Almost all PBXs are now computers and handle digitized voice, usually at the standard 64 kb/s. There are a few analog PBXs still around, but these are rapidly being replaced in most cases. Because of the processing power of the computer driving these modern PBXs, most products have an array of features that many users do not even know about. Some companies actually hire consultants to come in and tell the users about the features available on the PBX and how to access them. For instance, many PBXs have a *make busy* feature that allows users to set their office telephone to give a busy signal to all callers, even if the telephone is not is use. Good for getting some privacy to work undisturbed, but tempting when looking to goof off for a while. If the line's busy, the person must be working, right?

PBXs, which were invented long before the computers that now power them, provide two key service features to a company. First, a single person answers all calls and says something like, "Good morning, XYZ Corporation" (this might be a recording today, of course). The caller is greeted by the attendant and asked for the party he or she is trying to reach, perhaps Ms Smith, the caller's sales representative. If the caller knows Ms Smith's extension, he or she usually can dial this directly, but perhaps the caller has lost Ms Smith's business card and has looked XYZ Corporation's number up in the telephone directory (411 directory service is a good example of a per-use enhanced service). This single-point-of-contact feature is the major benefit of the PBX.

The second service feature is known as *revert* and is also a real benefit to all types of organization. Suppose Ms Smith happens to be out to lunch when the caller is directed to her extension. In many cases, after several rings with no answer, the call reverts to the attendant and rings at the central console again. The attendant can then take a message, find another sales representative to take the call, or do one of a number of different things. Even when the PBX system is linked to voice mail, there is typically an option to allow the caller to revert to the attendant. It does no good to leave voice mail for a person who has left for the day if the matter is urgent. More sophisticated features can link the PBX for messaging purposes to pagers or even cellular telephones.

Thus enhanced features are common on the business PBX and not as common, or as extensive, on residential systems. Consider business use for the moment. Most businesses have a LAN and router to link individual user PCs together, most often with a form of Ethernet LAN called *10base-T*. For data purposes, the router is the interface device used to link sites together. For voice purposes, the same role is played by the PBX. The links between PBXs are called *tie-lines*, but these are just 64-kb/s DS-0 leased private lines, just like the links between the routers. Figure 1-2 showed how the bandwidth on a DS-0 link could be shared by both data and voice, as long as the voice is packetized and compressed to look like data. This is, of course, exactly what VoIP does to voice.

But there is more. The problem is how to share the link between the PBX with interfaces and procedures appropriate for circuit-switching networks and the IP router with interfaces and procedures appropriate for packet-switched networks. Maybe the solution is to *put the PBX on the LAN*, just like the router. This makes sense when VoIP is used anyway. If the voice now looks just like data, why shouldn't the voice device look at and handle voice packets just like the data device looks at and handles the data packets?

The link handling voice and data would now look as shown in Figure 1-3. The PBX is just another type of server device on the LAN. A LAN server is just a PC used for a special function, such as storing all e-mail messages or holding central database records. Users have PCs that act as LAN clients, which are PCs used for general functions, such as reading and sending the user's own e-mail or accessing the database records of interest for a specific database query. Together, clients and servers are linked by the LAN, and the router links LANs at separate locations. The router can link LANs on different floors of the same building, or the router can link LANs around the world. Whatever the configuration of the network linking the LANs, the whole exists to allow any client to access any server in order to get the information the client needs to perform its functions. This philosophy is known as *client-server computing*, and related terms all convey the same basic idea. Thus *client-*

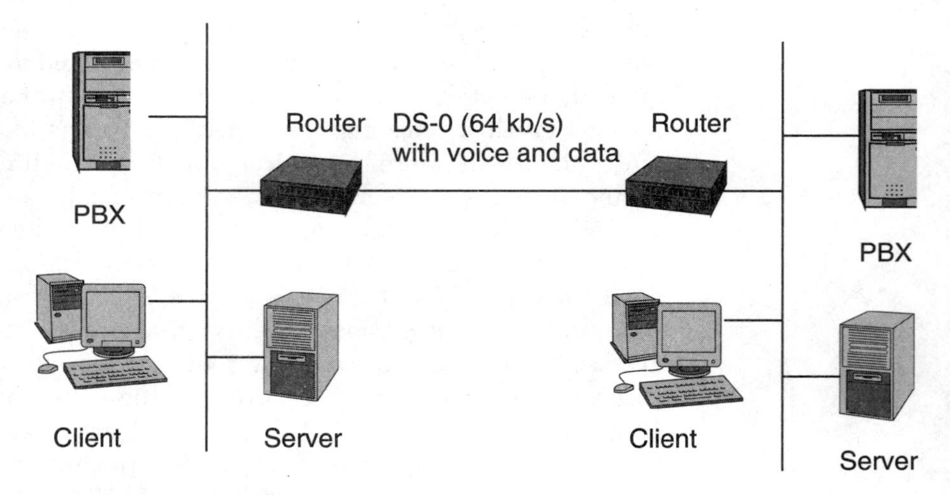

Figure 1-3
The PBX as a server on a LAN.

server applications are those applications which require a client version of the software to be used with a server version of the software, *client-server architectures* are those network architectures which expect all linked devices to be clients and servers, *client-server protocols* are those protocols which expect all devices to be clients or servers, and so on.

The role of the *Internet PBX* in VoIP and businesses will be explored more fully in Chapter 9. For now, it is enough to realize that VoIP allows a much closer relationship between traditional voice and traditional data devices. Not only does modern compressed and silence suppressed voice look like data, it can be treated like data in many cases. VoIP voice not only can be sent across data links, VoIP voice can be recorded, stored, processed, converted, and so forth, all by the same types of hardware and software that have been applied to data in the past.

There is one other related point that must be made regarding Figure 1-3 to avoid confusion with the many figures to come in this book. This involves the fact that the simple LAN as drawn and the LAN as actually implemented in most cases are really two different things. Network people tend to talk about *logical* LANs and *physical* LANs in general, but all that need be pointed out here is that the typical Ethernet-type 10Base-T LAN is a logical bus when it comes to operation but a physical star when it comes to cabling.

Ethernet and 10base-T LANs are usually drawn as a bus and represented as a central line (the bus) with other lines connecting the clients and servers and routers to this central bus. But such drawing is really just a convention based on the way the Ethernet and 10base-T network access protocol functions. It is a *logical* representation of the LAN. In reality, a 10base-T LAN is deployed as a star-shaped group of wires connecting the clients and servers and routers to a central hub (often called an *Ethernet hub*, although it is really only a 10base-T hub). Each wire occupies a port on the hub, and hubs can have as few as 4 or as many as several hundred ports. This hub and star are the *physical* representation of the LAN, but one that is seldom seen in literature. In this book, as in most, real 10base-T stars are always drawn as logical Ethernet buses. There is no hub to be found, but hubs are always present in a 10base-T configuration. Both are shown in Figure 1-4.

Both LANs in Figure 1-4 depict the exact same LAN in terms of number of clients, servers, and routers. But one is a logical bus, and the other is the physical star version of the same thing. This is worth pointing out, especially when voice people begin to explore VoIP and find that the wires in the wall look nothing like the pictures in the book.

Figure 1-4
The *(a)* logical bus and *(b)* physical star 10base-T LAN.

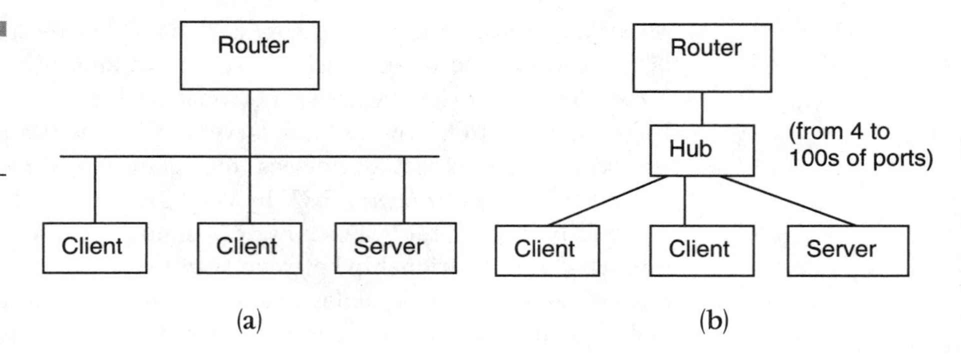

So VoIP server devices on a LAN can deliver many of the same enhanced services that traditional voice PBX devices do. Why should anyone care? The trend today is for closer computer telephony integration (CTI) than ever before. In one sense, VoIP is the ultimate conclusion for CTI. With CTI, and even without VoIP, a PBX with a LAN interface can forward the caller's number not only to Ms Smith's telephone but also to her PC. The client PC can query a LAN-attached database server using the caller's telephone number to put up on the PC screen not only the caller's number but also the caller's name, photograph, account information, credit history, spouse's name, favorite sports team, hobbies, and any other information someone in the organization has seen fit to track about the caller. And all this appears before the telephone on the desk has rung twice!

And since people are people, the theory is that customers are more likely to buy something from people they like. And people like people who seem to care about them, as evidenced by the fact that Ms Smith has taken the time and effort to "remember" all these details about the caller. It makes good business sense, and personal sense as well.

However, with CTI, the emphasis is on the computer, not the PBX itself. Consider another CTI enhanced service. Instead of actually picking up the telephone and dialing a number, Ms Smith can just open a directory window on her PC and scroll through her customer contact database as easily as scrolling through her e-mail directory (they might even be the same directory). All Ms Smith need do is point and click at the name, the number might never appear at all. The LAN conveys the number to the PBX, the PBX places the call, and the telephone basically rings at both ends at the same time. If the line is busy, this fact can be flashed on the screen without involving the desk telephone at all.

Needless to say, call waiting, call forwarding, and all the familiar PBX enhanced services can be handled the same way. Simple icons handle all incoming events, and a simple point and click does it all with regard to outgoing features. CTI can even be used to remind someone to call home before leaving the office (how often does this simple task get forgotten?) and even place the call for you.

But VoIP server devices on a LAN can deliver more and better enhanced services than PBXs alone, even with CTI. This is all due to the prevalence of Internet- and Web-related protocols and applications in the business environment of today. The explosive growth in the use of IP for LAN usage is examined more fully in the next section. It is enough to note here that when the Internet protocol suite (TCP/IP) is used on a private LAN, the result in most often called an *intranet* (sometimes seen with the capital *I* as *Intranet*). When two separate companies use the Internet protocol suite so that a client PC at one company is authorized to access information on a server at the other company, this is called an *extranet* (or *Extranet*, usage varies). Once the Internet and Web are wedded to the enhanced voice services that a PBX typically provides, a new level of effectiveness is reached.

Consider the case of a potential customer calling Ms Smith at XYZ Corporation today. In many cases, the potential customer has been to the company's Web site and has figured out exactly which color widget is needed or precisely which level of the service that XYZ Corporation fits best. In many cases, the product or service can be ordered directly through the Web site, an arrangement known as *electronic commerce*, or *e-commerce*. Some corporate Web sites are little more than electronic versions of a print catalog.

However, potential customers can find themselves quite frustrated with the Web ordering process. Some people simply will not buy anything unless they can talk to a human being to answer nagging questions about the product. For instance, the Web page says it comes in red, but the Web site only shows it in blue. Is this accurate? Pricing information is crucial. The Web page says 10 percent off the retail price this week only. When does the week start and end?

Ironically, customers sometimes have access to Web-based information that the company's sales representatives do not. The sales representatives have a telephone, and maybe a PC, but not often access to the company's Web page. Thus a customer gets the company's sales telephone number from the Web page, calls on a second line while still looking at the Web page, and asks the question about red or discounts to the person who answers the telephone. The sales representative often relies on the information that the customer reads to him or her over the telephone!

How much better would it be if there were a click box right on the page that said "talk to a sales representative" that anyone with a VoIP-enabled browser could press? The Web server would establish a VoIP connection from client (customer Web browser) to VoIP server, where the voice would be forwarded to the chosen sales representative. All talking would occur through the PCs, not the telephone. Now that's an enhanced service worth talking about.

The same arguments used above applied to a business's products and services can be extended to government agencies, educational institutions, medical centers, and any one of the many types of nonprofit organizations that maintain a Web presence today. VoIP is not just for businesses, but for any organization of any shape or size or purpose at all.

Most Efficient Use of IP

The mention of IP in the preceding section introduces the fourth and final benefit of VoIP that is always mentioned as an incentive to deploy VoIP equipment. That is, it is an IP world, and exclusively an IP world when it comes to the Internet and Web.

A brief history of the rise of the Internet and Web will be presented in later sections of this chapter. Chapter 5 details the TCP/IP stack itself. The rise of IP has coincided with the rise of the Internet and Web. And it all has to do with the positioning of the Internet protocol suite as the only set of protocols that can be used on the Internet.

Networking protocols have a long and rich history. Some of the protocols are considered to be *open*, whereas others traditionally have been considered closed, proprietary, or vendor-specific. Proprietary or vendor-specific network protocols are available only from the vendor that implements the protocol. For example, IBM controls the SNA network protocol, and Novell controls the NetWare LAN protocol. No one can buy SNA or NetWare from vendors other than IBM and Novell, with the notable exception of authorized resellers who buy from IBM and Novell and sell to others. Open network protocols include TCP/IP itself and numerous internationally standardized LAN protocols such as Ethernet (technically, as something called IEEE 802.3 LANs) or Token Ring. Open protocols are available from a variety of sources, and this competition helps keep the price of open protocols to a minimum.

Not so long ago, most organizations employed a number of network protocols to enable their clients and servers to talk to one another. More than one protocol was needed because one network protocol could not do everything. IBM's SNA was not intended for a LAN, Novell's NetWare ran poorly over a WAN, and so on. Also, whose network protocol would run best on a vendor's own equipment than the protocol the vendor specified, implemented, and supported on its own devices?

In order to allow all these protocols to be used on the increasingly common router-based network scheme of LANs connected by WAN links, many organizations built *enterprise networks*. Organizations had legacy mainframes, minicomputers, PCs on LANs, and so on all using networking to communicate. Even PCs were divided into the Windows-based PC platform group (perhaps sales), the MAC users (publishing), and UNIX-based workstation platforms (research and development). All had their own needs when it came to networking protocols. The cornerstone of the enterprise network was the *multiprotocol router*, a router that could understand and route packets structured according to a variety of protocols all at the same time. The multiprotocol router could not make a NetWare client attach to an IBM mainframe running SNA, but at least both SNA and NetWare packets could be sent through the same device over the same link between separate sites.

Three things were then needed to build an enterprise network:

1. Multiple sites
2. Multiple platforms
3. Multiple protocols

The multiprotocol router supported all the protocols needed by the multiple platforms. Of course, a *gateway* was needed somewhere if a client running one network protocol needed to interact with a server running another network protocol. Confusingly, early routers often were called *gateways*, but not in the same sense as the term *gateway* was used when it came to enterprise networks.

IP is the open protocol associated with the Internet and Web. By the mid-1990s, not only was it a good idea to allow users to access the Internet and Web to perform their daily tasks, it was pretty much mandatory. But the Internet, and the portion of the Internet known as the World Wide Web, is not a multiprotocol network at all. It is a single-protocol network, and the protocol is IP. Want to access a Web site? Does your client run IP? You're on. If not, you either need to get IP installed on your client, or you need to access a special gateway first. These special gateways were tried, but it proved better for a number of reasons to just put IP everywhere on the organization's clients and servers. And since IP was an open protocol, it was easy to find a vendor who made a version of IP for almost any platform, from Macs to mainframes.

Once this transition to the Internet protocol was made, the enterprise network became the *intranet*. Two of three ingredients were still present, but the third had disappeared. There were still multiple sites and multiple platforms, but they all ran IP. Since IP is assumed to be everywhere, only two things are needed to build an intranet:

1. Multiple sites
2. Multiple platforms

The preceding section ended with the realization that even a PBX could be put on a LAN using VoIP in order to support not only traditional enhanced voice services right to the desktop client PC but also more and better enhanced services. Why use IP to support the voice? Because it is there already. Many organizations plan to save money on building cable deployment alone because the voice would then require no special wiring of its own.

IP is typically bundled with every PC operating system sold today. This usually means Windows, but not always. UNIX and the Mac OS have not disappeared and are even enjoying a renaissance. Why buy anything else when every client and every server understands IP already? (This was not always the case. Before Windows 95 bundled it in, most Windows PC users had to buy IP software for about $100 in order to link Windows 3.1 directly to the Internet.)

So when it comes to the Internet and Web, IP is on the client and the server and everywhere in between. The multiprotocol router is just the single-protocol router, and the protocol supported is almost always IP and IP alone.

If nothing else, IP has proven to be very adaptable for uses other than data. Most users are familiar with Web sites that include audio and video content, as well as text-based data. Web pages routinely offer video about everything from product installation techniques to scenic panoramas. Radio broadcasts over the Web are quite popular as well. Customers accessing a corporate Web site often can speak to a sales representative or a technical support analyst just by clicking on a button on the Web page that says essentially "talk to me." It makes a lot more sense than picking up the telephone next to the user's PC, trying to find the number to call, and ringing a telephone next to an employee's PC. And, of course, the employee, given the traditional split between data networks and voice networks, might not have access to the same Web page that the user is looking at anyway.

The final benefit of VoIP, therefore, is that it makes the most efficient use of the IP protocol stack. If IP is everywhere, why not use it for voice as well?

Related Issues

The preceding section explored the four major reasons that VoIP is so attractive today. But within these four major categories of reduced costs, less bandwidth needed, better enhanced services, and efficient use of IP, there are related issues that should be examined a little more closely. Each of these areas is not different enough from the big four to deserve its own category. However, each of them is significant enough to deserve a few words of their own in this section.

Four issues are investigated here. These related VoIP issues are

1. International calling
2. Telemarketing
3. Call centers
4. The strange case of the fax

Each of these will be looked into a little further. The goal is to enable a comprehensive model of VoIP possibilities to be developed using all the benefits of VoIP for business and even residential use. These four related issues provide details into the operation of this proposed model. The VoIP model, in turn, will provide a framework for the rest of the chapters in this book, enabling the reader to relate each of the specific technology elements to the overall model presented at the end of this chapter.

International Calling

The VoIP benefit for long-distance calling has been mentioned already. But there is more to long-distance calling than just voice calls that span a whole country today. Emphasis on a global economy, spurred in no small part by the global Internet itself, means that the issue of long distance concerns not only domestic long distance but international calling as well.

The global economy means that businesses that were once firmly considered purely domestic enterprises can now market, sell, and support product offerings outside the country in which the company is headquartered. Not only does multicountry marketing better insulate a company from local economic swings, since many goods and services are routinely marketed globally by even modest-sized companies, a global presence is rapidly becoming a competitive necessity rather than a luxury.

In the past, a company seeking to extend itself beyond the borders of its home country had to worry about the cost of international telephone calls, which generally run about 10 times or more than calls of similar domestic duration, long distance or not. There are exceptions, of course, but if a way could be found to leverage the use of the global Internet not only for international data use (sales orders, e-mail, etc.) but also for international voice use, this could make the difference between opening a European sales office in Paris or not. And this, in turn, could make the difference between a company flourishing or languishing in the global marketplace.

Although this section so far has emphasized international calling in a business context, international calling for personal purposes should not be overlooked. It is not unusual today for families outside the United States to have one or more family members living and working in the United States. The social phenomenon of Central and South Americans with family members in the United States is but one example. The list could be extended easily to include Asians who have family members running businesses throughout Europe, Africans with family members attending universities worldwide, and so on.

The point is that these families need to stay in touch one way or another. In many cases, the medium of choice today is e-mail delivered over the global Internet. While Internet e-mail is a common enough means to achieve the goal of preserving family relationships, there are a number of drawbacks to using e-mail for this purpose. First, it requires both parties to have access to a computer. It is not even always a valid assumption that the person on the U.S. end of the e-mail exchange has a PC either. Many foreign families with PC access have a family member in the United States who does not have easy access to a PC and the e-mail application it represents due to mobility or lack of resources.

Second, e-mail remains for the most part a text-bound medium. There are ways to append voice, graphic, and even video attachments to Internet e-mail, naturally. But not only does the use of these attachments raise to bar in terms of end-user PC capability requirements, they also require a fair amount of sophistication on the part of the end users, especially when conversion is required. Thus the vast majority of e-mail remains just text.

However, so much more useful information is conveyed when voice is added to the text. Subtleties of diction and jest become apparent, as well as undertones of annoyance, anger, sarcasm, and so on. E-mail compensates for this by the clever use of "smileys" or "emoticons" to express the intentions of the user more closely. Most people routinely use the J or :-) symbol for "just kidding," since text always seems to come out harsher (or lighter) than the writer intended. (Video even adds a "smile when you say that" body language and facial expression element to voice nuances, but this is a separate issue.)

Voice communication is more personal and involves a higher layer of involvement than text alone. Telemarketing executives figured this out long ago, as annoying as the fruits of their labors are. So no wonder people prefer to phone home than to write. A quick survey of "10 minutes to Mexico for $10" phone card offerings shows how popular international calling has become. And VoIP benefits can be applied here too, whether the people talking have a PC or not.

Telemarketing

Selling goods and services by placing a telephone call to people and talking to them is big business. The proof need go no further than to recall the telemarketing calls that anyone receives in a given week. As annoying as these calls can be during dinner, they would not prosper if they did not generate the revenues needed to sustain them. This came as a surprise to some people at first. Why would people who routinely tossed a "junk mail" offering into the trash actually listen to—and buy from—a sales representative on the telephone? Mainly because people are used to paying more attention to audio communications, whether delivered in person or over the telephone. And sales is first and foremost an exercise in getting attention.

Telemarketing operations (more properly called *outbound-call centers*) vary greatly in terms of resources. Smaller efforts may just hire college students who make calls, sometimes from their dorm rooms, based on a list of telephone numbers and a printed script ("Hi! How are you today? My name is Valerie, and I'm calling to see if your chimney needs sweeping..."). These telemarketers usually get paid strictly for numbers of calls made and generally have a low success rate.

At the high end are telemarketing operations that use highly trained personnel. These telemarketers usually sit in a series of soundproof cubicles or rooms and rely on a PC and LAN to dial up one number after another automatically. They typically read from a professionally prepared script presented on the PC monitor, sometimes geared to the targeted customer's preferences and buying habits. Some are skillfully designed to appear as "surveys" and only make a subtle sales pitch near the end of what might be a 2- or 3-minute conversation only peripherally related to the product or service of primary interest to the telemarketing group.

The PC and LAN are used to adjust the telemarketing session during the conversation. The opening page ("Good morning, Ms Jones. I'd like a few minutes of your time to help us compile a survey on chimneys in your neighborhood.") usually has a series of choices available so that the telemarketer can decide how the conversation is to proceed based on initial response. There could be choices for "no chimney" or "hostile" or "had chimney cleaned last week by competitor." Choices are used to automatically update the telemarketing database in order to focus marketing efforts better in the future. Subsequent pages of the "sorry to have bothered you" or "thanks for your business" type are presented as required.

The point here is not to explore the psychology of telemarketing, as fascinating as this might be. Rather, the point is that professional telemarketing efforts rely heavily on the PC and LAN today (older operations from 10 years ago were based on the minicomputer). And while some integration between the data systems (the script and related information) and voice systems (the telephone directory and dialing) is common, it would be even more efficient to combine the data and voice sides of a complete telemarketing operation even further. Instead of having the PC relay the next number to be dialed to a power dialer on a PBX, why not use the PC and LAN for everything? Why talk on a telephone headset (it saves the arm and leaves the fingers free to enter responses to change script pages on the fly) when a PC can just as easily have a microphone stand in front of it?

Telemarketing is the business side of using VoIP for outgoing telephone calls in a residential setting. The basic architecture of the telemarketing center for a large corporation in the United States today is shown in Figure 1-5.

Call Centers

Call centers are the reverse of telemarketing in the business voice environment. Telemarketing handles huge numbers of outgoing calls in an effort to offer goods and services or generate business. Call centers handle huge numbers of *incoming* calls in an effort to sell goods and services

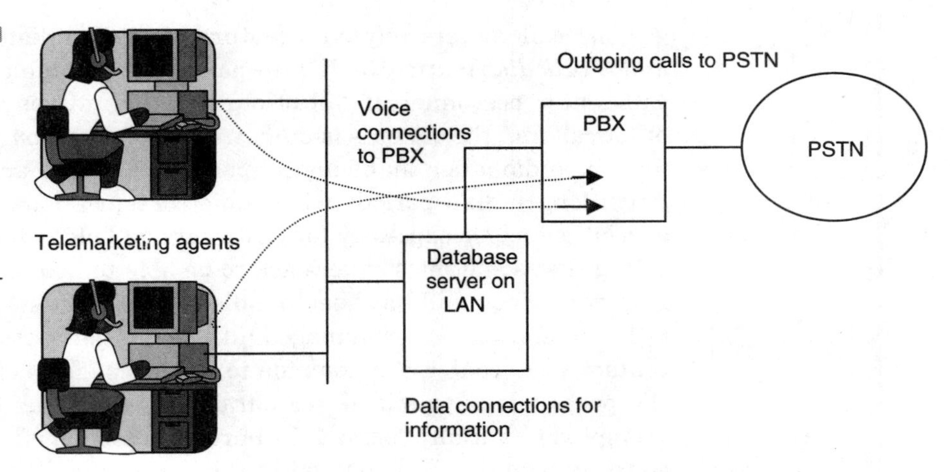

Figure 1-5 A modern telemarketing operation. Every location needs both voice and data network connections.

or retain business. Call centers are more properly called *inbound-call centers* to distinguish them from telemarketing's outbound-call centers. However, when the term *call center* appears alone, it is usually the inbound version that is meant.

Businesses that rely on the telephone to sell goods and services have been using call centers for years. Goods sold by catalog companies rely on the telephone number listed in the catalog to generate sales. The catalog could be in print or on the Web, but the preferred contact method is still often the telephone. And service companies such as travel agencies, airlines, hotel chains, and so on all rely on call centers to handle their sales.

Most people are familiar with the call center in the form of the 800 number. This is often called a "toll free" call, and any call made to an 800 number is indeed free to the caller. The recipient usually pays a bulk-rate discount on calls made to the 800 number, but the call is not free to the recipient by any means. However, it is usually a toll or long-distance call.

Many businesses found that there was a higher level of customer satisfaction with a product if there was a way of reaching the company before or after the purchase was made. And the customer satisfaction level was even higher if the call was free to the customer. Therefore, companies routinely began to put 800 numbers on product brochures and instruction manuals. Whether a customer has a question about a flight on an airline, or a setting for a new video recorder, or how to plug in the new modem, he or she usually can find an 800 number to call to get assistance.

Most call centers rely on a feature of the familiar PBX, called *automatic call distribution* (ACD), to parcel out incoming calls to available call center personnel. If "all of our representatives are busy assisting other callers," the caller is usually treated to *music on hold* or some other form of audio other than silence (periods of silence are perceived to last much longer than periods with audio content). Some companies record special messages expressly for customers on hold, which can be so interesting that sometimes callers ask to be able to hear the rest of the message when their call has been handled to their satisfaction!

In an effort to better manage all the potential reasons someone might contact a call center, it is common to have some form of directed access to the proper department or section of concern. Thus the caller may be prompted to initially "press 1 for hardware problems" or "press 2 for software problems" when contacting the call center. It is best not to have these options nested too deeply, since callers often do not have the patience to wade through four levels of menu choices just to be put on hold for 15 minutes, no matter how interesting the choice of audio content might be.

Call centers today are as dependent on the PC and LAN as their telemarketing counterparts. For sales, call center personnel have access to product availability, customer records, financial information, and much more through the PC and LAN. Call centers for customer support have access to installation guides, exception reports, trouble histories, and related information, also through the PC and LAN.

Modern call centers also are integrated in terms of voice PBX and data LAN. By the time the voice call has been routed to an agent, the caller's records may be on the agent's screen already. Sometimes the caller is required to enter an order number or customer code to allow this to happen, which leads to mistakes and repeated efforts. How much more efficient would the operation be if all the voice communication could be handled through the PC and LAN as well, perhaps by using the caller's IP address, with VoIP? The basic architecture of the call center for a large corporation in the United States today is shown in Figure 1-6.

The Strange Case of the Fax

The persistence of facsimile (universally called *fax* for short) services and equipment in the world of e-mail and the Internet has come as a surprise to some. It should not. Faxing has several advantages over e-mail and file transfers attempting to do the same things. Before exploring the rea-

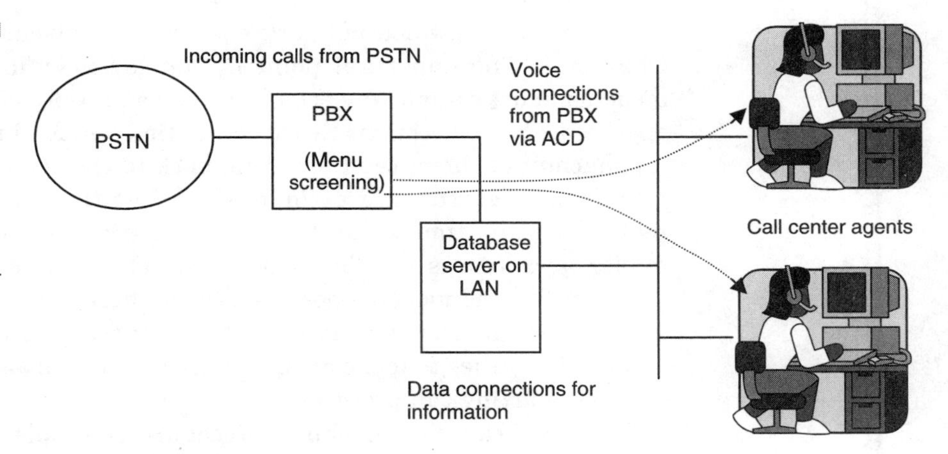

Figure 1-6 A modern call center operation. Every location needs both voice and data network connections.

sons for the persistence of fax in the e-mail world, however, a few words of background are in order.

Fax services more closely resemble voice services than data services. That is, with few exceptions, a fax is sent over the same telephone lines as those used for analog voice conversations. Many fax machines have a regular telephone handset attached and can be used to make voice telephone calls. Telephone companies have no easy way, and even few sophisticated ways, to distinguish a call made to send a fax from a call made to speak to someone. A fax call looks just like a voice call to the PSTN.

This is not to say that there are no differences at all between a connection used to send a fax and a telephone call. There are. Voice calls made from person to person have silences of varying length in both directions as people alternately talk and listen. Fax calls are predominantly "all talk" in one direction, from sender to receiver. Voice can be very tolerant of PSTN errors because people do their own error recovery, usually without even noticing ("What was that you just said?"). Faxing is less tolerant of errors because the end result can be just too grainy to read. So fax is sometimes referred to as *error-sensitive voice*.

However, nothing special need be done to the existing global public voice network to accommodate fax traffic. There is one exception to this rule, but the exception is now so universally followed that it is more or less taken for granted today. The PSTN uses special devices known as *echo cancelers* when the delay on a voice circuit exceeds a certain amount. Echo on a voice circuit can be quite annoying to human speakers if the echo is significantly out of synch with the speaker.

The details of voice network echo will be discussed more fully in Chapter 3. The important point is that fax machines, just as analog modems, must disable any echo cancelers on a voice circuit if they are to operate in both directions at the same time on the line. Now, receiving fax machines seldom send anything back to the sender, with the exception of brief status and error reports, but these can be sent by the receiver at any time while the sender is still sending a fax. This full-duplex operation is possible only if any echo cancelers present are disabled for fax and modem operation. Since there is no way of knowing for sure if echo cancelers are used or not on a particular connection, the fax standard specifies that the disabling procedure is always to be performed before the faxing can proceed.

This echo canceler disabling procedure is simple enough. A 200-Hz tone must be sent in each direction for at least 1.5 s before anything else happens on the connection. All telephone company echo cancelers are able to detect this tone and turn themselves off for the duration of the fax or modem call. This familiar modem "whistle" is evident whenever anyone dials the Internet from a PC or accidentally dials a fax number instead of a voice telephone number. The "whistle" identifies a fax or modem call, but little telephone network equipment outside the echo cancelers can detect this tone or cares about it at all.

Thus fax calls look just like voice telephone calls to the PSTN. Modem calls also look like voice telephone calls to some extent, but fax calls are generally of much shorter duration than Web sessions. Depending on what source is followed, anywhere from 30 to 40 percent of "voice" calls from the United States to Europe are actually faxes.

However, none of this explains why faxing still exists in a world of e-mail and file transfers. Why fax something to Europe when you could send it in an Internet e-mail? Even if the information is more complex than simple text, why not just transfer the file over the Internet? There are actually six good reasons why fax machines still exist and will continue to exist in the Internet world. These are

1. *Faxes usually have a higher legal status than e-mail.* A signed fax is usually admissible in court. E-mail, which is difficult to sign effectively (but it can be done), seldom enjoys the same status. Also, the date and time stamp on a fax is generally more respected than e-mail timestamps.
2. *Not everything is in a computer yet.* It may be hard to believe, but there is lots of useful and needed information that is not inside a computer somewhere. And the older the information, the more

likely it is that this is true. This information still needs to be sent to others, and faxing is the most efficient way to do it.

3. *A fax is, by definition, a copy of what was sent.* A facsimile is just what it claims to be: A copy of what was put into the fax machine at one end pops out at the other, like a remote copy machine with the PSTN in the middle. There is no concern about graphics format conversion, line breaks, or any of the other annoying problems with e-mail and file transfers.
4. *A fax is something tangible.* A fax can be further copied, filed, read, or distributed right away. No processing is required! A lot of e-mail is printed and filed before anything else. Why? Because if the computer crashes and takes the hard drive content with it, the message(s) are potentially lost forever. Now, a fax can be scanned into a PC for backup purposes, but a fax is "real" right now.
5. *Not all languages in the world are equally suitable for e-mail.* Many languages around the world are what are known as *logogram* languages (often called *pictogram* languages as well). Languages such as Chinese and Japanese are written with thousands of special signs that stand for words. Even English and other Western languages have these logograms; they form most of the upper row of a keyboard. So % means and is pronounced "percent" and never "circle-slash-circle," as one might expect. Some languages that change letter forms depending on words, such as Arabic, can be challenging for word processors and e-mail systems. In many countries, therefore, the preferred method of electronic communication is not e-mail but fax.
6. *All those millions of fax machines are not going away.* E-mail requires the use of a computer on each end of the communication. Fax machines are needed for faxing, of course, but it is much easier to "borrow" a fax machine to send a fax than to "borrow" a PC to check e-mail. Faxes can be sent to places all over the world where the presence of a PC on even adequate Internet access is not always a given.

What has all this talk about faxing to do with VoIP? Just this. Since faxing is more like voice than data, and since faxing is not going to go away soon, perhaps VoIP could handle faxing as well as pure voice applications. The savings on fax expenses could be considerable.

None of these four related issues of international calling, telemarketing, call centers, and faxing are particularly esoteric. Many organizations do all four, although telemarketing can be farmed out easily to firms specializing in telemarketing efforts. However, it is hard to monitor results and maintain tight control when the process is outsourced. It all depends on the type of product and service.

But then the incentive to support these activities with VoIP shifts to the telemarketing house. It does not subtract from the allure of VoIP.

The Lure of IP Telephony

The attraction of IP telephony, therefore, is not only in the promise of monetary savings, although this incentive is very powerful. IP telephony, as embodied in VoIP, offers tighter integration between voice and data networks than was ever possible or even dreamed of before. People can live and work more efficiently, although much of this efficiency translates directly into cost savings as well.

Three related developments throughout the 1990s have made VoIP an idea whose time has definitely come. These developments are the rise of the global economy and global Internet, the rise of the PC, and the rise of the World Wide Web (more often now just the Web). As these three concepts become more intertwined and indispensable for education, recreation, professions, and many other aspects of day-to-day living, it is tempting to try to integrate all of them into one seamless network for voice, data, and even video. Each of these trends is worth a short investigation to examine the role it plays in VoIP.

Global Economy, Global Internet

It is no longer the exception that world markets rise and fall as one, as exemplified by the global depression that threw the world economies into the doldrums in the 1930s. The economies of the world's nations are so interdependent today that a currency crisis in Southeast Asia or a banking disaster in Japan causes instant ripples across the stock markets of the United States and Europe. Such firmly "domestic" companies as McDonald's and Coca-Cola now derive so much of their revenue overseas that a fluctuation in currency exchange rates can drastically affect the stock prices of these and other firms.

Despite the risks inherent in diversifying into the world marketplace, companies are always eager to do so. The whole world economy should not become depressed all at once. If Asian operations are struggling, perhaps the European market is booming. How much more of a risk is it to remain in a tight corner of the marketplace when local disaster strikes?

The rise of the multinational corporation has been going on for a while, but the easy global connectivity provided by the Internet has increased the pace of this trend. In the United States, it is hard to find people who realize that Shell Oil Company is not a U.S. company but a Dutch company. Nestle is a Swiss-based company, not a "native" American firm founded in the United States. However, even the most nationalistic citizen of the United States would not pass up a Shell station to fill up with Exxon when needing gasoline or seek out a Hershey bar when a Nestle's Crunch is around. Shell, Nestle, and many other companies have such a global presence and are so firmly entrenched in the countries in which they do business that no one considers them a "foreign" company.

Products today are manufactured in Malaysia, warehoused in Mexico, and marketed in Spain by a company in the United States, and such products could be shipped all over the world. Along with this spreading out of the corporate structure into the wide world has come the need to network everything together to keep it flowing and to figure out how things are doing. As recently as the mid-1980s, a worldwide computer company was monitoring its Asian manufacturing operation simply by calling the manager, leaving a message, and waiting for a return call. The news was always good, until the end of the quarter when red ink was spilling all over the corporation's profit and loss statements. Clearly, some method was needed to monitor economic health on a day-to-day basis, not just during quarterly spreadsheet consolidation.

However, running private lines all over the world to network far-flung operations was too expensive a proposition to even contemplate in most cases. There were exceptions, of course, or else Shell and Nestle would never have penetrated global markets to the extent that they have. But these pioneers have been quick to adapt to changing network technologies to solidify their global positions. Shell, for example, uses a satellite network to validate credit cards right at the pump. But terrestrial leased lines paid for by the mile are not the way to go when it comes to building a global network to control a sprawling corporate structure.

The Internet has changed the whole economy of global networking finance. Just rent some office space almost anywhere in the world, get local access to the Internet, and start exchanging e-mail, transferring spreadsheets, and accessing remote databases with the home office.

Local telephone line charges may be high, but these are nothing compared with the cost of an international circuit.

The Internet's growth has been highlighted many times and need not be rehashed here. One example is enough, and mostly because it deals with an area of the world, Africa, that is often slighted when it comes to discussions of telecommunications infrastructure, sometimes by the nations in the region itself. With no disrespect intended, the fact is that there are more telephone access lines in New York City than there are in the whole continent of Africa. Yet the Internet flourishes in Africa.

This was not always true. As recently as 1996, Internet connectivity throughout Africa was spotty at best, and many countries had simple, text-only e-mail capabilities alone. Today, only one African country remains totally Internet-free (Somalia), and almost all countries have full Internet access, including the Web. One of the things that helped popularize use of the Internet in Africa was sitting the collected ministers of communication down at a conference and letting them use the Web for 2 hours. Like almost everyone else who sees and uses the Web, they came away impressed enough to make the Internet happen where they could.

Thus the global economy is served by the global Internet. The more useful the Internet is, the better it is for the economy.

The Rise of the PC

A famous quotation attributed to John von Neumann, primary inventor of the modern stored program computing architecture, has the legendary genius declaring something along the lines of "There is a world market for about six commercial computers." The quote is always good for a laugh that such a computer pioneer could be so wrong. The only problem is, von Neumann was more right than he was wrong.

The occasion was the introduction of the world's first commercial computer, the Sperry Univac in 1951. After the computer had been used successfully in government projects from computing artillery tables to modeling atomic bomb implosions, it was time to see if the same technology could make it in the business world. The trouble was that the early computers cost about $4 million and had less processing power than a singing birthday card has today. And these were 1951 dollars, when the cost of a new car was about $1000, fully loaded. Thus a company buying a computer was faced with parting with funds worth about 4000 new cars. Today, a car costs about $20,000, and this is just for the basics in many cases. Four thousand cars are now worth about $80 million. If computers cost this much today, selling all six would still be an accomplishment. The computer also filled the whole floor of a building and was mainly powered for computing purposes by a myriad of vacuum tubes that failed consistently and required constant replacement.

Naturally, the price of computers has come down rapidly, while at the same time the processing power they represent has skyrocketed. For the first 30 years of commercial computing, *computer* meant a large mainframe, usually from IBM, housed in and completely filling an entire controlled-environment room often with glass windows (so everyone could see and appreciate how progressively corporate funds were being spent). Most of the early buyers of these monster computers where in the utility industry, such as power companies and telephone companies, firms that struggled mightily just to mail a correct bill to each of their potentially millions of customers each and every month. Computers, then as now, excelled at record keeping above all else.

By the late 1970s, computer technology had advanced to the point where a minicomputer cost about $100,000 and could do almost everything a mainframe could. The invention of the integrated circuit based on semiconductor technology in the 1960s fueled this triumph of lower costs and miniaturization. The first computer networks appeared around the same time in the mid-1960s to allow remote users to access the central computer. The next logical step was to give each user a computer of his or her very own.

This step toward individual, or personal, computing began at several research institutions, most notably Xerox's Palo Alto Research Center (PARC) in California. In a frenzy of creativity rarely seen anywhere in history, a few short years in the late 1970s and early 1980s saw the birth of the personal computer (PC), "windows" on monitors, graphic user interfaces (GUIs), the mouse, the LAN, and the first network protocols and operating systems designed for LAN instead of WAN use.

Most of these ideas grew out of a preceding one in a natural progression. The early PCs, which could be built for about $10,000 by the early 1980s, were as powerful as a mainframe that could run 16 programs all at once. But there was only one monitor to watch all these programs execute. So add a "window" for each one. But how to get simple text-based monitors to display windows of varying sizes? Shrink or expand the graphics (invent the GUI) to fit the window instead of fixing the letter *A* to always be the same size and appearance. But there is only one keyboard. How can the keyboard quickly be associated with one window of another? (Today we say "give the window keyboard focus.") Add a mouse that can be used to point at a window and click to essentially say, "I want the keyboard to be associated with *this* window for now." (Special keystroke sequences were tried, but they needed a window of focus of their own, and if anyone terminated the keystroke program, the computer was effectively hung.) PC users were isolated from other users, even though they were only a cubicle away. So invent the LAN to tie them all together.

Books have been written about how Xerox, through PARC, managed to "fumble the future." The Xerox corporate structure, on the East Coast, could never figure out how what was happening at the research and development arm on the West Coast would help to sell copiers. Even the inventor of the laser printer (yes, that was Xerox too) was "banished" to PARC at one point. Part of the problem was the issue of trying to manage the research and development (R&D) group from across the country. Perhaps if the Internet had been as useful then as it is today, people would all be buying Xerox PCs and Xerox Windows and so on. Contrast IBM, where the R&D groups in Yorktown Heights and Hawthorne are about 5 or 10 minutes away from the IBM headquarters in Armonk, New York.

All of this is more than idle musings about days gone by. The point is that when an idea comes along at the right time, products come to market and thrive with breathtaking suddenness. Everyone that saw a PC wanted a PC. Not so much because of the package, but because the user was in total control. The PC came with spreadsheet and word-processing software, a rarity in the mainframe world at that point. Perhaps VoIP will follow the same trajectory for the same reasons.

In any case, by the mid-1980s, IBM had made the PC respectable by offering its own version of what to that point had been more of a curiosity than a useful business tool. Many companies would buy no computing equipment other than IBM equipment, so an IBM endorsement of the PC was a major step on the road to the desktop.

Another famous quote from around this time is attributed to Ken Olsen, founder of Digital Equipment Corporation (DEC), the company behind much of the popularity of Ethernet LANs, UNIX, and related computing methods. Supposedly, Olsen claimed that he could not see why anyone would want a PC in their home. Just like von Neumann's quote, this one is always sure to generate a chuckle. And just like von Neumann, Olsen was more right than wrong at the time.

By the mid-1980s, the PC had proven itself to be a valuable business tool. No one would seriously argue this point, even the mainframe and minicomputer vendors. However, PCs were still prohibitively expensive to own and operate. Most full-featured PCs cost close to $10,000 or more.

TABLE 1-1

The PC, 1981-1999

Feature	1981	1988	1992	1996	1999
RAM	64 kB	256 kB	1 MB	16 MB	128 MB
CPU speed	8 MHz	16 MHz	33 MHz	133 MHz	450 MHz
Hard disk	5 (MB if present)	40 MB	80 MB	1.5 GB (1500 MB)	16 GB (16,000 MB)
OS size	360 kB	1.2 M	7 MB	170 MB	600 MB
Base price	$6000	$5000	3000	$2000	$1500

In fact, in November 1991, DEC was advertising an Intel 486 processor running at 66 MHz with 16 MB of RAM and a 500-MB hard drive and a 14-in color monitor for $11,999. A double-speed CD-ROM (still rare at the time) and set of small speakers added $1000 to the price. For networking, a 9600-b/s Hayes external modem listed for $799.

No wonder no one seriously considered home PCs. A new PC cost about as much as the family car. Today, of course, some 20 percent of PCs sold cost less than $500. More than half of all PCs sold to home users are second PCs, not first PCs. All PCs come prepared for Internet access, of course. And soon PCs will be given away, as long as the "buyer" agrees to sign up for an Internet service package for 2 or 3 years.

The role of the PC in education cannot be underestimated. Basic instruction in PC skills begins in grade school, and it is safe to say that no one can graduate from high school without the assistance of a PC and the Internet. Many colleges and universities require incoming freshmen to have laptops, and some schools include this cost in the price of tuition.

As a final word on the rise of the PC, consider Table 1-1. This is a compilation with minor variations of one of the authors' desktop systems used for business and personal activities going back to the typical IBM PC in 1981. Note the rise in power and the fall in price and the fact that the useful life of the PC in terms of years before a new PC becomes absolutely necessary also has fallen. Also note the size growth of the operating system (but this is another story).

The Rise of the World Wide Web

A few words have already been said about the rise of the Internet earlier in this chapter. There is no need to detail all the landmarks along the way as the Internet made its way from a research and development network used mainly by colleges and universities doing U.S. government research to the global force it has become today. Such histories are readily available on the Internet itself, although the variation of interpretation of events and claims of who did what when is astonishing.

Instead, this section will briefly outline this rise of the Web as the most significant portion of the Internet. It is safe to say that without the Web, the penetration of the Internet into every business, every household, and every corner of peoples' lives would not be as much of a given as it appears to be.

Children used to ask their parents, "What did you do before there was television?" and now they ask, "What did you do before the Web?" The Web is a relatively new phenomenon in the world of networking, arriving in most places on the Internet in 1993 or 1994. The Web demanded powerful PCs and a common network (the Internet) to link everything in the universe together as a form of *hypertext* (although the term *hypermedia* is more applicable today).

The basic ideas behind presenting information not in a linear fashion, as in a book such as this one, but with embedded links that can be followed according to the whims and needs of the user (reader) have been around for many years. The term *hypertext* can be traced back to Ted Nelson in 1981, who in his book, *Literary Machines,* proposed the then radical idea of using computers not for number crunching or record keeping but for storing books and other forms of printed material. Nelson was inspired by Vannevar Bush. Bush was a key scientist and educator supporting computing projects and the general application of pure science to practical problems facing the world. In 1945, Bush designed a computerized database he called the *memex* to allow users to read documents and follow the trail of information from place to place as they liked.

Far beyond the capabilities of any computer of the time, hypertext got a real boost from the Apple Macintosh in the form of an application bundled with the Mac in 1987 called *HyperCard*. HyperCards could be linked not only between others of their kind with text but also with other media such as graphic artwork and sounds, giving hyper*media*, not just hyper*text*, links. Windows users are perhaps more familiar with Windows Help files, where green highlighted and underlined words and phrases can be "clicked on" to link to more in-depth or related information.

The step from HyperCard and Windows Help to the Web was a small one, and once made, it was amazing that it took so long to happen. All that was needed to make the leap to the Web was to allow the links to be made not from file to file within the same computer but across a network. Naturally, the key here is that all user PCs (the clients running the Web browser) and all the information on the Web pages they access (the servers running the Web site) are accessible on the same network. This was the role that the Internet played in the development of the Web.

Oddly, the Web actually came into existence not by extending hypertext to Web pages on a network but by linking separate pages with hypertext. In 1989, a software engineer at CERN, the European Particle Physics Institute in Geneva, Switzerland, named Timothy Berners-Lee became frustrated at the difficulty in tracking down physics papers and experimental results scattered all over Europe. The Internet was around then, of course. Using the Internet in those days, however, meant running the e-mail client application to send messages to colleagues, running the telnet client application to remote login to where material of interest was located, running the ftp client application to fetch a copy of the file, and so on.

Berners-Lee, and others like him, wanted two things added to the Internet to make life easier. First was a type of "universal client" that could send e-mail, perform file transfers, display images, and do many other things, all from the same application. Second was embedded hypertext links that could be followed almost anywhere using a common form of addressing and naming now known as the *universal resource locator* (URL). He got the second before he got the first. The universal client (which is now the familiar *Web browser*) was difficult because of the processing power demanded from a single network application that could do it all.

In any case, Berners-Lee proposed a method for linking text documents using "networked hypertext" in March of 1989. He teamed with Robert Cailliau to produce a more detailed design document that appeared in November 1990, more than a year and a half later. The document noted that such a system would be "a web of nodes in which the user can browse at will." The term *World Wide Web* appeared here first, although without the capital letters. Berners-Lee and Cailliau also developed a HyperText Markup Language (HTML) to embed the links, based on the Standard Generalized Markup Language (SGML) for formatting electronic documents.

Lacking a formal browser and concentrating on the server side first, the first Web site used a modified version of the telnet remote login application to access the "Web page," such as it was in those days. A slightly edited view of the Web "home" page in 1991 is shown in Figure 1-7.

The rest, as they say, is history. Berners-Lee formally presented the Web to the general public in December 1991 at the Hypertext '91 Conference in San Antonio, Texas. Then as now, everyone who saw the Web wanted the Web. The original telnet-based (called *line-mode*)

Figure 1-7 The World Wide Web in 1991.

Welcome to the World Wide Web
THE WORLD WIDE WEB

For more information, select by number:

A list of available W3 client programs [1]
Everything about the W3 project [2]
Places to start exploring [3]
The first international WWW Conference [4]

This telnet service is provided by the WWW team at CERN [5]
1-5, Up, Quit, or Help

browser was made available over the Internet in January of 1992, almost 3 years after the initial inspiration.

The dream of a universal client remained elusive, however. The line-mode browser was no one's idea of something easy to use and navigate. By 1992, the popularity of Windows for DOS-based PCs and X-Windows for UNIX-based workstations, coupled with the "horsepower" advances of desktop systems, made the use of a GUI windows approach and mouse click interface more feasible. An associate at the University of California at Berkeley named Pei Wei developed the first modern-looking Web browser, Viola, in July 1992. Viola was an interesting project for Pei Wei, who soon moved on to other things. A graduate student at the National Center for Supercomputing Applications (NCSA) on the University of Illinois in Urbana/Champaign (UIUC) campus, Mark Andreesen, along with Eric Bina, released a more full-featured Web browser for UNIX in February of 1993. The browser was called Mosaic and was freely available over the Internet in UNIX, Windows, and Macintosh versions by the fall of 1993. Andreesen, of course, later made Mosaic over into Netscape.

With the introduction of Mosaic in 1993, the Web exploded onto the Internet. Throughout 1994, Web sites popped up all over the Internet, from an estimated 5000 sites at the end of 1993 to a probable 50,000 or so by the end of 1994 (even then, no one was quite sure about the actual numbers). Today there are perhaps hundreds of thousands of Web sites, serving up millions of Web pages with embedded text links, graphics, audio, video, and almost anything one can think up. The integration of all these forms of information was the big payoff of the Web.

Except voice. Common, everyday telephony never was included in the vast array of content available on the Web. People cruised the Web to find information but still picked up the telephone to have a real conversation. Why not? The Web addressed limitations in the way that information was handled and presented by the Internet. There were few perceived limitations in the way that voice was handled and presented by the PSTN.

The Web grew out of a data network intended for text (the Internet) that added more than text content over the years. Maybe the same trick could be done with the global voice network. If the Internet could integrate video and audio to data with the Web, maybe the PSTN could integrate video and data to audio with something else.

Great idea. It just proved difficult to accomplish. This book is all about the integrated delivery of voice over IP-based networks such as the Internet and the Web portion of the Internet. But there have been attempts in the recent past to deliver integrated data services (and other services) over the PSTN. Since the technologies concerning voice that have been developed for these methods still have a role to play in VoIP, a few words about these integration efforts are in order.

Early Integration Efforts: ISDN and B-ISDN

The key to this section is that integration of voice and video and data did not have to occur on the Internet with the Web and VoIP. This integration of voice and video and data could have happened on the PSTN with ISDN and ATM. But it didn't happen on the PSTN, and this section explains why.

ISDN stands for *Integrated Services Digital Network*. ISDN is an international standard that has been around since the early 1980s and was intended to bring to the global telephone network many of the features and services that everyone now enjoys on the Web and Internet. There are two parts to ISDN: the integrated services part and the digital network part. The *digital network* aspect of ISDN means that all information flowing on an ISDN must be represented in digital form, from data to video to voice. This is not a problem at all today, since data are already in digital format, video is rapidly transitioning from familiar analog TV services to digital, high-definition TV (mostly driven by government conversion mandates), and voice is digitized the minute it enters the core telephone network. The only remaining large-scale analog infrastructure outside the analog TV arena is the analog local loop (access line) used for residential voice services. With ISDN, even this local loop is converted to deliver digital content for all services.

Thus the digital part of ISDN is more or less a prerequisite for the first part of ISDN: the integrated services. The *integrated services* defined for ISDN include digital voice (of course) as well as digital video, telemetry services (low-speed digital information such as burglar alarms and heat sensors), and (most importantly) data services in the form of a "high speed" packet-switching network. In other words, just about the same types of things that people do routinely over the Web today.

But "high speed" in 1984 is not the same as "high speed" today. The original ISDN specification for residential integrated services ran at 144 kb/s in the form of two *bearer* channels (B-channels) at 64 kb/s for user information and one *delta* channel (D-channel) at 16 kb/s for signaling and low-speed data purposes. Today, the two B-channels usually can be *bonded* to yield 128 kb/s in both directions at the same time. This 2B+D channel structure is still sold by telephone companies not so much to access the ISDN integrated services but to provide higher-than-analog modem speed access to the Internet, where the integrated services that were supposed to be a part of ISDN really are today.

However, there were no real "integrated" services with ISDN at all. ISDN really just integrated *access* to a number of different network services and allowed a number of different devices in people's homes to share this access. This is readily apparent in the architecture of ISDN for residential services, which usually was shown as it appears in Figure 1-8, although some possible premises equipment and networks are illustrated.

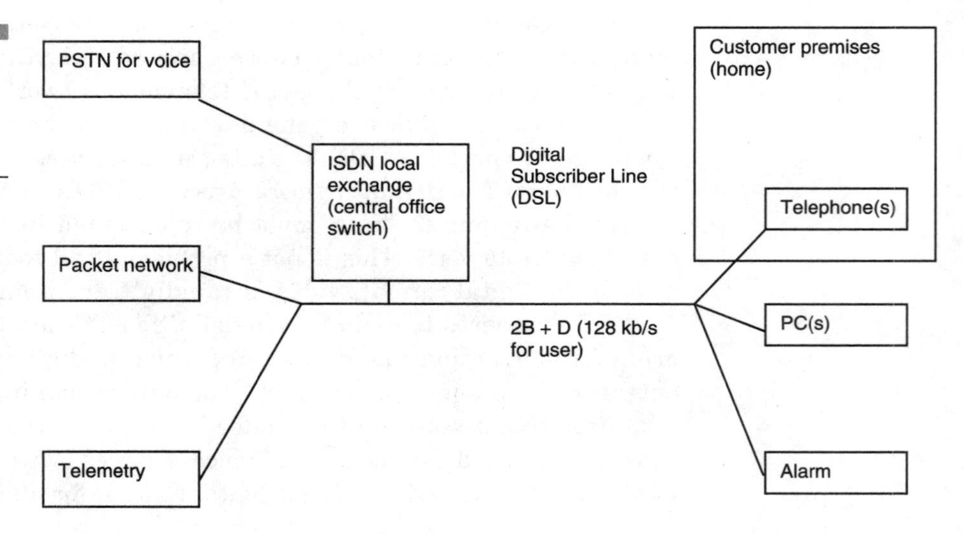

Figure 1-8 The ISDN architecture for residential services.

It did not take potential ISDN customers long to look at the drawing, scratch their heads, and say, "Okay, so where are the integrated services?" There are still many separate networks on the service provider end of ISDN. This was all right, the rationale went, because there were no devices in the home that could access more than one of these network services at a time anyway. ISDN benefited the local telephone companies more than it benefited the average residential user at the time. The telephone company could handle many traffic streams on the integrated access line without having to build, supply, and maintain a separate line for each device. Without ISDN, if people needed to talk, access the Internet, and send and receive faxes at the same time, they needed a separate line for each one. With ISDN, access to all these devices could be integrated, but the networks remain separate.

By the 1990s, this idea that there were no devices in the home that could access more than one of these network services at the same time was no longer true. The device had arrived: the PC. The PC could be used to access a data network like the Internet, as people who were eager to enjoy the Web quickly discovered. PCs also began to come bundled with speakers and microphones and voice software. The same telephone line that was used to access the Internet also could be used to make and receive telephone calls. This often caused astonishment to users who plugged in the telephone line to the PC modem only to have a small telephone icon on the PC desktop ring to announce an incoming call. The call could be completed with a simple mouse click, speaking into the microphone and listening through the PC speakers. More advanced PC models included an address book for speed dialing, and the PC could even be used as the world's most expensive answering machine. Even faxing could be handled with bundled fax software that was compatible with stand-alone fax machines at the other end. PCs even began to appear with cable TV tuning cards in them. Just plug in the cable TV coaxial cable, and 16 channels could be watched at the same time, most as small but full-motion icons on the desktop (only one could be listened to, however).

To the credit of the telephone companies, they all saw this coming by the mid-1980s, shortly after the basic ISDN architecture was completed. The problem was that computing power was growing faster than standards organizations could keep up. Integrated access, which was what ISDN provided, was not enough anymore. A way had to be found to integrate the network services that the customer accessed over the digital link.

At the same time, the telephone companies wanted to increase the bandwidth available to the home as well. After all, if the PC could be telephone, Web browser, fax machine, and even television all rolled into one, why not integrate voice, data, and video services and deliver them all over the same telephone line that ISDN used?

This was not possible mainly because the speeds defined for ISDN could not handle the digital speeds needed to deliver all these services at once, especially digital video. ISDN did include some support for *video telephony*, but the quality was nowhere near what people were used to from the cable TV networks. ISDN functioned at *narrowband* speeds, defined as lower than about 2 Mb/s. And true network service integration would require major changes to the way that voice, video, and data were delivered to users through the switching office and over the telephone lines. The existing networks were just too different in design to enable video over the telephone network or allow voice over a cable TV network without changing both radically.

The solution, which was ready to go in outline in 1988, was to invent a new version of ISDN that ran at *broadband* speeds (B-ISDN). B-ISDN did two major things to ISDN. First, it increased the speeds on the access lines to a whopping 155 Mb/s to start with and extended to 622 Mb/s. This could easily handle the most demanding digital video services. Second, B-ISDN replaced every existing network service from voice to video with a new technology called *Asynchronous Transfer Mode* (ATM). An ATM network, which packaged everything into new units called *cells*, could handle voice but was not specifically a voice network. ATM also could handle video but was not specifically a video network. And so on, for almost any service one could imagine delivering over a network in the first place. How B-ISDN changed the basic ISDN architecture is shown in Figure 1-9.

B-ISDN was intended to be an *evolutionary* standard, meaning that no one should try to change everything overnight. The International Telecommunications Union (ITU), which developed the standard for ATM and B-ISDN, guessed in 1988 that it would take anywhere from 20 to 30 years for B-ISDN to be available almost everywhere. The problem

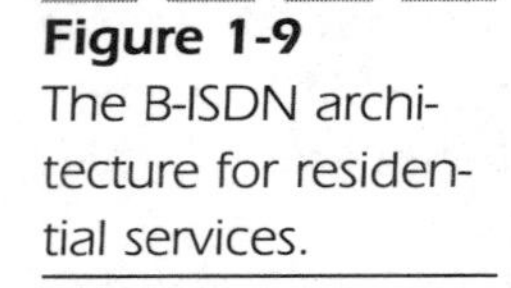

Figure 1-9
The B-ISDN architecture for residential services.

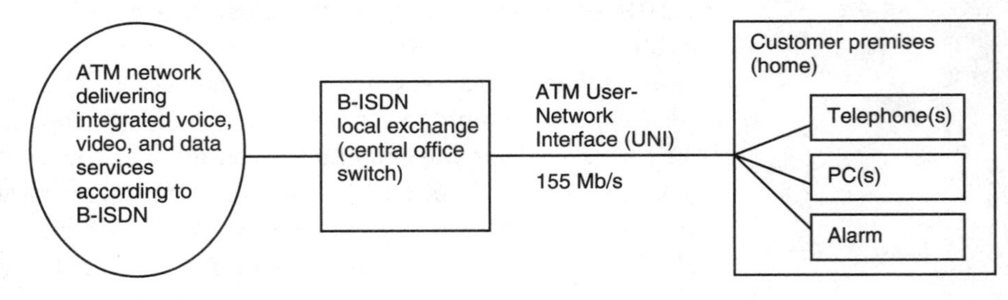

was that integration at the premises in the form of the PC and Web occurred much faster than deployment of the B-ISDN services and ATM network necessary to deliver the services.

Something else was needed right away to integrate network services right to the user desktop PC, whether B-ISDN came to town or not. That something else turned out to be the Internet. And VoIP is the way to integrate the voice network with the Internet and the PC or even a traditional telephone handset. Also, as it turned out, modern video digitization techniques need nothing close to 155 Mb/s, or even 100 Mb/s, to deliver video as good as seen and heard in a movie theater. Thus B-ISDN is left today as a solution in search of a problem.

Today, talk of B-ISDN is rarely heard. ATM exists only to provide those who can afford it faster Internet access and backbone links. However, several protocols originally developed for B-ISDN live on in VoIP implementations (such as H.323, discussed in Chapter 7). ATM even has a role to play in most VoIP implementations, although this may be a temporary situation. The debate rages over whether ATM makes any sense at all in the IP-Internet-Web world of today, but this also will be discussed in greater detail in a later chapter.

A Model for IP Telephony

All the pieces are now in place to develop a model of VoIP, or IP telephony. This model incorporates important aspects of each of the issues examined so far in this chapter, from international calling to telemarketing to call centers and beyond. One of the reasons that so much effort was expended describing these issues was to make the model more concrete and understandable when first encountered.

The model addresses four areas of interest, using business and residential use as the source and destination of the voice call. Keep in mind that this is only a model, so nothing will prevent a person at work from contacting a business's help desk or call center. From the point of view of this model, therefore, this is still a residential-to-business use of IP telephony.

The model for IP telephony used in this book appears in Figure 1-10. This model will be used as a framework for all subsequent discussions of VoIP equipment, applications, software, and so on. It should be noted that there is nothing standard or special about this model of IP telephony. Many other models are possible, but this is the one used in this book.

Looking at each compartment in turn shows how VoIP can be used to

Source	Destination	
	Business	Residential
Business	Sector: business to business Use: faxing, tie-line replacement Need: Better QoS for IP, managed IP network? Outlook: Do it now	Sector: business to residential Use: telemarking Need: IP-enabled PBX, ISP to PSTN gateways Outlook: Do it carefully
Residential	Sector: residential to business Use: call centers, catalog sales Need: voice-enabled Web site, IP-enabled ACD Outlook: Do it carefully	Sector: residential to residential Use: long distance replacement Need: Many PSTN gateways, basic voice QoS Outlook: long distance now, local perhaps later

Figure 1-10 A model for IP telephony. Source is the originator of a call, and destination is the receiver.

create a total IP telephony system that integrates voice capabilities for almost anything into the Internet and Web. Each category sets out aspects of VoIP application use, the most pressing need at the moment to make it happen, and the outlook for VoIP to occur in that particular segment.

Start with the use of VoIP from one business to another business, the upper left-hand box. VoIP can be used in this sector to send faxes from one business to another and to replace tie-lines between business PBXs, which are just point-to-point 64-kb/s voice circuits paid for by the mile. There are other potential business uses for VoIP, but these are the ones mentioned most often and that have the most potential to save a business some money. However, if VoIP is to be used for these purposes, there are some needs that must be addressed to make VoIP a worthy vehicle for these uses. First and foremost, the network delivering the VoIP, probably the Internet if any business-to-any business connectivity is the goal, must have adequate quality of service (QoS) to deliver voice that people will use. Of course, the Internet only supplies connectivity, and no basic bandwidth, delay, or other QoS guarantees at all. This QoS concern may mean that business-to-business VoIP will be confined to *managed IP* services for the time being. Managed IP services place business traffic onto a special portion of the service provider network in an attempt to "manage" QoS through limiting traffic to maximize customer bandwidth, minimizing hops (routers) to limit delay, and so on. The Internet is still accessible through managed IP service, but not with the QoS guarantees available for traffic placed on the special backbone. Finally, the outlook for business-to-business use of VoIP or IP telephony is very upbeat. For even modest-sized businesses with limited budgets, VoIP for business faxing and tie-line replacement should be done now—the sooner the better.

Next, look at business-to-residential use of VoIP, the upper right-hand box. The major use here is telemarketing (outgoing-call centers). The major need right now is for IP-enabled PBXs with links to both the corporate LAN and the PSTN through the Internet service provider (ISP). These ISP-to-PSTN gateways are the crucial piece for this IP telephony application, since most residential customers will only receive calls today over the PSTN and not the Internet. These devices exist and will be explored in Chapter 8. However, the current outlook is to do this very carefully for now. In other words, it would be rash and risky for any business to retool its whole telemarketing department for IP telephony and then find that the PSTN gateway is down more than it is up and running. A more cautious and phased approach is the best for now.

The lower boxes examine the residential use of IP telephony to contact businesses and even other people at home. This is usually true IP telephony, since in a residential environment it can be assumed safely for now that people are using a telephone attached to the PSTN for their end of the call.

For residential-to-business use of IP telephony, the lower left-hand box, the major applications are call centers (incoming-call centers) and catalog sales departments. As much as this activity has transitioned to the Web, there are those who will not buy unless they speak to a human and will not be happy fetching up documents on their own unless someone can reassure them that this set of instructions will solve their problem. What is most urgently needed here is a voice-enabled Web site with a "talk to me" button to click on and an IP-enabled automatic call distribution (ACD) system that acts like the IP-enabled PBX for telemarketing, but in reverse. Both devices exist today, of course, and although not mentioned in this context, PSTN-to-ISP gateways are needed as well, just as in the business-to-residential direction. And just as with IP-based telemarketing, the outlook here is for careful deployment of VoIP and IP telephony technology and equipment and methods.

The final category, in the lower right-hand box, considers pure residential-to-residential calling using VoIP and IP telephony. This is the same application targeted by 1010XXX (technically, it is 101XXXX) advertising and is essentially an alternative to using the PSTN for long-distance calling. The big attraction here is to replace PSTN-based international calling, but any long-distance calling can be replaced cost-effectively with VoIP and IP telephony as long as the ISP-to-PSTN gateways are in proper locations (i.e., in each country). In fact, this is the pressing need here. There have to be many PSTN gateways, at least one in each country. Naturally, the IP portion of the network needs to deliver basic voice QoS, which may not be a given if a major portion of *all* telephone calls, not just call center calls, find their way onto an IP network such as the Internet. The outlook for service providers of residential service is to do long-distance, especially international, calls right away with IP telephony. Local calling perhaps may follow later, but only once IP telephony and VoIP become more common than any other way of making a telephone call.

This model for IP telephony will be used to give form and substance to the detailed discussion of hardware and software, equipment and vendors, services and capabilities, and many other aspects of the whole field throughout the rest of this book.

Technical, Economic, and Social Factors

The model for IP telephony just presented is not necessarily a reliable road map for the future. There are three factors that ultimately decide whether some scheme or another becomes the next television, which people are astonished to realize was not always around, or the next Pony Express, which despite the romance and legend surrounding it, lasted all of 18 months until the telegraph and stagecoach forced it out of business. So which is VoIP, television or Pony Express?

The three factors are the technical feasibility, the economic feasibility, and the social acceptance of the new proposed service or product. All three factors play a role, and any one can be fatal to a new technology. In some sense, VoIP is like a three-legged stool, with each leg representing one of the technical, economic, and social factors involved with the new technology. If one leg is much shorter or longer than the others, nobody can sit comfortably on the VoIP stool, and they might even fall off.

Technical feasibility, which is just a decision as to whether some technology can be implemented or not, is not always a given despite all the advances in computing and processing power people take for granted today. For example, the market for robots that could perform everyday household tasks such as vacuuming, making beds, or washing cars would be tremendous. However, even after 30 or so years of intensive research into computerized mechanics and especially in the field of artificial intelligence, such robots remain a dream. No one would even buy stock in a company that announced a "house robot" product without first determining if the robot was a person dressed to play the part. Powerful processors can beat the world's best at chess but cannot be made to understand simple commands that 3-year-olds follow without question, such as "stay right here" and "follow me."

A robot is just not technically feasible at this time. However, even if a house robot could be made, is it possible for a company to make money manufacturing and selling such robots at a price people can afford? This is where the economic feasibility of a new product or service based on a particular technology comes in. No one will make and market something unless they think they can make money doing so. Initial prices can be high, but if they do not fall rapidly enough, the market remains small, and the technology withers. Color televisions, when first introduced in the 1950s in the United States, cost about $1500. This does not sound like much today, but the price of a basic 1957 Ford Fairlane, a popular model at the time, also was $1500. How many families can afford to buy a color television instead of a needed new car? Of course, the price of television sets dropped as car prices climbed, and this made the color television a fixture in most homes. A similar issue involves the new digital television sets, initially priced between $10,000 and $60,000. If prices do not fall rapidly into the ranges where most people with the average household incomes in the United States of $30,000 or so can afford them, the optimistic rollout schedule for digital television may be compromised.

If a new technology is feasible but is not accepted in the marketplace, the price is most likely set wrong. If the price is set too low, no one can make money selling the service or product, as many ISPs found out when Internet access market factors drove prices below $20 per month. If the price is set too high, the market remains small, and it is hard for suppliers to prosper, as might happen with digital televisions.

VoIP is feasible both technically and economically for service providers, although there are still arguments about exactly how to make money offering VoIP and IP telephony services. However, even if the economic issues are addressed adequately, there is still the third issue—

social acceptance. And this often proves to be the trickiest of all to overcome regardless of capability and price.

Take video-on-demand (VOD) services as an example. This plan for switched digital video (SDV), championed by the local telephone companies in the early 1990s, would deliver pay-per-view movies over the same copper wire as used for telephony. Telephone companies wanted to use VOD and SDV to fight back against the cable TV companies, which were seen as a potential threat to telephony revenues because the cable TV companies were beginning to offer "voice over cable" services. To counteract this possible erosion of their customer base and revenues, the local telephone companies developed SDV systems and brought them to market trials in many areas, notably in Bell Atlantic territory. Selected people were able to view digital versions of movies such as *Top Gun* while still making telephone calls, all delivered without the need for a cable TV connection.

The technology was rock solid and, in fact, became the basis for asymmetrical digital subscriber line (ADSL) systems that are now being deployed to counteract the use of cable modems for high-speed Internet access. The economics appeared sound: Movies cost about the same as they did to rent at the local video store, and use of the family car was not required. The movies in some systems could even be paused, "rewound," and otherwise manipulated.

But SDV never caught on, mainly for social reasons. People could not tape the digital movies the way they could tape cable TV movies (there are *still* no digital video recorders for the mass market). People *liked* going to the local video store, sometimes just to get out of the house for a while. People did not like using pay-per-view for movies, and pay-per-view has only succeeded for certain high-visibility and specialized sports events such as boxing matches. In addition, the digital processing needed to convert "regular" analog television signals to digital signals at the time ruled out live broadcasts (even today, there is a noticeable delay in live digital broadcasts).

The real problem, however, was so simple that no one expected it: The SDV remote control unit did not work the way people expected it to. Some people can watch 10 channels (or more) at a time. They just constantly flip through the channels on a broadcast TV or cable TV with the remote (when people had to get up off the couch and move to change channels, this was an unknown phenomenon). While driving others around them to distraction, this "channel surfing" works because all the channels are always there; the remote just tunes the TV to the channel of interest at the moment (literally).

However, the SDV remote worked differently. There was only *one* channel on the SDV TV. There was not enough bandwidth on the local loop to deliver more than 100 channels, as some cable TV systems could. So a click of the remote could change the channel, but only about a half second later after the request made its way up to the SDV "head end" and the new channel made its way back down to the subscriber. Naturally, people used to instant channel changing usually pushed the button three or four times because "nothing happened" in the time frame in which they expected it to. As a result, the SDV system jumped up three or four channels instead of one. This in turn elicited more clicks to bring it back, usually missing the channel in the opposite direction. Eventually, frustrated users turned off the SDV and turned the cable TV back on.

Where does VoIP stand when it comes to technical, economic, and social factors? On a very level stool, as it turns out. The technology is there for the asking, and the economics appear to be sound for international calling, faxing, and related applications at present. The social issue may be solved by the fact that much of IP telephony is based on the Internet. People like the Internet, and this liking often leads them to overlook obvious and potentially serious glitches such as unavailable local Internet access, Web page delays, and e-mail viruses. The more VoIP can be tied to the Internet, the more certain its social acceptance will be.

Regulation

New services and products may be perched on the three-legged stool of technology, economics, and social acceptance, but government regulation can be a saw that cuts off a few inches here and there from one or the other. The final result can be a more stable and level stool or a wildly tilting platform that will support nothing. Thus regulation can encourage or discourage, according to the economic or political goals of the regulators.

There is no need to explore all the ins and outs of regulation when it comes to IP telephony. Such a discussion could easily fill a book of its own. Rather, the intent here is to broadly outline the role that regulation has and will play when it comes to VoIP. At the risk of being too broad, this section will divide the whole world into two sections: the United States and the rest of the world. Although there may be objections that the role of regulation around the world varies widely in scope and intent, there are some overall points that can be made in the way of examples when it comes to IP telephony.

In the United States, the role of regulation at the state and federal levels has been lately to preserve the profits for service providers that must often make huge capital investments before earning a dollar from a service. Many businesses fall into this category, from railroads to airlines to cable TV to telephone companies. Historically, this regulation has taken the form of establishing monopolies for service providers in certain areas. Lately, the trend is more toward the structured introduction of selected (certified) competitors into previously restricted markets, although debate rages as to just how successful or unsuccessful this approach has been.

Many technologies are encouraged at the federal level in the form of direct subsidies or exemptions from charges that would otherwise apply. Sometimes this practice is also attacked, but the principle is firmly entrenched in the U.S. Constitution, which charges the federal government with encouraging practices that improve the quality of life for U.S. citizens. The principle has been applied many times, and the telegraph was a direct beneficiary of this practice. Samuel Morse, the telegraph inventor, was not a businessperson but an art professor at a major university. Without federal money, the first telegraph links would not have been built. Use was encouraged by direct subsidies that ensured free use until the telegraph could stand on its own and pay its own way. The Internet prospered with the constant aid of federal money for more than 20 years before ISPs, faced with having to turn a profit, busily began signing up ordinary citizens for Internet access. Fortunately, by then, they had the Web to sell.

However, ISPs are still exempt from many of the access and related charges, generally known as *settlements*, that other communications companies routinely pay to each other for completing calls that originate with one service provider (who bills the customer) and terminate with another service provider (who would otherwise get nothing). This is slowly changing, and the important point is not whether such an "Internet tax," which is how the ISPs generally perceive the issue, is a good idea or not, but how regulation in the United States could encourage or discourage the use of VoIP.

There are special rules for ISPs that deliver data over telephone lines. But the question is whether VoIP is voice or data. VoIP travels over the same Internet routers and links as data, is inside IP packets, and looks just like any other stream of bits on the network. How could there be different rules for data and VoIP anyway? How could IP packets with voice be distinguished from IP packets with data in the first place?

People have certainly tried. Early makers of VoIP software were threatened with classification as "telephone companies" subject to all the certification rules and access charges that pertained to the traditional circuit voice carriers. This effort failed, and interest in similar proposals to firmly regulate VoIP efforts diminished as the interest of traditional voice carriers in IP telephony increased, although this may be too cynical a position.

However, it is always worth remembering that regulations can change at any time. The current crop of IP telephony developers, vendors, and service providers is skating on the thin ice of a benevolent state and federal regulatory environment. A sudden warming could change the ice to slush, however.

Outside the United States, things are even more complex. Regulation in many parts of the world exists to generate revenues from telecommunications services for the central government. Telephony in many countries is run by a government bureau, just like the post office. This practice began when telephony was seen as just another form of messaging, like letters or telegrams. Often, telephone call price structures were kept artificially high to maximize these revenues, which sometimes provided the largest contribution to the national budget. Needless to say, there is little to no competition in either services or equipment.

The issue is now that when the Internet and later the Web came to many countries around the world, the central government embraced the Internet as a sign that the country was progressive and enlightened and always seeking to improve the lifestyle of its citizens. There were exceptions, of course. China shunned the Internet because of the lack of control it represented. France, which still has fewer Web sites per capita than any industrial nation, thought it had something just as good or better in the form of the Minitel system and so just shrugged until recently. The countries that embraced the Internet generally did so by exempting Internet companies from all the regulations and charges that would apply to "normal" telecommunications companies doing business within their borders.

So far so good. But what happens when VoIP comes to town? When used for international calling, IP telephony is much less expensive than the domestic alternative. This is free-market enterprise at its most basic: When competition comes to town, prices must fall. And this is exactly what happened when early and primitive but inexpensive VoIP was offered in many countries outside the United States. For example, the price of an international telephone call from South Korea to the United States was $1.00 per minute before VoIP was available to place the same call. By early 1998, with competition from VoIP, the price had dropped to $0.31 per minute. There are many more examples of how VoIP competes with international calling, but two more will suffice. The price of a 10-minute telephone call to Australia in April of 1998 was $10.90 with AT&T, $10.89 with MCI, but only $1.80 with VoIP provider IDT. Naturally, AT&T and MCI can only lower their rates so far because they have to pay access charges to the Australian telephone companies, while ISPs generally do not. From Japan, a call to the United States cost $1.71 per minute but only $0.10 over the Internet in July of 1997.

Sometimes, this use of IP telephony is called *Internet bypass* to reflect the fact that the Internet is used to "bypass" the higher price structure charged by the incumbent carrier. The overall architecture of an international call made from the United States to another country is shown in Figure 1-11. Note the key position occupied by the IP and PSTN gateways.

In many of these cases, the lower VoIP prices took revenue away not only from the national carrier (12 percent in the case of Japan) but also from the central government. Some have fought back, and IP telephony is now illegal in at least two countries. (But how do they find it mixed in with Internet traffic?) Some countries have stopped fighting the future. In South Korea, for example, people can now dial 00727 before making a call to make an IP telephony call or 001 to use the PSTN.

Deregulation initiatives all around the world have been in place for years. However, progress has been either fast (as in most of western Europe) or slow (as in many other parts of the world) as governments struggle to replace steady telecommunications revenue streams with alternatives. As always, regulation remains a volatile environment where things can change rapidly as political and economic agendas shift. The current regulatory IP telephony environment will be more fully explored in Chapters 11 and 14.

Figure 1-11 International calls using the Internet and VoIP.

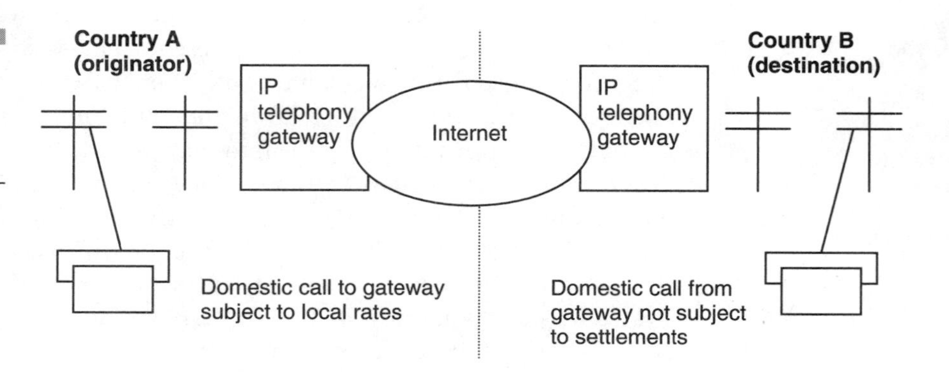

CHAPTER 2

Digital Voice Fundamentals

Voice over IP (VoIP) takes a speaker's voice, digitizes it, places it inside the same Internet Protocol (IP) packets used for Internet and Web communication, and sends the digital voice over an IP network. The IP network can be a local-area network (LAN) such as an intranet, a private wide-area network (WAN) running IP, a managed IP mixture of public Internet and private extranet logical networks (such as a virtual private network, or VPN), a frame relay or Asynchronous Transfer Mode (ATM) network, or the global public Internet all on its own. It makes no difference to the VoIP. IP is IP. This is one of the great attractions of IP for network service providers: There are almost no limitations to the type of network on which IP can run. And IP runs on anything without major changes to the IP packet structure and operation. IP is a general-purpose network protocol that can be tuned to one specific type of network or another but requires no special versions to function properly from LAN to WAN and back.

Digital voice is just another form of IP packet content, just like a file transfer or e-mail message. However, the zeros and ones inside a VoIP packet represent human speech. This chapter examines all the details involved in taking analog human speech and turning the analog voice into a digital stream of bits. Along the way, all the latest voice digitization techniques will be explored, from compression to silence suppression and beyond. The whole process must begin by being very clear about the precise differences between analog and digital information and the analog and digital links used to carry that information.

Analog Voice and Digital Data

Most scientists agree that what makes human beings radically different from all other mammals is that we talk with each other. There has not been enough time for evolution to make human brains that much different from apes, and DNA research has reinforced this belief. Human speech, however, learned rapidly after birth, appears to drive many of the differences in brain organization that allow people to carry on a conversation with their parents but not with their dogs. This came as a surprise to researchers, since conventional wisdom attributed speech to human brain organization and not the other way around. People got bigger and better brains because they needed to express bigger and better thoughts vocally. Between the ages of 5 and 11 years, the average person learns about two or three new words *a day*, because it is necessary to do so.

Be that as it may, only humans appear to use the voice as the primary means of deep social interaction. And as human society spread throughout the world, the need to keep in touch vocally led to the development of the worldwide telephone system and now drives the need to add voice to the global Internet.

Voice or human speech is generated by air from the lungs vibrating special vocal cords in the throat and also to a lesser extent the tongue and lips to form sounds, known technically as *phonemes*. The vibrations occur at a range of frequencies (number of vibrations per second) that do not vary much from culture to culture. However, the exact array of sounds made varies from culture to culture, and studies have shown that if a certain sound is not learned properly by the age of 7 years or so, the ability to generate that sound correctly is lost forever. This is one reason that the later a person learns a foreign language, the more likely he or she is to always speak with a pronounced accent. The sounds are arranged into voice units called *words*, words are organized into *phrases*, phrases make *sentences*, and so on.

The distance at which the human voice can be heard also varies. This range depends on weather factors, as well as the power of the sound. Some sounds are more powerful than others, a feature of voice that will be important later on in this chapter. Voice is a pressure wave generated in the air that travels from mouth to ear. Voice pressure waves are *acoustical* waves, as are the sounds made be musical instruments, animal growls, bird calls, and so forth. Pressure waves are fundamentally different than the type of electrical (technically, electromagnetic) waves that operate in power systems, wireless cellular telephone networks, TV systems, and data networks. Acoustical waves rely on air to carry them. The thinner the air, such as at higher altitudes, the less power is in the voice. Water, being very dense, carries sound waves so well that ears, normally capable of determining the direction of a sound source with ease, cannot discern where sounds come from; they seem to come from all directions at once.

Voice, carried on these acoustical waves, is a form of analog information. This means that the amplitude of the signal can vary from a maximum to a minimum and take on any value at all in between. Voice sounds can be described in one of several ways. There are frictives and labials, for instance. But when it comes to understanding voice digitization techniques, it is easiest to describe voice as a mixture of high-amplitude sounds called *voiced sounds* and low-amplitude sounds called *unvoiced sounds*. In the English language, voiced sounds include the vowel sounds such as *a*, *e*, *i*, *o*, and *u*. Voiced sounds are formed in the throat and are determined directly by vocal chord positioning. Unvoiced sounds are the consonants such as *b*, *t*, *v*, and so on. Unvoiced sounds are formed by the tongue and lips as air is passed directly through the vocal cords. Voiced sounds have about 20 times more amplitude than unvoiced sounds and carry most of the power of human speech. However, the unvoiced sounds are what ears rely on to distinguish between words having the same vowel sounds at their core, like the difference between *belt*, *felt*, and *melt*.

Figure 2-1 shows the relative amplitude of voiced and unvoiced sounds in an analog acoustical wave. The figure represents a pressure wave as a time waveform, a common enough practice. Other languages use other sounds, but all are formed with either the throat or mouth playing the major role. This limited range of possible sounds is an important feature of analog voice that comes into play in modern voice digitization techniques. However, the sound should never be confused with the letter or letters of the alphabet that represents it. The explosive *k* in the English word *skate* is quite different from the *k* sound in the word *kite*.

Voiced sounds, as illustrated by Figure 2-1, have fairly regular wave-

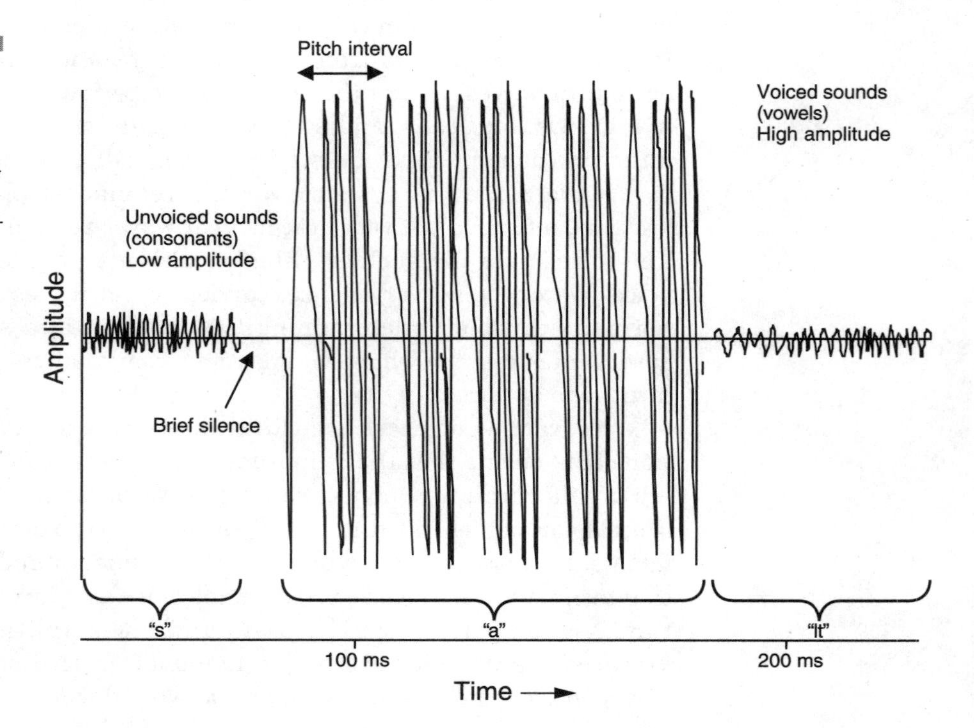

Figure 2-1
The English word *salt* as an acoustical wave representing analog voice.

forms, since the vibrations in the throat are very stable. On the other hand, unvoiced sounds have a more random and unpredictable waveform pattern because the position of the mouth can vary during sound generation. These observations also will become important in modern voice digitization techniques.

Voiced waveforms tend to repeat in what is known as the *pitch interval*. The length of the pitch interval can vary, most notably between men and women (this is not a sexist comment, just a report on the findings of acoustical engineers). When men speak, the pitch interval lasts between 5 and 20 ms (thousandths of a second), and when women speak, the pitch interval typically lasts 2.5 to 10 ms. Voiced sounds last from 100 to 125 ms. Thus a single voiced sound can have anywhere from 5 to 50 pitch intervals that repeat during a single sound. Oddly, not all the pitch intervals are needed for understandable speech. People can understand speech faster than people can generate speech. Some television commercials, in a quest to jam as much talk into a 20-s spot as possible, routinely remove some of the repeated pitch intervals in voiced sounds to speed up the voice without distorting it. The result sounds somehow strange to viewers, but few can say exactly what the difference is.

One further aspect of human voice should be discussed before moving on to the invention of the telephone. This is the fact that *silence* plays a large role in human speech communication. The listener at the moment is almost totally silent during a conversation, save for short sounds generated to let the speaker know that he or she is still paying attention ("uh-huh") and understanding the speaker ("Go on"). These feedback sounds are most frequent when two people are conversing and almost totally absent when one person is addressing a group of listeners (thankfully).

Moreover, silence occurs in the speaker from the need to draw a breath at the end of a sentence or long phrase. Silences of shorter duration occur within phrases, between words, and even within words themselves, usually words consisting of multiple syllables (a *syllable* is a consonant and vowel combination treated as a unit of speech).

During a typical conversation between two people, active voice is generated by one person or the other only about 40 percent of the time. Fully half the time comes from one person listening, and the other 10 percent comes from pauses between sentences, words, and syllables. During a telephone call, the listener also pays attention to background noise, not only obvious noise sources such as TVs or radios but a persistent low-level hum. This persistent background noise is known as *ambient sound*. The presence of ambient sound even in the absence of speech allows the listener to realize that the line is still "live" and has not failed abruptly.

Thus human speech is characterized by three major features:

1. A mixture of high-amplitude voiced sounds and low-amplitude unvoiced sounds
2. A mixture of more regular and predictable voiced waveforms and more random unvoiced waveforms
3. Silence nearly 60 percent of the time during a two-way conversation

Any voice digitization technique should take all these voice features into account. And if one voice digitization technique acknowledges only one characteristic and another takes into account all three, then it can be said with confidence that the second voice digitization method is better than the first.

The Analog Telephone Network

Direct voice communication is possible only when people are in close proximity. Shouting from the mountaintop increases distance but usually decreases understandability. And one can hardly shout across the Atlantic Ocean. So a method of voice telecommunication (which just means "remote communication") remained a goal that was only achieved around 1875 with invention of the telephone.

Other forms of telecommunication had existed prior to the telephone, of course. Signal fires, mirrors, and even the telegraph had allowed people separated by greater or lesser distances to communicate. But none of these methods allowed people to just chat away as they pleased. The telephone not only was voice communication, it also was *interactive* communication in "real time," as it happened. Most people are familiar with the basics of the story of how Alexander Graham Bell, a teacher of the deaf in Boston, Massachusetts, and so very familiar with precisely how voice and hearing worked, happened to invent the "harmonic telegraph" that carried voice over a wire using electricity. Fewer people know that Bell set out to invent a way to multiplex (combine) several telegraph messages at the same time over a single telegraph wire as a way to make money to marry the girl of his dreams, who came from a relatively well-off family. The telephone was just a happy accident, and only Bell's highly trained ear was able to realize that in the hisses and squeals of the multiplexed telegraph were all the elements of voice, just jumbled a bit.

It took years to work out all the details, but what Bell essentially had done was to mirror the variations of the human speech acoustical wave with an electrical wave. Bell converted analog information (the voice acoustical wave) into an analog signal (the electrical wave). All the rises and falls in amplitude of the voice wave were mimicked by the electrical wave. The electricity was provided by a simple battery yet could reach miles. At each end, a device converted analog voice into analog electrical waves (the transmitter), and another device converted the analog electrical wave back into analog voice delivered to the ear (the receiver). In combination, the telephone *transceiver* formed the interface between human and network.

Bell discovered that the entire frequency range of the human voice, from about 100 Hz (cycles per second) to about 10,000 Hz, was not needed to make the voice understandable over the telephone. In fact, only a small fraction of that almost 10-kHz range was required. Little power was generated above 4000 Hz, and it turned out that some 80 percent of the entire power of the human voice lies within the 300- to 3300-Hz range.

Figure 2-2 shows what is called the *power spectral density* of voice. This is just another way of looking at the voice waveform. Now, however, frequency forms the horizontal axis, not time, and power forms the vertical access, not amplitude. Technically, the figure shows the *long-term* power spectral density, because at short enough intervals, speech power may not necessarily look like the figure. The power units are in decibels (dB), the standard unit of power gain and loss in a communications system.

Thus there is no need to build an analog network to carry 10 kHz, or even 4 kHz. This is good because the less bandwidth (frequency range)

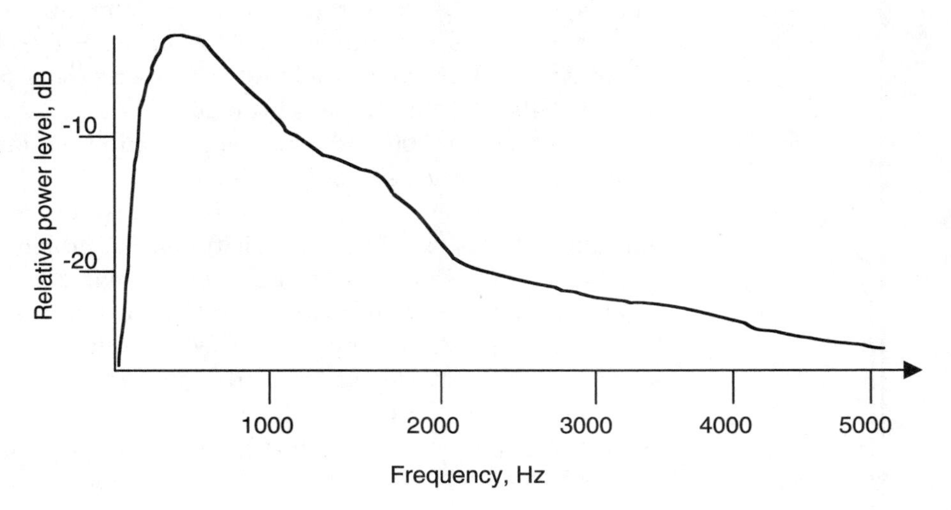

Figure 2-2 The power spectral density of voice.

that a network has to be engineered to carry without drastic power losses, the less expensive the network will be. And more separate voice channels can be multiplexed onto a voice backbone link of a given bandwidth. How far can this process go? In Figure 2-2, the voice power peaks around 500 Hz. Why not just build a network to carry anything from about 100 to 1000 Hz and save even more?

Because eventually voice quality will suffer as the bandwidth is restricted much above 300 Hz and much below 3300 Hz. The voice quality does not deteriorate all at once, and the effect is more pronounced as the high-frequency components are chopped off the voice *passband* (the range of frequencies the network will carry). For example, as the high-frequency passband limit moves downward, voices tend to sound more mechanical and eventually become indistinguishable. Think of the intercom at the local fast-food restaurant, which generally has a much smaller passband than a telephone network. Most people can identify the voices of friends over the telephone. But this might not be true if the voice passband were from 300 to 2700 Hz, and it is in some older and smaller telephone networks. Even with a voice passband from 300 to 3300 Hz, the Bell System standard, if TV show audio or a stereo is heard in the background, the audio sounds distorted. Music over the telephone sounds odd, as most people who have experienced music on hold know. There is no bass (which is far below 300 Hz) and no high-frequency components to singers and strings above 3300 Hz.

What about *expanding* the voice passband to improve voice and audio quality and give everyone more bandwidth? This makes no sense for voice, because any perceived improvements above 3300 Hz are marginal, and there is not far to go between 300 and 0 Hz. Moreover, the 300- to 3300-Hz voice passband is so entrenched in the Public Switched Telephone Network (PSTN) in the United States (in many other parts of the world it is 300 to 3400 Hz) that there is little hope of changing this. Thus any analog signal between 300 and 3300 Hz gets into (and out of) the voice network, and nothing outside this range will.

The telephone operates in a *full-duplex* fashion, meaning both people can talk at the same time. Electricity had no problem traveling in both directions at the same time in these early analog systems, and the electrical waveforms just looked different when two people were speaking. The talker heard his or her own words over the telephone line, in a form of *echo*, but this turned out to be all right. When the receiver was placed over one ear, half of a person's hearing was impaired. The person reacted by speaking louder to compensate, and early telephone conversations quickly turned into shouting matches (which distorts what is said). The speaker echo in early telephone systems provided a cure for this hearing loss, and as long as the echo was not noticeably delayed compared with the speaker's voice, which could be very annoying, everything was fine. Later systems added circuitry to try to eliminate echo (analog echo suppressors and later digital echo cancelers) and added *sidetone* to the telephone handset to pipe some sound directly from transmitter to receiver.

Oddly, rumors were circulating about the invention of the telephone before the telephone actually was invented. Newspapers had to print retractions from one inventor or another saying, "No, I haven't invented the telephone yet." People wanted and needed the telephone badly. Inventors tried things like placing telegraph wires and metal plates directly into the mouths of volunteers and telling them to talk, fortunately without serious injuries. Nothing worked until Bell accidentally spilled the sulfuric acid from an early transmitter on his lap and cried out loudly "Watson, come here! I want you!" It is probably fortunate that Bell was a man a great reserve, or the first words of telephony may have remained unprintable.

Once the telephone was invented for real, therefore, it really took off. People, it seemed, liked to talk. A lot. All of this astonished Bell, who imagined that the telephone would be a useful enough tool for businesses to call suppliers and hospitals to call pharmacies but hardly dreamed that people would want one in their homes so they could call friends and family instead of writing letters to them. Bell was wrong about that, but he did become rich and did marry the girl of his dreams.

The telephone company that Bell founded eventually became American Telephone and Telegraph (AT&T). Until 1984, when a reorganization called *divestiture* occurred, AT&T was really a huge agglomeration of local telephone companies linked by a nationwide long-distance carrier called AT&T Long Lines. But most people thought of AT&T Long Lines as being AT&T itself. This was so because each of the more than 20 local telephone companies that were part of AT&T had individual names, mostly tied to states, such as New York Telephone or Illinois Bell. Because of this naming convention, the term *Bell System* came to be applied to all the local telephone companies (the *Bell Operating Companies*, or BOCs) including AT&T (really AT&T Long Lines).

More details about the rise and fall of the Bell System in 1984, and what it all means to VoIP, will be given in Chapter 3. It is enough to point out here that not every telephone in the United States belonged to AT&T. There were many *independent* telephone companies that offered services mainly in rural and less populated areas, since the Bell System deployed first in major metropolitan areas of the northeastern United States. Chief among the independents were GTE and, later, portions of Sprint. Sprint can trace its roots as a local telephone company back to the Brown Telephone Company in Kansas in 1899. Just like AT&T, GTE and Sprint had their own long-distance networks independent of AT&T Long Lines. But many independents continue to operate with as few as 200 or 300 subscribers. All independents, large and small alike, are now interconnected into one huge national network grid, if not directly, then through long-distance carriers such as AT&T, MCI, and Sprint. In fact, the last independent without links to AT&T went out of business in rural New Jersey in 1945.

Digital Computer Networks

The growth of the analog voice network in the United States was rapid by the standards of the day. By 1941, even after the hard times of the global Depression in the 1930s, more than 40 percent of homes in the United States had telephones. Almost everyone who wanted a telephone could get a telephone, although even local service could be expensive in some cases. Perhaps surprisingly, use of the telephone for routine business purposes was slow to catch on. Businesses generally had an office telephone, but usually only in the manager's office. It was not until the 1950s that an office telephone for each and every worker regardless of job function became a fixture.

By 1962 or so, though, analog telephones linked to the analog links and switches that comprised the PSTN were available to almost everyone in the United States. The analog voice network connected all the analog devices (the telephones). However, in the middle and late 1960s, a new type of device that was not an analog telephone began to be linked to the PSTN. This was, of course, the digital computer.

Today, the term *digital computer* seems redundant, but it is worth pointing out that there are *analog* computers as well as digital computers. Analog computers are mostly specialized devices today, odd machines with gears and dials that show the phases of the moon or the height of the tide. When used alone, the term *computer* means *digital computer* unless more information is given.

Digital computers can only handle zeros and ones in the form of binary digits, or bits. In contrast to analog information, with any valid value between maximum or minimum, digital information takes on only one of two values, 0 or 1. The power of digital information lies in the fact that long strings of 0s and 1s can represent almost anything, as users of computers that show movies and play audio CDs can testify.

The first practical digital computers were developed as part of the war effort in the United States in the 1940s. The first commercial digital computers became available in 1951 and were popular with utility companies that had to handle large numbers of accounts, customer bills, payments, and so on. Many companies were already familiar with IBM business machines designed for this magnitude of record keeping, so when IBM introduced its own line of computers in the 1950s, the market was already there.

The idea of using a network to connect computers, however, took a while to catch on. The government computers had no general need to talk to each other, although computers used for military purposes had to and did. But as for making ordinary commercial computers within a company talk to each other, that was a different matter. It was the airline industry in the United States that led to the rise of the first successful network of digital computers in the world: the SABRE system.

The end of the war brought prosperity to the United States, which alone among the heavily industrialized nations of the world remained unscathed by the world war ending in 1945. Not only was industrial production geared up for the war effort, but the people of the rest of the world had few places to turn to buy the goods they needed to rebuild their own shattered economies. Thus the U.S. economic sectors boomed in heavy equipment, manufacturing in general, housing for the hordes of new workers, and most important for computer networks, travel for the newly mobile population with money to spend on work and play.

The rapidly increasing pace of commerce and life demanded faster methods of travel than the pokey railroad could provide. Steam locomotives, common throughout the 1950s, required stops for water every 50 miles and for coal every 100 miles. Overnight trips from New York to Chicago and 3 days to San Francisco were too slow for the needs of the times. Where the railroads failed, the commercial passenger airlines stepped in. In 1926, a year that saw the railroads carry about 1 billion passengers, the commercial airline industry, with 20 planes and 300 seats, carried only 5800 paying customers. By 1956, however, the automobile had so eaten into rail travel that the airlines, carrying some 40 million passengers, actually handled more travelers than the struggling rail industry. Most people could see it coming. Even in 1946, the 16-hour railroad ride to Chicago from New York by train took only 4 hours by air.

The problem was that handling the reservations required for this deluge of air passengers was overwhelming the airlines by the late 1950s. Overbooking led to irate passengers and was hardly a way to encourage an industry (as late as 1971, fewer than 50 percent of people in the

United States had ever flown on an airplane). Computers seemed like the answer. As early as 1952, American Airlines used a computer called the *Magnetronic Reservisor* to help sort out its reservations mess.

In 1953, C. R. Smith, president of American Airlines, and R. Blaire Smith, an IBM senior sales representative, met on a flight from Los Angeles to New York. It took a while, but by 1959, American Airlines and IBM had agreed to develop the Semi-Automated Business Research Environment (SABRE) system to bring computer power to the airline reservation system. The first SABRE computer, installed in Briarcliff, New York, near IBM headquarters, in 1963 handled 84,000 telephone calls a day. But the plan was to eventually put SABRE computers all over the United States and network them together, which was done during 1964.

Once networked, the SABRE computers could handle flights that originated in New York and stopped in Chicago and Kansas City before heading for Los Angeles by way of Phoenix. People got on and off at each stop, of course, and a networked system was needed to coordinate it all but still allow tickets to be sold locally. Once SABRE could display more than just American Airline flights (United soon had its own APOLLO system as well), and the travel agencies got onto the system in May 1976, computer networking became relatively common and accepted. Most people's first contact with a computer network was at the local travel agency.

Analog Links and Digital Links

The SABRE system used ordinary analog telephone lines to transfer digital information between computers. Since telephone lines normally expected to carry analog information (voice), new techniques had to be developed to allow digital information to cross the analog links. Since only analog signals between 300 and 3330 Hz (the voice passband) could enter the analog telephone network, something was needed to "make the data look like voice." In other words, the only way that digital 0s and 1s could cross the voice network was as *sound*.

Strangely enough, around the same time that the SABRE system was using analog lines for digital information, the telephone companies were starting to add *digital* links to the PSTN. Digital links, unlike analog links that allow a signal of any value between a maximum and minimum, only allow a limited number of signal values. The discrete signal values on the line represent digital values. In its simplest form, a digital

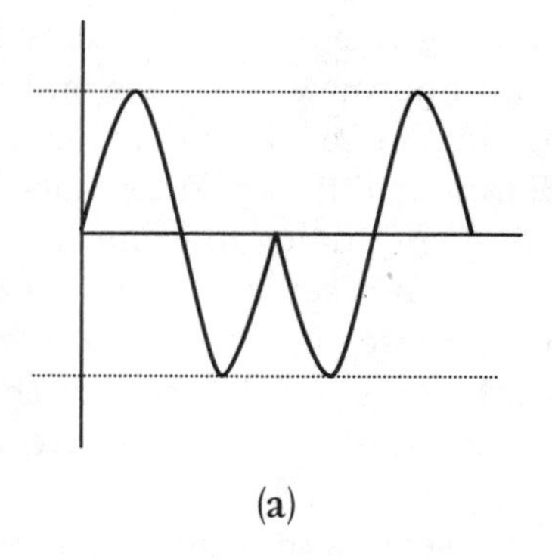

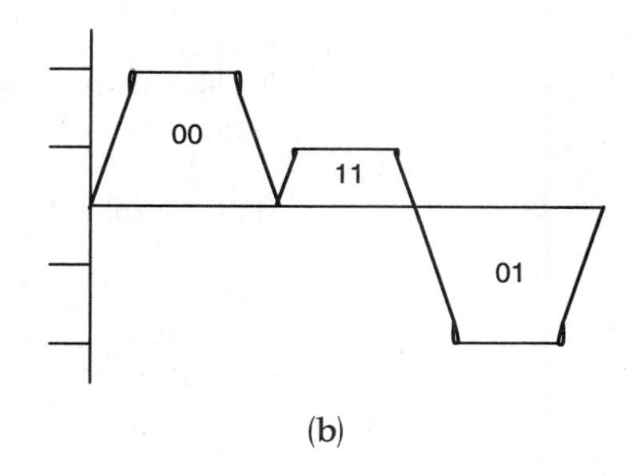

(a) (b)

Figure 2-3
(a) Analog link where any value in a range is valid.
(b) Digital link where signal values represent 0s and 1s.

link used one signal value (line state) to represent a 0 bit and another signal value to represent a 1 bit. More advanced techniques can represent a whole string of 0s and 1s, perhaps 8 or more, with only one signaling value or line state. The difference between an analog link and a digital link, using amplitude alone to distinguish digital signals, is shown in Figure 2-3.

For the most part, however, during SABRE's development, the portions of the telephone network used by customers to access the digital links remained analog. Thus, despite the presence of digital links deep within the telephone network, access was still by means of an analog line.

Ironically, the interface device for computer networks that allowed digital information to become analog sound and get it back again already existed. It was called the *modulator/demodulator*, or *modem* for short.

The Modem

Legend has it that during the early days of the last world war, in 1941, Bell Laboratories, then the research and development (R&D) arm of the AT&T Bell System, wanted to demonstrate the latest in digital computing technology at a conference at Dartmouth College in New Hampshire. However, shipping the computer on a truck from New Jersey posed several problems. The machine was big and bulky, and there were real concerns about the risk of saboteurs and spies along the proposed route.

What Bell Labs came up with was a way to ship only a teletypewriter terminal with a keyboard for input and roll of paper for output. Over a regular analog AT&T telephone line, the terminal could use a device that Bell Labs called a *modem* to modulate the digital information into quasi-analog signals over the analog line and demodulate arriving quasi-analog signals back into digital information for use on the terminal. Modem signals are *quasi-analog*, in the sense that modem signals are analog sounds, to be sure, but distinctly different from voice because they represent digital information and not analog information (voice). Anyone who has picked up a telephone while someone else was on-line has heard these quasi-analog signals. However, use of the *quasi-* prefix has been dropped over the years. Modems at each end of the analog link were needed. In this way, the computer remained secure while remote access was allowed.

The number of times per second that an analog link can change line conditions per second is called the *baud rate*, for Baudot, an early telecommunications pioneer. Emile Baudot, in France, invented the 5-bit *Baudot code* for early teletype machines in 1875. In Great Britain, this is also called *Murray code*.

The line conditions that can vary to represent digital line states are only three in number. These are the amplitude, phase, and frequency of the electromagnetic wave on the wire itself. In practice, one of these line conditions is held constant, while two of them vary to represent the 0s and 1s. For example, it is common today to generate a carrier wave at a certain center frequency and vary both the amplitude and phase to represent 0s and 1s. Almost all modems operate in this fashion today.

The modem languished until computer networks became more common. Today, no one would even consider buying a PC without making sure the PC came with the latest and greatest in modem technology. Early modems in the 1960s were the size of a modern microwave oven and ran at only 110 b/s (bits per second). As recently as 1991, state-of-the-art modem speed was 9600 b/s (9.6 kb/s) and not 56 kb/s, as is the norm today. And the 1991 modem cost about $800.

The Codec

Thus the rise of computer networks in the 1960s made the modem a common device for computer networking. However, the 1960s also saw the first large-scale efforts to digitize voice in order to send the digital voice over a digital communications link. Digital links used digital signals instead of analog signals. But what about the information sent across the digital links? The information could be analog (voice) or digital (from a computer). How could analog voice be sent over a digital link?

Enter Bell Labs, the inventors of the modem. Scientists and engineers now came up with a *coder/decoder*, or *codec*, a kind of modem in reverse. That is, instead of digital information to or from an analog line, the codec took analog information (voice) to or from a digital line.

Codecs were used to digitize many voice calls and send them over two pairs of unshielded twisted-pair telephone wire, forming a digital link and so added multiplexing to the mix. The device that included the codecs and also multiplexed was called a *channel bank*. (Since multiplexers also demultiplexed incoming channels, it is a wonder, but a relief, that they were not called *muxdemuxes*.) This was the first application of digitized voice in the PSTN.

More details on this new addition of digital links to the PSTN will be presented in Chapter 3. One other possible combination needs to be mentioned here, however. What is needed when a digital computer generating digital information is linked to another digital computer over a digital telephone link? This is where the channel service unit/data service unit (CSU/DSU) comes in.

The CSU/DSU

Digital links allow only a limited number of line conditions, or signals, to represent information flowing on a link. If analog information is to be sent on a digital link, this is where the codec comes in to convert from analog to digital and back again.

But what about when the digital information that emerges from the serial port on a PC (for example) needs to be sent not over an analog link but over a digital link such as an Integrated Services Digital Network (ISDN) line? At first glance, it would seem that no interface device would be necessary at all. But this is not the case. Without going into all the technical details, which are readily accessible in many basic data communications books and Web tutorials, the fact is that the way 0s and 1s are represented as they emerge from the serial port on a PC differs in many ways from the way that 0s and 1s are represented on an ISDN line. Now, there are several ways to represent bits at the serial port and several ways to represent bits on a telecommunications link. But all the various ways acknowledge a couple of basic facts. First and foremost, the serial port bits are not intended to be sent for many miles, but less than 100 ft or so. Second, the serial port bits do not have to worry much about signal loss over such a short distance. Third, there is not much risk of interference due to noise on serial port bits. There are also some other differences, but these are the major ones.

When all these things are considered, it just makes sense to use one type of digital coding for bits on a serial port and another type altogether for bits on a WAN wire. Both bits streams are still serial, but they are coded differently.

What device converts from digital to digital and back again? Awkwardly, the device is called the *channel service unit/data service unit* (CSU/DSU). The terms began when the two devices actually were separate boxes and have stuck ever since. Usually the CSU/DSU (sometimes seen as DSU/CSU, the order hardly matters anymore) is all one device. In the original packaging at AT&T, the DSU handled the conversion of the analog voice channels to digitized voice and also handled the conversion of the 0s and 1s to a form suited to the digital link. The CSU was a device that boosted signals, provided a place to loop back the signal from the network, and allowed related "housekeeping" functions.

Today, the CSU/DSU is one device that interfaces with the network interface device (NID) that forms the demarcation point between network and user device. Thus the CSU/DSU is customer premises equipment (CPE). The most familiar corporate CSU/DSU package is the T-1 multiplexer used to send and receive digital information at 1.5 Mb/s. The CSU/DSU hooks up to a serial port or some other form of data termination equipment (DTE). Residential and Internet users are familiar with the ISDN "modem," which is actually a CSU/DSU designed for use on an ISDN digital residential line (ISDN BRI). The typical functions of the CSU/DSU are shown in Figure 2-4.

It should be noted that the CSU loopback is typically to the network side of the device, through the NID. The DSU loopback is usually to the DTE (PC serial port). However, there are exceptions.

Figure 2-4
The CSU/DSU.

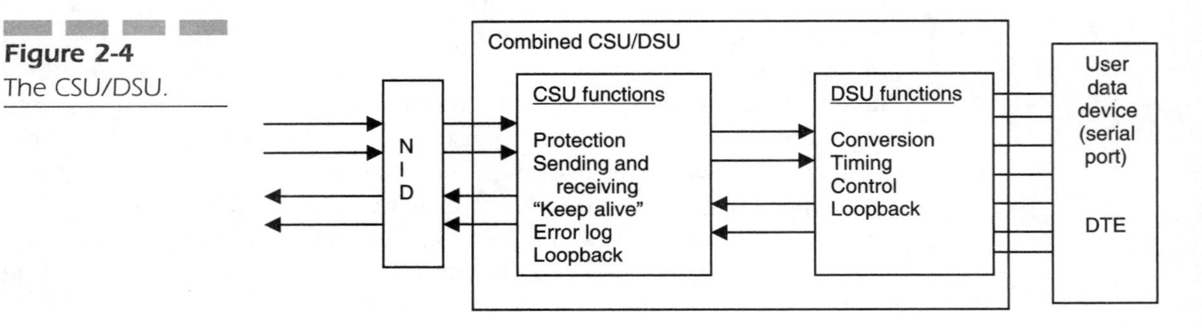

Putting It All Together

Modems, codecs, CSU/DSUs. When to use what? Figure 2-5 shows how it all fits together. There is only analog and digital information (voice and data). There are only analog and digital lines. Today, almost all analog links are the access lines used in residential telephone services, except for many cable TV systems, which are mostly one-way anyway.

This exercise actually has been quite important to understanding VoIP. This is so because most people use analog lines not only to talk on but also to link their digital PC to the Internet with a modem. However, what if people begin to hook their telephones to the PC and then to the Internet? Then the modem has to employ a type of codec behind a modem on the PC side to encode the analog voice as bits in the first place, except when the access line is digital, as with ISDN BRI. Then the codec feeds a CSU/DSU. Without a basic feel for the terms, the discussion can quickly break down.

The main point goes beyond terminology, though. Why bother to digitize voice if the analog voice works just fine at home to begin with? As it turns out, there are sound reasons (pardon the pun) to digitize voice in the first place, and few of them have anything at all to do with better Web and Internet integration.

Figure 2-5
Interface devices.

Line is	Information is	
	Analog	Digital
Analog	Network example: PSTN voice Interface device: *Transceiver*	Network example: PC Internet access Interface device: *Modem*
Digital	Network example: T-1 voice Interface device: *Codec*	Network example: T-1 data, ISDN Interface device: *CSU/DSU*

Why Digitize Voice?

Analog voice sent across the PSTN began to be digitized (sent on digital links) in the 1960s, although the technique had been around since the late 1930s. The amount of effort to perform this digitization given the state and price of computer processing power was considerable. Today, people are used to having a small number of microprocessor "chips" that can be held in the palm of a hand do almost anything. These *chipsets* form the basis for all network technologies today. There were no chipsets back then. There were no real microprocessors, simply because the foundation of that whole industry, called *very large scale integration* (VLSI), did not occur until 1967 or so (accounts vary). There were miniprocessors and macroprocessors, however. These clunky boxes, about 7 ft tall and a couple of feet wide and long, allowed digitized voice to happen in the first place.

Today, digital voice is taken for granted. Back then, the decision to digitize voice was not a given, and powerful arguments had to be marshaled to convince everyone to spend good money on voice digitization. Bell Labs put together an exhaustive list of nine benefits of the digitization of voice. It is a good idea to review them here not only because they still apply to VoIP but also because digital voice has become so common that few people ponder all the reasons it was done in the first place.

Here is the list. When voice is digitized:

1. Multiplexing is easier.
2. Signaling is easier.
3. PCs and other computers can be used.
4. The voice switch becomes a big computer, and all lines are digital.
5. Lines are not as noisy.
6. Line quality can be monitored closely.
7. New services can be provided.
8. Lines are more tolerant of noise.
9. Encryption is possible.

Each of these points needs a few words of explanation. Some and even all of these points appear in many other books and articles on digital telephony, but usually in more technical terms. This section adds some of the detail and examples often missing in those other accounts.

Multiplexing Is Easier

This reason for digitizing voice comes first for good reason. Multiplexing is just the art of sending multiple signals along the same physical path. Telephony started with efforts to multiplex telegraph signals on the same line. Obviously, there is a financial incentive for combining signals, as long as the equipment at each end required to multiplex the signals is less than the cost of installing new links.

This is almost always the case, since construction costs are always rising, or at best staying about the same, while the costs of digital end electronics are constantly and rapidly falling. Currently, the price to install new physical plant in the PSTN, most often in the form of fiberoptic cable, is about $40,000 per mile, or about $7.50 per foot.

Analog voice channels have been multiplexed for years over analog circuits and cables (analog *carrier systems*). But when analog voice is digitized, multiplexing is easier to accomplish. Why? In contrast to digital electronics, analog electronics (which admittedly are less common today) prices are more stable over time, falling only slowly. The bottom line is that digital multiplexing equipment is less expensive than analog multiplexing equipment and can be used as long as the voice being multiplexed is in digital form.

Signaling Is Easier

In this context, the term *signaling* means call control messages. Call control messages set up and take down telephone calls, determining the number that was dialed and how to route the call through the PSTN to set up the connection. All telephone calls on the PSTN are connection-oriented; a stable path must be established through the network from source to destination before communication can take place.

Thus signaling can be the *off-hook* signaling "message" generated when an analog telephone is picked up and draws a dial tone. Another signaling "message" is the *on-hook* or hang-up signal that terminates a call. Each dialed number is even a little signaling "message" all its own. The term *message* appears in quotes because each of the examples, when used on an analog telephone line, is not really a message at all in the sense of a data packet. Analog signaling is done with either electric current flowing (on-hook, off-hook) or sounds (*touch-tone* dialing). Even the older *pulse dialing* (click-click-click-click-click) was just a broken stream of electric current. Whole banks of equipment in the telephone company central office were needed to interpret these analog signals to complete calls and bill the customer correctly.

When voice is digitized, however, so is the signaling. On-hook, off-hook, dialed numbers, and so on are now strings of 0s and 1s. In fact, the signaling on a digital link is often placed into packets, just like Internet data. Even if the digital signaling is not packetized, it is much easier to deal with signaling messages that are in digital form. This is so because all the switches and other equipment in the telephone company central office are now just big computers that can handle digital signaling much more effectively than their analog counterparts.

PCs and Other Computers Can Be Used

Digitizing voice makes the stream of 0s and 1s representing speech look the same to the network as computer data. Thus specialized equipment is not always needed to handle voice, as it was in the days when all voice was analog on the network. Any general-purpose PC platform will do, or UNIX/LINUX workstation, as long as the "horsepower" is there to handle the voice traffic. And without voice digitization techniques, the common PC could not be used as a fax machine, answering machine, or speed-dialing system with the same ease as today.

Digital voice allows the PC to act as a telephone today. And instead of adding video or data capabilities to a telephone, as ISDN and other schemes planned, it is much easier to add voice capabilities to the PC. There are still some tricky issues to resolve. For instance, PC use could be easier for people without technical backgrounds, but major efforts are always underway to make the PC more accessible to this portion of the population.

The Voice Switch Becomes a Big Computer, and All Lines Are Digital

ISDN ushered in a whole new era of digital central offices that were basically big computers. By the late 1980s, a complete, but tiny, ISDN central office switch could be implemented on a PC platform. The dozen or so lines on the ISDN PC switch were all digital as well, using simple serial interfaces, and the package was positioned for small telephone companies.

For years in the 1980s, the workhorse ISDN voice "switching system" of the AT&T system and former pieces thereof was based on a 3B20 central processor. It could have easily been called a "digital voice computer," but AT&T could not market a computer because of regulation restrictions and was reluctant to even label anything a computer. The 3B20 was a minicomputer (no PCs had appeared in mass yet) and very powerful. The 3B20 ran the UNIX operating system, an AT&T invention, and used C language (not yet C++) to implement all voice functions, from call routing to interfacing with billing systems to multiplexing. In other incarnations, the 3B20 provided programmers with a timeshare development platform and package. When applied to voice, however, it became the "electronic switching system."

Needless to say, installing a "central office software package" on a PC or minicomputer is a lot more cost-effective than spending millions on a huge, special-purpose hunk of hardware that is dedicated to only one task. Once voice is digitized, the advances in digital computing translate directly to the telephone system.

Lines Are Not as Noisy

Digital voice was invented originally to solve the problem of noise on analog telephone lines. Analog signals of all kinds, not just voice, are much more susceptible to noise than their digital counterparts. This is so because an analog signal's values are valid anywhere between a minimum and maximum. If noise, which is just an unwanted signal, not sent by the originator, is added to the analog signal, then a receiver or amplifier on the line has no way to distinguish between the original waveform and the waveform with the noise added. Thus the noise is received or amplified right along with the information-carrying signal. It is said, therefore, that noise on an analog line is *cumulative* or *additive*.

Digital links operate differently. Only a limited number of line states are permitted, and these always represent 0s and 1s. Noise still occurs on digital links, of course, but the effects can be minimized because the noise is limited to less than half the difference between the line states. For example, if an amplitude of +3 V represents the bit combination 01 and an amplitude of +1.5 V represents the bit combination 11, then impulse noise of less than about 0.75 V plus or minus will still allow the receiver to say something like, "Oh, the signal is +2.0 V, but that's close enough to +1.5 V to call this a 11."

Digital links do not properly have amplifiers but rather *repeaters* or *regenerators* that "clean up" the digital signal before sending it on its way. In the example given, the repeater would send out the digital signal in that bit slot as +1.5 V, thus eliminating the noise that would otherwise accumulate on an analog link until the noise totally swamped the original signal.

Most people became aware of the superior quality (clarity) of digital voice through a series of brilliant television advertisements from Sprint. The "Hear a pin drop" campaign was very successful, and some spots even featured shots of exploding microwave towers, a mildly bogus point because digital microwave was replacing analog microwave carrier systems with many voice channels even as the spots were airing.

Line Quality Can Be Monitored Closely

This feature of digital voice is a consequence of the last feature. Analog voice quality delivered over a telephone network is a purely subjective measurement. Now to a large extent this is still true of digital telephony: How does the voice *sound* to the ear? There are several standards, all of which ultimately revolve around people in a room listening to "telephone calls."

The advantage of digital voice in this regard is that noise that exceeds the line-state limits is manifested as bit errors. And bit errors can be precisely measured and monitored by all components of the digital network. Thus particularly noise links or equipment can be isolated more easily and replaced or repaired. And a system that reproduces 99 percent of the digital voice bits correctly at the receiving end is, by definition, better than a system that reproduces only 97 percent of the bits correctly (actually, both systems are pretty horrible). The point is that digital error information can be used independently of subjective methods to determine system voice quality. This idea also can be used as a design tool (built for 99.99 percent accuracy), a management tool (use the best links for this connection), a repair service tool (preventive maintenance on links with rising error rates before complaints start), as well as a monitoring system for contract or regulatory compliance.

New Services Can Be Provided

Many of the types of services that digital voice systems can provide are difficult, if not impossible, to implement with analog voice systems. For instance, a digital voice switch can provide voice mailboxes for subscribers. Analog voice systems cannot, unless someone actually would consider supplying an analog cassette tape like those found in home answering machines to record messages for customers. Once digitized, voice is just another stream of bits that can be stored on a computer hard drive like any other file. If the voice switch is a "digital voice computer," this new service becomes just a matter of adding enough storage space for the mailboxes.

However, regulation has hampered the deployment of many of the services possible with digital voice. Telephone companies traditionally have been limited to simple "carriage" of information from point A to point B without processing or altering the information in any way in between. This is one reason telephone companies have been called *carriers*.

New services provided on digital links, however, can be based on digital information in general, not just digital voice. In one sense, Internet access to an Internet service provider (ISP) over an ISDN link is one example of a new service that can be provided easily by digital telephone lines designed for digitized voice. Color faxing, video telephony, and related services are all further examples of the types of new services envisioned for digital voice systems. What regulation limits have done is allow many of the anticipated services to appear on the Internet, which continues to be unregulated at present.

Lines Are More Tolerant of Noise

This is just another aspect of the line-monitoring feature of digital voice mentioned earlier. Noise levels that would reduce an analog signal to gibberish can be tolerated on digital links. And the noise on an analog system is most noticeable during periods of silence. This is so because analog silence is transmitted at a much lower power level than active speech. Thus noise levels on analog systems must be kept very low.

On digital links, however, silence is just coded up and sent as a string of 0s and 1s, just like active voice. Thus the silence power level is the same as the active speech power level when digital voice is used. This means that noise has less of a chance of disrupting the conversation, even at relatively high levels.

More will be said later about silence suppression, which eliminates periods of silence from the bit stream. For now, it is assumed that every moment of a conversation is expressed as a stream of bits, even extended periods of silence at one or both ends of the conversation.

Encryption Is Possible

Security and privacy have become more and more of a concern in all networking situations. Purely analog voice systems relied on special voice *scramblers* to try to jumble up the sound at the sending end so that eavesdropping became more difficult between source and destination. But scrambling could only operate with very restrictive parameters, or else the receiver could not descramble the analog voice enough to make it understandable. The same analog noise limitations applied to intentional scrambling techniques. Thus, with enough effort, most analog scrambling could be understood by an eavesdropper.

True voice encryption is possible once voice is digitized, however. All the 0s and 1s can be jumbled as much and however the sender pleases, as long as the sender has enough processing power to implement the encryption technique and the receiver knows the method used (and has enough processing power as well). The same advanced methods that protect electronic commerce (e-commerce) on the Internet can now be applied to voice.

The Digital Signal Processor

All the advantages of digital voice over analog voice detailed earlier are implemented today using the *digital signal processor* (DSP). These chipsets (groups of microprocessors bonded to a common board) enable advanced signal processing to take place more cost-effectively than ever before. Signal processing is just the act of transforming or enhancing a signal, and DSPs operate on digital signals, naturally, but signal processing in general can be used in many analog applications, such as modems.

Signal processors can be used to modulate a signal in the form of an analog or digital waveform and also to amplify, filter (selectively pass or block certain frequency ranges), or perform equalization ("smoothing out" certain electrical characteristics). DSPs use digital logic to operate on digitized waveforms. In some cases, an analog signal is digitized just to allow a DSP to act on it (to remove noise, for example) and then converted back to an analog waveform.

Since DSP chipsets are in a sense little computers all by themselves, DSP chipsets have their own memory (storage) and processing unit and run programs to make the DSP perform the required tasks. This programmability is a major attraction for DSP applications. Now the DSP program code is loaded into a DSP chipset like a program into a PC. DSP *microcode* (programs appropriate for microprocessor chips) is "burned" into the chip during the programming process and generally is not changed. This makes the DSP very versatile and adaptable to many applications, such as VoIP.

The main applications of DSPs today are in four major areas, the last of which is of direct concern to VoIP. First, DSPs are used today in *echo cancelers*, which are special electronics that eliminate the electrical "echoes" that circuits pick up mainly from differences in electrical impedance from one part of a telecommunications link to another (e.g., where a local loop or access line enters a voice switch). Next, DSPs are used to interpret the dual-tone multifrequency (DTMF) "touch tones" that are generated whenever a nonrotary, pushbutton telephone is dialed. These DSPs act once the analog DTMF sounds are converted to digits, which are what the telephone numbers are, after all. Third, DSPs are used in all high-speed modems today. All modern modems do their own equalization and signal conditioning (minimizing the effects of line characteristics such as *envelop distortion*). Before DSPs became powerful enough to do so, the line itself had to be conditioned to allow reliable "high speed" operation (9600 b/s in those days). Finally, DSPs are used to implement the *low-bit-rate speech encoding* that is a characteristic of VoIP operation. In many cases, the DSPs operate with 64-kb/s PCM (pulse code modulation) voice as input and produce voice at 8 kb/s or even lower rates without periods of silence ready to place inside a string of IP packets.

Thus a lot of 8-kb/s voice starts out as plain old 64-kb/s PCM voice. To understand how the VoIP DSP converts 64-kb/s PCM voice to compressed, silence suppressed 8 kb/s or less, a brief look at where 64-kb/s PCM voice comes from is needed.

Three Steps to Digits: Sample, Quantize, and Code

The voice digitization process that produces 64-kb/s PCM voice takes place in three basic steps. The steps are to *sample* the input analog waveform frequently enough to accurately reproduce the signal, *quantize* the samples to produce the 0s and 1s that represent the analog waveform as a string of bits, and finally *code* the quantization bits into a form appropriate for transmission over a link that might span many miles.

Actually, there are two main ways to produce digital voice. The first way, and for a long time the only satisfactory way, to produce digital voice was to model the analog waveform as closely as possible with a string of 0s and 1s. This is the 64-kb/s PCM approach. But this is not the only way to produce digital voice. It is possible to look at the analog waveform as a whole and attempt to code not arbitrary waveforms but only those waveforms which occur naturally during speech. Devices that model speech are called *voice coders*, or just *vocoders*. This section will deal mainly with waveform modeling, and a later section in this chapter will detail voice coding methods, which are the ones mainly used in VoIP. But both methods should be understood.

Of course, the three steps in sequence produce 64-kb/s PCM voice from analog voice at the sender. But analog-to-digital (A/D) conversion is only half the digital voice story. At the receiver, a digital-to-analog (D/A) process converts the arriving bits back into sound. However, the complexity of digital voice is on the A/D end, at the sender. It is a relatively straightforward process, although by no means trivial, to take the arriving bit stream and generate a sound that corresponds to a given sequence of bits.

The details of these three digitization steps have been determined by mathematicians and electrical engineers in a series of basic and complex formulas that is the common language of math and engineering. Many digital telephony texts from this point forward consist of graphs, tables, and formulas showing precisely how each can be implemented to minimize distortion of the voice signal. This treatment need not be as detailed or mathematically rigorous. However, there is a lot that can be said about each step to illuminate how voice digitization works without turning the balance of this chapter in an engineering textbook requiring at least some higher mathematics and engineering concepts to understand completely.

The approach here is more in the form of analogy and common-sense interpretations of engineering concepts. This sacrifice of rigor in the name of accessibility should be forgivable, however, for the purposes and intended audience of this book.

Sampling

In order to model an arbitrary waveform, the first thing that needs to be done is to establish a set of times when the input analog waveform is to be sampled. Typically, the sample intervals are spaced regularly at precise intervals, although there is no mathematical reason that this uniform sampling time must be fixed. It is just easier this way. It is a proven mathematical theorem that if the sampling of the analog waveform takes place often enough, the input analog waveform can be recovered completely at the other end of the link. All that need be done is to use an electronic circuit called a *low-pass filter* to "smooth out" the digital samples generated during the sampling process.

The sampling process generates what are known as *pulse amplitude modulation (PAM) samples*. Although more needs to be done to send the PAM samples over a WAN link, the PAM samples are all that are needed to reproduce the input waveform, as long as there are enough PAM samples to "connect the dots" at the receiver. The basic idea is shown in Figure 2-6.

The term *PAM* means that the sampling pulse train of constant amplitude is modulated to model the analog waveform. Therefore, this is pulse amplitude modulation.

In 1933, Harry Nyquist established the minimum sampling rate needed to reproduce an analog waveform. This *Nyquist sampling rate* is set at twice the highest-frequency component of the analog input waveform, sometimes expressed as twice the bandwidth, where *bandwidth* is just a measure of the frequency range of the input signal. Thus, if an analog voice signal reaching up to 3400 Hz is to be sampled at the Nyquist rate, the sampling frequency must be at least twice that, or 6800 Hz, or samples per second.

Sampling does not have to be done at the Nyquist rate. The Nyquist rate is a minimal requirement to reproduce the input waveform, but sampling can be done at rates higher or lower than the Nyquist rate. If sampling takes place at rates lower than the Nyquist rate, the result is distortion of the waveform known as *aliasing*. Aliasing just means that there is more than one output waveform that fits the "connect the dots"

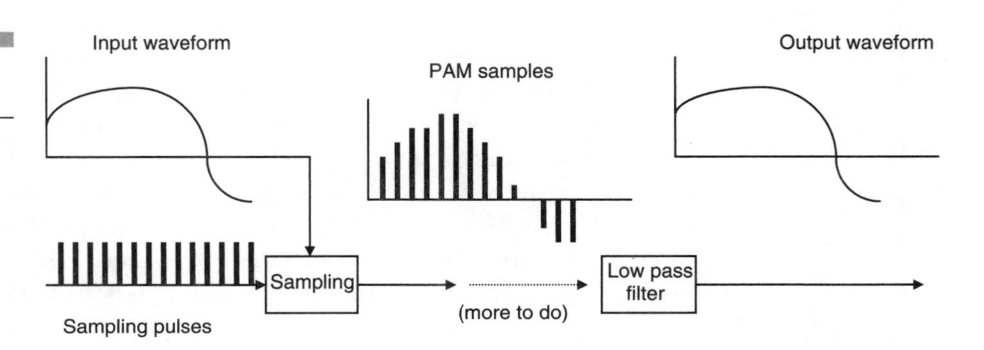

Figure 2-6
PAM samples.

pattern of the samples. There is no aliasing at the Nyquist rate and above.

Why sample above the Nyquist rate? Because then there is *more* information than needed at the receiver to reproduce the input waveform. Thus, if some samples are corrupted by errors, the samples can be discarded if detected and the analog waveform reproduced anyway. Digitization of analog waveforms, whether for voice or for music, that are sampled above the Nyquist rate are said to be *oversampled*.

For analog voice, the sampling rate is set at 8000 Hz (or 8000 samples per second). This sets the voice bandwidth at 4 kHz, which is larger than strictly required. However, a 4-kHz bandwidth for voice digitization has several advantages, one of which is the fact that 4 is a power of 2, and most digital processors are based on the binary (base 2) numbering system. Thus 64-kb/s PCM voice generates PAM samples once every 8000th of a second, or 125 μs (microseconds).

PAM samples conceivably could be put directly on a link and sent to a receiver. However, in practice, this is rarely done, except for short distances. PAM samples do not travel long distances without significant distortion. However, some PBX systems have used PAM samples directly.

It should be noted that PAM samples are not really digital. PAM samples, like analog waveforms, can take on any valid value between a minimum and a maximum. A PAM sample can just as easily be +6.5473 as +6.5472, or many values in between. There is nothing inherently digital about PAM samples. Thus the digitization process actually has just begun with PAM samples. It is still necessary to produce a string of 0s and 1s to represent the PAM samples. In PCM voice, this is the job of the quantization step.

Quantizing

In reality, the analog signals that produce 64-kb/s analog voice are digitized twice. The PAM samples are first converted to a string of 0s and 1s through the quantizing process, and then they are converted once again into a form suited for long-distance transmission over a digital telecommunications link. It is also possible to quantize and then use a modem to send digitized voice over an analog link, but this possibility is not considered here. Not only is this rarely done today, but 64-kb/s PCM voice was developed for digital links called *DS-0 links*. DS-0 stands for *digital signal level 0* and forms the basic building block of all digital voice multiplexing systems.

This section deals with the quantization step. It is the quantization step that actually produces the pulse code modulation (PCM) "words" that express analog PAM samples as PCM sets of 0s and 1s. The quantization step digitally encodes the PAM samples and so forms the true A/D conversion step of the whole process.

Each PAM sample is used to produce a string of 0s and 1s. There are only so many 0s and 1s that can be used to represent a given PAM sample, however. If PAM samples are generated 8000 times a second, then the bits used to represent the PAM sample must be sent in 125 μs, or 1/8000 of a second. This means that if each PAM sample were represented by 16 bits, the digital link would have to run at 128 kb/s (8000 × 16) just to get rid of them all. Obviously, since 64-kb/s PCM voice runs at 64 kb/s, each PAM sample is represented by exactly 8 bits. This is nice because 8 is also a power of 2, 8 bits make a byte or octet, and the early DSPs used to digitize voice were 8-bit processors, meaning that they always grouped 8 bits together to act on anyway.

However, consider using 3 bits to represent PAM samples, as shown in Figure 2-7. Something strange seems to be happening. First of all, it

Figure 2-7
Quantizing PAM samples.

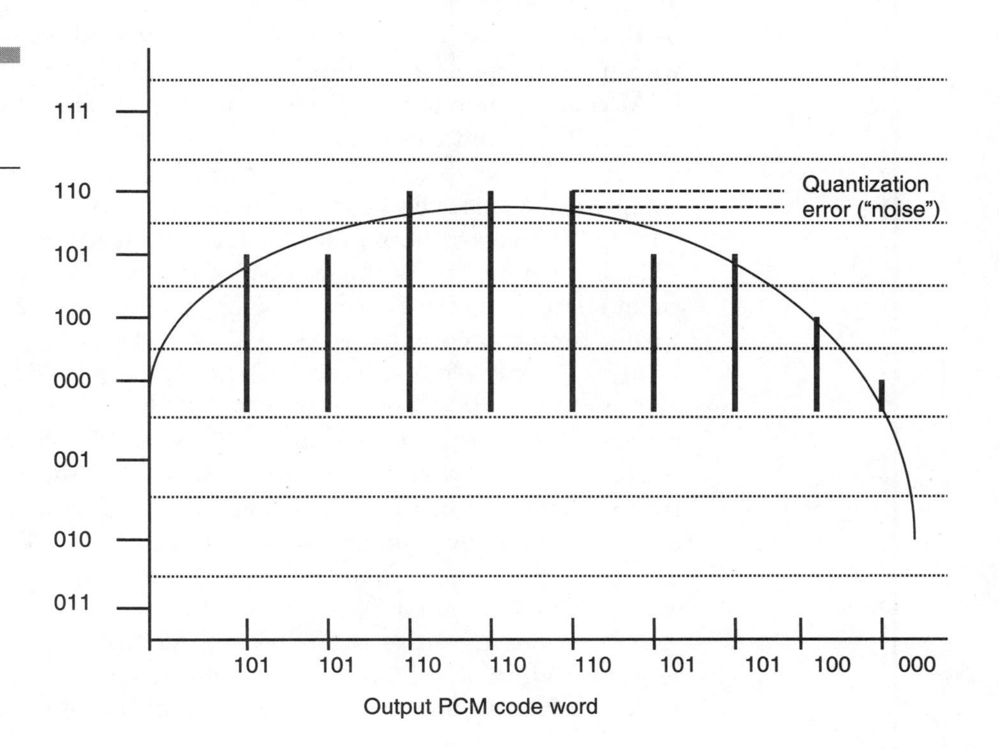

should be noted that the digital "steps" used to encode the PAM samples must cover the whole range of analog values, from maximum to minimum. Since there are only 3 bits per sample, there are only 8 levels or steps in the entire PAM range. In the figure, the PAM samples are not shown, but the input analog waveform is shown, for the sake of clarity. Note also that the values on the vertical axis are not in numerical order. This is done intentionally. It is best to arrange the levels so that only one bit position differs in value between adjacent steps. This is so that single-bit errors on the transmission link will only cause an error of one step at the receiver, which is not too bad considering that each PCM word only represents 1/8000 of a whole second. Therefore, there is no numerical relationship between PAM sample and PCM word other than that this particular PAM sample produces that particular PCM word.

The main point of the figure, however, is not really how the quantization steps are arranged. The real point is the fact that the constantly varying analog input, or rather the PAM samples that represent it, produces a series of PCM code words that are the same value. In other words, many PAM samples will generate the same PCM word, even though the PAM samples have different values. This is a natural consequence of having a limited number of quantization steps as determined by the number of bits used in the PCM word. All sample values falling within the range of the quantization interval must be represented by the PCM value located at the center of the step, as shown in the figure.

Thus the quantization step must introduce some distortion or error into the PAM sample series. This error is known as *quantization noise*. Quantization noise can be minimized by increasing the number of steps, but this increases the number of bits required for each PAM sample. This, in turn, increases the number of bits per second the link must get rid of just to keep up. However, as long as the quantization noise is kept within limits, the digital voice quality will be acceptable. For 64-kb/s PCM voice, only 8 bits are available to represent PAM samples.

Sometimes the irony of sampling at the Nyquist rate to precisely reproduce an analog waveform only to introduce quantization noise at the A/D step that prevents exact reproduction of the analog waveform at the receiving end of the link is pointed out. This error is just a consequence of changing from analog (any value valid) to digital (only certain values valid) formats. Quantization noise can be minimized (at the cost of more bits) but cannot be avoided.

Before moving on to the final step of the PCM voice digitization process, coding, one final point on quantization should be made. This point revolves around the observation that the voice digitization process is not concerned with arbitrary waveforms but with the analog waveforms that represent speech. Minimizing quantization error for speech requires that some allowance be made for the fact that human voice contains a mixture of low-amplitude unvoiced sounds and high-amplitude voiced sounds and not much in between. Therefore, PCM voice digitization systems use a procedure called *companding* to take this high- and low-amplitude speech reality into account.

Companding refers to how this practice *comp*resses some parts of the voice amplitude range and ex*pands* them again at the receiver. The problem with establishing equal regions to quantize PAM samples (called *uniform quantization*), as shown in Figure 2-7, is that when applied to "real" speech, there are too few levels for low-amplitude unvoiced sounds and more than enough levels for high-amplitude voiced sounds. The high levels do not need the fine "granularity" that the lower amplitudes require, so the loss of some levels for high-amplitude sounds can be tolerated. Companding addresses this issue by establishing fewer quantization levels for high amplitudes and creating more quantization levels for low amplitudes.

Companding can be illustrated in principle quite easily. Instead of dividing the amplitude range into a number of ranges that are equally spaced, companding divides the amplitude range into smaller and larger ranges. This principle is illustrated in Figure 2-8. The coding levels are in numerical order for clarity. The left-hand portion of the figure shows a linear, equal relationship between amplitude and quantization coding. The right-hand portion of the figure shows how companding (compression/expansion) of these equal regions works. Six companding regions are shown, and now there are more coding levels available at lower amplitudes and fewer at higher levels, which is exactly the desired result.

Note that the relationship between companding regions in Figure 2-8 just divides the remaining amplitude in half, and then half again, and so on. Actually, companding methods are more complex than this simple "half" rule, but not much more complex. The two major companding methods are called *A-law* and *mu-law* (sometimes written with the Greek letter mu: μ-law) companding. A-law companding is used in international situations and in the vast majority of countries around the world. Mu-law, which is also called $\mu = 256$ companding, is used mainly in North America and the United States. In other words, A-law is used in E-carrier and mu-law is used in T-carrier, just to name the two most common ways of multiplexing voice in each environment.

In the most common voice companding methods, eight ranges are established on each side of the horizontal axis. Thus there are eight

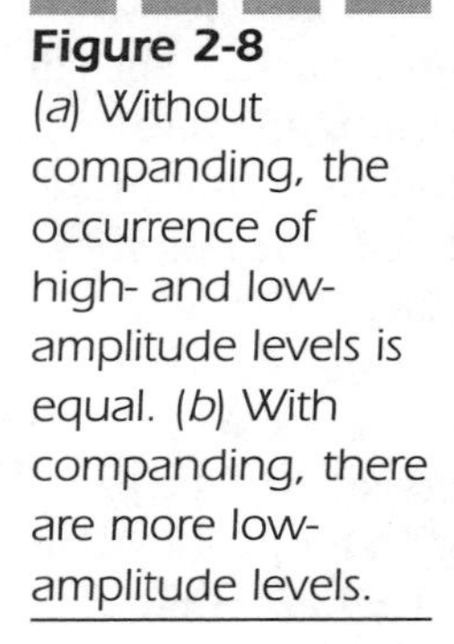

Figure 2-8 *(a)* Without companding, the occurrence of high- and low-amplitude levels is equal. *(b)* With companding, there are more low-amplitude levels.

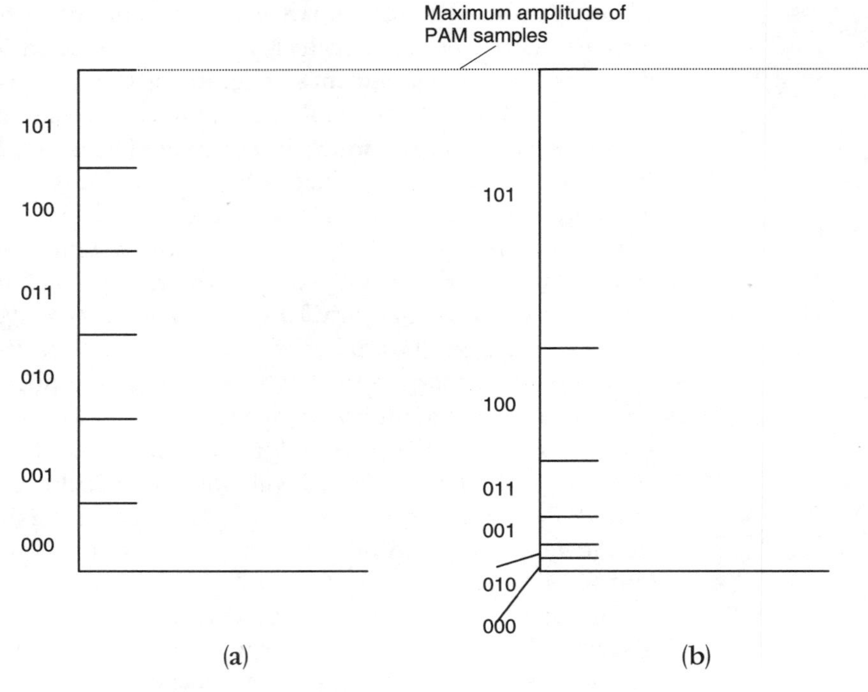

Figure 2-9 The 8-bit PCM word.

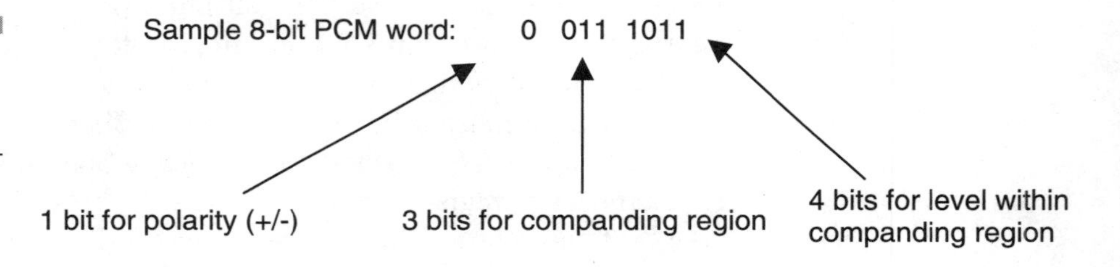

ranges for both positive and negative amplitudes. Each of the eight companding regions can be indicated by 3 bits. A fourth bit is used to indicate polarity (+ or −), giving 4 bits in all. These 4 bits always determine which of the 16 companding regions into which the PAM sample falls.

However, 64-kb/s PCM voice results in 8 bits to represent each PAM sample, not only 4 bits. The other 4 bits are used to divide each companding region up into 16 equal steps for quantizing the PAM samples. Four bits are required to represent these 16 levels within each companding region, making 8 bits altogether for the whole 8-bit PCM word. The structure of the entire 8-bit PCM word is shown in Figure 2-9.

Both A-law and mu-law companding produce 8-bit PCM words, but of course, they are not interoperable. Thus an A-law encoder cannot be used with a mu-law decoder. All codecs are either A-law or mu-law, but they cannot be both at the same time. Fortunately, the conversion from one form of companding to the other is simple and quick, amounting to little more than a lookup table. In the "old days," it actually was quicker (and less expensive) to "reanalog" the digital voice and reencode it using one form or the other than to try to convert between A-law and mu-law directly.

Sometimes there are still debates about whether A-law or mu-law companding is really "better." Human speech, as it turns out, varies in terms of high amplitudes and low amplitudes and other characteristics depending on whether the speaker is a man or woman. Some languages rely on differences in pitch as well as sound to determine word meaning. Which method best addresses each individual issue is still a topic of some discussion. But both methods work well, and for the foreseeable future, 64-kb/s PCM voice will continue to use mu-law in North America and A-law everywhere else.

Coding

The last step to producing digital voice ready to be sent over a WAN link is the coding step. The coding step produces the *line code*, or 0s and 1s suitable for transmission over long distances. This might sound odd, since the output of the quantizing step is an 8-bit PCM word that is already digitized and in the form of 0s and 1s. However, sending digitized PAM samples in the form of PCM words is hardly ever done except over short distances. Many other line codes are available that are more suitable and efficient than trying to send PCM words directly on a line.

The line code most often used in the United States to send PCM words is called *bipolar alternate mark inversion* (bipolar AMI), or just AMI, since all AMI is bipolar. AMI was first applied to digital telephony in the T-carrier systems of the 1960s at the first level of the T-carrier multiplexing hierarchy, T-1 (sometimes seen as T1). Although still in use, most T-1 links do not use AMI but a coding method based on AMI called *binary 8 zero substitution* (B8ZS). The ways that B8ZS enhances plain AMI are not of concern here. The point is just that AMI is used to send the 8-bit PCM words over the digital T-1 link.

Technically, AMI is *bipolar coding*, but use of the term *AMI* is universal outside voice digitization textbooks. Bipolar coding uses alternate mark inversion to represent 1 bits on the line. A *mark* is just a term taken from the telegraph system and used to mean a voltage pulse on the line. Voltage pulses can be either all positive or negative, which is a form of direct-current (dc) operation, or alternate from positive to negative and back, which is a form of alternating-current (ac) operation. When successive 1 bits are represented by pulses that alternate between positive pulses and negative pulses, this is the essence of AMI operation. Any 0 bits are represented by a total lack of voltage on the line for the duration of the bit. Thus bipolar coding appears as shown in Figure 2-10 for an arbitrary string of 0s and 1s.

Bipolar coding solved several nagging problems on early digital links. If enough dc pulses build up on a telecommunications line, it stops working. The use of ac pulses prevents this, and ac pulses travel much farther at the same power levels than basic dc pulses. The lack of a "dc component" helps with telecommunications systems because dc power has to be supplied directly to a line. Power from an ac circuit does not require direct electrical connections, so ac telecommunication circuits are easier to isolate and protect.

Bipolar coding is not the only digital line code in use. B8ZS is a popular variation, although opinion differs as to whether B8ZS is still AMI or deserves a category of its own. B8ZS is used to address a problem with the original AMI line code, which codes long strings of 0 bits as just flat periods on the line without pulses. This absence of activity can cause synchronization problems between sender and receiver (Was that just 14 consecutive 0s between pulses or 15?). B8ZS inserts intentional *bipolar violations* on the line to indicate the presence of 8 consecutive 0 bits. Bipolar violations, which are errors in pure AMI, represent consecutive 1s by pulses (marks) in the same direction, positive or negative, and not alternating. Other line codes besides AMI/B8ZS are used in the United States as well but are beyond the scope of this discussion.

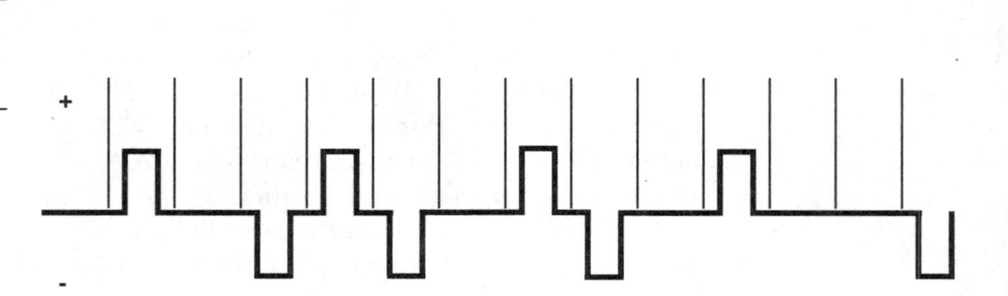

Figure 2-10 The bipolar AMI line code.

Outside the United States, neither AMI nor B8ZS is often used. The rest of the world more commonly uses high-density bipolar 3 (HDB3) for E-1 (2.048 Mb/s), E-2 (8.448 Mb/s), and E-3 (34.368 Mb/s) level digital line coding and uses *coded mark inversion* (CMI) for E-4 (139.264 Mb/s). HDB3 (sometimes called *B4ZS*) is a bipolar coding with violations for three consecutive 0 bits. CMI is not really a bipolar code at all but more closely resembles the type of bit representation seen on serial ports, but adapted for WAN transmission needs. All these E-carrier levels can multiplex many digital voice channels together for efficiency of transmission. In each case, the basic digital voice rate remains 64 kb/s.

Once the analog voice has been sampled, quantized, and coded, the digital voice that results can be sent across a WAN link and converted back to an analog waveform simply by translating the PCM words back into a very close approximation of the analog waveform that began the whole process. Quantization noise always prevents an exact duplication of the waveform, however.

A Basic System: 64-kb/s PCM Voice

The three waveform digitization steps of sample, quantize, and code that produce 64-kb/s PCM voice have been embodied in the United States in the T-carrier digital hierarchy and outside of North America in the E-carrier digital hierarchy (Japan uses a hierarchy, often called *J-carrier*, that is slightly different). Both T-carrier and E-carrier establish a multiplexing scheme for combining the basic 64-kb/s DS-0 digital voice channels for transmission over a single physical medium. Despite the terminology, the single physical medium could be two pairs of unshielded twisted-pair copper wire or two strands of fiberoptic cable, and so on. The key is that many DS-0 voice channels are combined onto a higher-speed transmission path using time division multiplexing (TDM).

The first level of the multiplexing hierarchy is T-1 or E-1. In the United States, the term *DS-1* is more properly used to describe the bit stream sent on the physical T-1 links, but everyone says "T-1" when they should say "DS-1" anyway.

To feed DS-0 signals into a T-1 or E-1, a simple multiplexing unit is needed to take 8 bits from each input channel and combine them into a T-1 or E-1 transmission frame. This is the T-1 or E-1 multiplexer itself. If the input stream is voice, which is always analog at the source (the human), then the voice is digitized before it is multiplexed. This is done either by a built-in codec or by taking voice that is already digitized, e.g., from a digital PBX, and putting the digital voice directly into the multiplexer (code conversion may be required).

Bits from a PC or other type of computer can be sent directly to a channel on the multiplexer as long as a few things are done. First, the multiplexer must be able to understand the coding used on the computer port, and this must be matching at the output port. Second, since this is just straight TDM, the bit rate from the computer serial port cannot exceed 64 kb/s (or 56 kb/s in some cases). In fact, the bit rate must always be *exactly* 64 kb/s (or 56 kb/s) from a computer port to DS-0. If there are no active data to send, then the serial port must generate enough bits to fill up the line rate.

The general multiplexing process for a T-1 and an E-1 is shown in Figure 2-11. Note that T-1 takes 24 DS-0s and E-1 takes 30 DS-0s. Also, T-1 uses mu-law companding, and E-1 uses A-law companding for digitized voice. T-1 is a North American standard and E-1 is the international standard, so when a DS-0 crosses a border to a country outside of North America, the E-1 format applies to the link. There are some other differences as well, just enough to make interfacing T-1 and E-1 interesting.

DS-0s, T-1s, and E-1s form the basic building blocks of the global PSTN. More details of the structure of the PSTN are explored in Chapter 3.

Breaking the 64-kb/s Barrier

The PCM voice digitization process is a waveform process that seeks to closely reproduce an arbitrary analog waveform of almost any shape. The standard PCM results in 64-kb/s digitized voice by generating 8-bit PCM

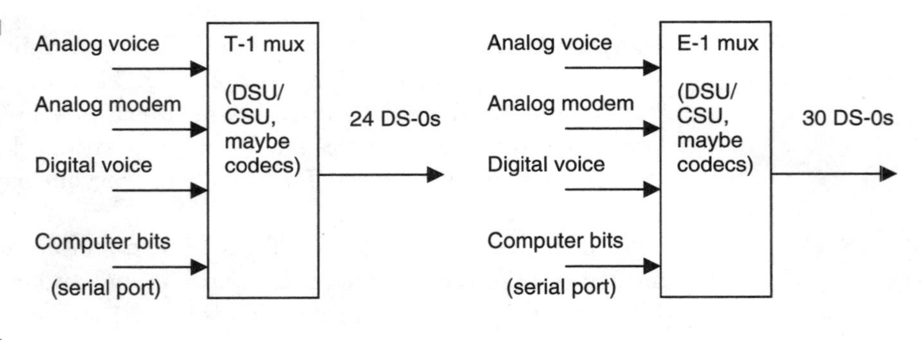

Figure 2-11 T-1 and E-1 digital multiplexing: *(a)* 24 input ports of almost any combination; *(b)* 30 input ports of almost any combination.

words 8000 times per second. Naturally, if more voice channels can be multiplexed onto a transmission link, this would be more efficient for telephone carriers in terms of users supported, which is always an incentive worth pursuing. There are two main ways of adding this traffic-bearing capability: increase the bandwidth of the multiplexed links or decrease the bandwidth requirement of the digital voice. This section details the second approach.

There was a related incentive to decrease the digital voice bandwidth requirement. In the 1980s, cellular voice systems and digital wireless systems became popular. For regulatory reasons, adding bandwidth to a wireless system is more difficult to accomplish than in land-line systems. Transmitters cannot just decide to increase the bandwidth used on a wireless link; there are licensing issues involved in most cases. Most of the frequency ranges used for wireless voice systems are tightly controlled and regulated, in some cases actually being auctioned off to the highest bidder. For wireless systems, decreasing the voice bandwidth below 64 kb/s is the most attractive way to increase voice-carrying capacity.

The big question is how to break the 64-kb/s barrier and maintain voice quality (also tightly controlled and regulated) at the same time. There are three major approaches to breaking this barrier. The first is to develop more efficient *waveform coding*. The second is to use a different voice digitization method known as *source coding*. Source coding takes into account that the source of the analog signal is not an arbitrary waveform but human speech. However, source coding is not particularly well-suited to encoding analog modem "sounds" or fax machine traffic. Thus the third approach is to combine the best aspects of waveform and source coding into *hybrid coding*.

Whether more efficient waveform coding is used, or source or hybrid methods are employed, all three seek to more effectively deal with redundancies, or repeated and so extraneous information, that are always present in voice communications.

Removing Redundancies

The easiest way to visualize the redundancies present in voice is to examine two consecutive PAM samples before they are quantized. Suppose the value of the first PAM sample on some arbitrary scale is +6.334 and the value of the second is +7.769. If the sender quantizes and sends the +6.334, then the receiver *already knows this* when the second sample arrives. In other words, only the +1.435 is new information. All the receiver need be told is "the waveform is up +1.435 from last time." This value of +1.435 is a lot less than the full +7.769 and so will take fewer bits to represent, since the waveform for human speech can only rise and fall so fast.

There are two main ways to accomplish this removal of redundant information and still use waveform coding. The first method is called *differential pulse code modulation* (DPCM), and the second is called *delta modulation* (DM). DPCM just encodes the *differences* (differential) between the PCM code words after they are quantized. Delta modulation is applied directly to the waveform with oversampling and in its simplest form generates a simple 1 bit for an "up one unit" or a 0 bit for a "down one unit" stream of bits. These principles are shown in Figure 2-12.

Early DPCM and DM voice digitization systems suffered from a phenomenon known as *slope overload* that degraded the quality of the digital voice that resulted from the application of these methods. Slope overload occurs when a waveform changes so rapidly that excessive noise is introduced. This noise is added because the increments established in straight DPCM (can only code ± 1.7, for example), and DM (called *sigma delta modulation* when the "up" and "down" unit size is fixed) cannot accurately reflect the rapidly changing waveform.

To deal with the problem of slope overload, both DPCM and DM now routinely adjust or adapt the "step size" if the waveform is increasing or decreasing rapidly. In other words, after three successive codings of current maximum increments, DPCM can increase the maximum increment indicated by 1101 to twice its present value. As long as the DPCM receiver knows this "rule," there is never any confusion at the decoder side of the codec.

This adaptive operation to compensate for the effects of slope overload makes adaptive differential PCM (ADPCM) and adaptive delta modula-

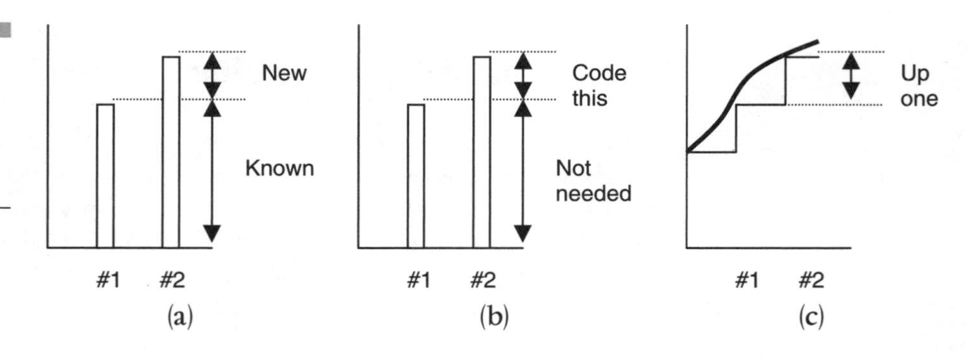

Figure 2-12
(a) Redundancies, *(b)* DPCM approach, and *(c)* DM approach.

tion (ADM) the preferred mode of operation for these techniques. ADPCM is an international standard all by itself and operates at 32 kb/s, half the normal PCM rate. ADM also usually operates at 32 kb/s, although some ADPCM and ADM implementations operate at other bit rates entirely. For example, some ADPCM systems have operated at 16 kb/s for voice.

Waveform coding cannot cut digital bit rate much below 16 kb/s (or even 32 kb/s in some cases) and still be useful for the wide range of services that telephone companies provide. Every drop in bandwidth sacrifices something in terms of voice quality due to the effects of quantizing noise. And voice digitization is not just a matter of analog voice in all cases. Analog modem signals need to be quantized for T-1 and E-1 links, as well as for direct serial port input from computers. Obviously, if a "voice channel" operates at 16 kb/s and not 64 kb/s, then the serial port input is limited to 16 kb/s. Thus voice efficiencies do not translate directly into data efficiencies, but quite the opposite. Analog modem waveforms (and fax machine waveforms also) are vastly different from the voiced/unvoiced pattern of speech, and going below 64 kb/s often limits the data rate of these modems to 9.6 kb/s or less. This was less of a problem in the days when state-of-the-art modems speeds were 9.6 kb/s or less, but the newest 56-kb/s V.90 modems all firmly accept 64-kb/s voice channels and in fact will not function properly below 64 kb/s on the digital end.

Thus, breaking the 64 kb/s has to be done carefully and is most common today in private voice networks or wireless systems. Ironically, the very techniques that make low-bit-rate voice possible on cellular systems and thus bring the price of voice down also conspire to limit data throughput when the wireless system is used for serial data bit transport. Land-line Internet access at 56 kb/s or faster is taken for granted, but throughput is limited to 9.6 kb/s or at most 19.2 kb/s in many cases on digital cellular systems.

Predicting Voice

Waveform coding is best when the input analog waveform can be modem signals or fax machines or just plain voice. The more forms, and the more arbitrary they are, the more it makes sense to mirror the input waveform as closely as possible within the limits of quantizing noise.

Suppose, however, the input analog waveform is intentionally limited to human speech. Are there more redundancies that can be removed to lower the required digital voice bit rate even further? As it turns out, there are many more repeated characteristics in voice that can be used to digitize speech as low as 2.4 kb/s.

Vocoders Most of these methods are called *source coding* techniques because they make specific assumptions about the source characteristics of the analog waveform. For instance, not only is human voice composed almost exclusively of low-amplitude unvoiced "hisses" and high-amplitude voiced "buzzes," but each buzz and hiss has a characteristic duration. There are never many hisses or buzzes in a row. The voiced buzz has repeated patterns called *pitch intervals*, and so on.

These characteristics make it possible to *predict* voice. In other words, if a speaker is currently buzzing, he or she will start hissing quite soon. Once the first pitch interval is detected, it is safe to predict three or four repetitions. This predictive operation is the essence of source and hybrid voice codings today, which are the most common forms used in VoIP. Devices that employ source or hybrid coding are called *voice coders*, or *vocoders*. The term *vocoder* is used to indicate that the method is "tuned" to voice digitization, not arbitrary analog waveform reproduction, as is true of a codec.

In practice, predictive coding describes voice waveforms as a *model* with several parameters, often as many as 10. The source and destination vocoders both implement and run the same prediction algorithm, the set of rules that embodies that particular form of predictive coding. All that needs to be sent between sender and receiver are the *differences* between the predicted waveform and the actual waveform seen at the sender ("They didn't stop hissing when we thought they would; keep hissing!"). This information is expressed as the value of a number of parameters and can be sent with very few bits, as long as the predictor is a good model of the speaker's voice patterns.

If source coding is so good at optimizing network capacity, why wasn't it used years ago? Because the benefit of saving bits is balanced by the cost of processing power. Waveform coding is simple and quick. Source coding is neither. The input waveform must be analyzed and the parameters adjusted and sent. The predictor that forms the basis for the operation of the sender and receiver must be running constantly and synchronized between sender and receiver. All this requires a lot of memory and processing power in the vocoder chipsets and makes vocoder software quite complex.

Until the late 1980s, vocoder technology was expensive and primitive. Voice produced by vocoders was mechanical sounding and synthetic, although it was clearly understandable. The old Texas Instruments "Speak and Spell" toy had a chipset vocoder that expressed bits stored in memory as speech at the press of a button. But no one would seriously consider talking on a telephone that way unless they had to.

There are several forms of vocoder. All assume that speech is produced by a *linear* system (i.e., a system where each output is determined by the sum of the inputs), the human vocal tract. This linear system is periodically *excited* by a series of impulses, determined by the pitch interval if the sound is voiced. Unvoiced sounds are excited by a random-noise sequence.

All linear systems model the vocal tract and the parameters that relate to this tract using different techniques. It is the differences in techniques that make up the families of vocoders. But they all attempt to produce a bit stream that results in what sounds like the original speech without worrying about how closely the output waveform resembles the input waveform (which is the reason vocoders can sound so artificial but still be understood perfectly). Transmitters analyze the input speech to determine the model parameters and excitation. Receivers synthesize the speech.

The poor quality of vocoders is due to the simple nature of the algorithms used. All sound is either high or low amplitude, with nothing allowed in between at all. Even worse, the ear can be very sensitive to pitch during voiced sounds, but all vocoders struggle with pitch intervals, a problem that has never been solved to everyone's satisfaction. Vocoders are very sensitive to errors, errors due to problems calculating the vocal tract modeling parameters as well as bit errors on the line.

Vocoders are used in voice and music synthesizers, but this discussion is confined to telephony application. The earliest vocoders were simple *channel vocoders*, conceived in 1939, about the same time that digital voice was invented. Noting the ear's insensitivity to what is called *short-time phase*, channel vocoders chop up speech into 20-ms segments and only worry about the magnitude of the segment to produce voice at 2.4 kb/s. An improvement was the *homomorphic vocoder*, which added pitch information to the magnitude, but at the price of functioning at around 4 kb/s. Once chipset processing power took off, *formant vocoders* could be used to analyze the voice in many ways, theoretically resulting in voice at around 1 kb/s or below. *Formants* are voice characteristics such as pitch intervals and the like. However, practical problems with determining the voice formants precisely severely curtailed the formant vocoder's popularity.

Many vocoders in the early 1980s used *linear predictive coding* (LPC) to achieve bit rates as low as 2.4 kb/s. Each speech sample is assumed to be a linear combination of the preceding samples, giving LPC its name. A block of about 20 ms of speech is stored and analyzed by the sender. The predictor coefficients are determined, quantized, and sent to the receiver. The arriving "speech" is sent through a series of circuits to produce the *prediction error*, or *residual*. This information is used to produce the actual sound heard at the receiver.

Even today, voice produced by traditional source code vocoders sounds very artificial. This is so because the operation of pure source coders is very dependent on how closely the vocal tract model used corresponds to the input voice signal. As it turned out, it was very hard to come up with a predictive model that applied equally well to men's and women's voices, and old and young alike. There also was a problem with making one method work with both the more guttural languages (with many rasping consonants such as in German) and pitched languages (many Hawaiian words consist only of vowels) in use around the world.

Hybrid Coding Most modern codecs employ a *hybrid coding* method to address the issues of young and old, language differences, and so on. Hybrid coding is used to combine aspects of waveform coding and source coding into a single package. Hybrid coding breaks the 16-kb/s waveform barrier and at the same time yields acceptable voice quality all the way down to about 2.4 kb/s (1-kb/s voice is promised soon). Pure source vocoders produced understandable speech at these low bit rates but could not produce natural-sounding speech at any bit rate.

Today, several forms of hybrid codecs exist (hybrid methods use the term *codec* by preference, but they are also sometimes called *vocoders*). However, the most successful and widely used are categorized as time-domain *analysis by synthesis* (AbS) codecs. These codecs use the same linear predictive model of the vocal tract as LPC vocoders. But instead of just dividing everything into high/low and voice/unvoiced, AbS codecs use the excitation signal to better reflect the true nature of the original waveform.

AbS codecs appeared in 1982 as the *multi-pulse excited* (MPE) codec. Later implementations were the *regular-pulse excited* (RPE) codec and, very popular today for VoIP, the *code-excited linear predictive* (CELP) codec family. MPE gives good voice quality at about 10 kb/s, while RPE operates at 13 kb/s. RPE is used in the pan-European wireless Global System Mobile (GSM) digital network.

All AbS codecs usually start with the same 20-ms chunk of speech, called a *frame*, that vocoders do, although this frame size can vary. Then the AbS encoder performs the familiar analysis of frame parameters and then excitation signal, just like a vocoder. But then the analysis by synthesis comes in. The encoder at the sender actually synthesizes many different representations of the input frame and then compares each excitation signal to the input frame by acting as its own receiver. Naturally, this requires a lot of processing power to perform in real time, but this "closed loop" encoder operation is the distinctive feature of AbS codecs.

The potential time and effort the chipset requires to pass every possible excitation signal through the synthesis filter (a process called *error weighting*) are enormous. Usually some means of cutting down on the complexity must be found, but this either means raising the required bit rate, sacrificing some voice quality, or both. This is the reason that many of these techniques have to hedge a bit with phrases like "around 10 kb/s."

CELP appeared in 1985 and was quite sophisticated right from the start. In fact, it was too sophisticated to run on most 1985 processors. The original, full CELP algorithm required 125 s to produce 1 s of digital speech—on a Cray-1 supercomputer. Reductions in complexity of CELP coupled with the rise in processing power since 1985 (a modern video game is almost as powerful as the Cray-1 was back then) means that CELP versions are readily available on DSP chips. Several standards have been based on CELP running at bit rates from 16 to 4.8 kb/s. At 16 kb/s, CELP voice is as good as 64-kb/s PCM, but of course, CELP voice devices are generally more expensive than PCM components.

CELP codecs have been produced that run below 4.8 kb/s, but these are relatively new. The current goal is to break the 2.4-kb/s barrier. Most new methods classify the input speech frames into voiced, unvoiced, and transition categories. Each type has its own dedicated encoder with its own set of rules. For example, unvoiced frames, without a reliable pitch interval, use no "long term" predictor. One of the newest CELP techniques is known as *multi-band excitation* (MBE) codecs. A related method called *waveform interpolation* (WI) codecs is used with CELP but is not directly related to the AbS and CELP family.

Today, the CELP family includes ACELP (algebraic, code-excited linear predictive coding), MP-MLQ (multipulse, maximum-likelihood quantization), LD-CELP (low-delay CELP, used to cut down on the processing time needed for CELP digitization), and CS-CELP (constant-structure CELP). All are refinements of the basic CELP process. For example, ACELP and MP-MLQ usually employ a 30-ms frame with a 7.5-ms "look ahead" for prediction comparison, but the precise details of them all are not of concern here.

A Note on Voice Quality

Digitizing voice does not fundamentally change what voice is used for. People still have to listen to the speech represented by the 0s and 1s. Voice is an analog phenomenon. Digitization began as an effort to minimize the effects of analog noise on voice transmission systems.

For this reason, the quality of a digital network is easy to measure. Simply put a known string of 0s and 1s through the network and compare what went in with what came out. Any network that matches the test bit string with 99 percent accuracy has better quality than a network that matches the bit string with only 95 percent accuracy (actually, both are terrible, but this is just an example). The quality of analog information and analog networks is not so absolute. The only way to compare 32-kb/s ADPCM voice with 64-kb/s PCM voice is just to "reanalog" the sound and ask a number of people to listen to it, usually through a regular telephone handset.

In actual practice, the participants are asked to rate the voice quality on a scale of 1 to 5, where 1 is "What was that?" and 5 is "Like the person is right next to me." Various levels between the integers have been added to this rating system until it is sort of like rating a new dance tune: "I liked the beat, so I gave it a 4.1." There has to be a statistically significant cross section of the customer population (old and young, male and female, etc.) and a minimum of 40 people (who are supposed to be "naive" about the method being tested). The results are tabulated to compute the average (arithmetic mean), and this gives the *mean opinion score* of the voice quality, the MOS.

The MOS scale for analog voice quality has been used for a long time. MOS is well known and relatively easy to carry out in test environments. Officially, the ratings are 5 for *excellent*, 4 for *good*, 3 for *fair*, 2 for *poor*, and 1 for *bad*.

Over time, the 5 rating usually has been considered to be *local quality*, where both parties are serviced by the same central office or two central offices linked by a few trunks. The 4 rating (or above) is known as *toll quality*, and most long-distance calls fall into this category. Ratings between 3 and 4 are typically considered to be *cellular quality*, ratings in the 2s are down to almost the fast-food intercom level ("You want fries with that?"), and ratings between 1 and 2 are barely understandable and require constant repetition of phrases prompted by the listener. Most regulators expect to see consistent MOS ratings of 4 or better for landline services and 3 or better (sometimes 3.5 or better) for wireless telephony. Naturally, a service provider with a 4.3 will always claim "better voice quality" than a service provider with a 4.0, even though both essentially provide toll-quality voice.

Thus service providers quickly discovered that while 64-kb/s PCM voice consistently gave an MOS of 4.5, 32-kb/s ADPCM voice had an MOS of about 3.0. Dropping PCM voice with more advanced ADPCM methods to 16 kb/s produced an MOS of about 1.5.

What has all this to do with the current rise of voice below 64 kb/s? The new DSP chips have dropped in price and increased in power dramatically since the original ADPCM broke the 64-kb/s barrier. Even more important, the new LPC algorithms, sometimes called *voice compression*, have become sophisticated enough to "flatten the slope" of the MOS dropoff due to using fewer bits to represent voice. So voice at 16 kb/s no longer is condemned to an MOS of 2 or lower. Almost all the techniques described in this chapter can be used to give an MOS of 3 or higher and in some cases 3.5 or better.

Other analog voice quality measurements are used besides MOS. The most common are the *diagnostic rhyme test* (DRT) and the *diagnostic acceptability measure* (DAM) systems, which measure voice according to other, finer scales. For example, an MOS of 4.0 translates to about a 95 on the DRT and about a 70 on the DAM scales. DRT and DAM are more sophisticated than the simpler MOS. DAM, for instance, allows for the fact that listeners may disagree about speech quality but typically agree on the amount of background noise present. DAM listeners separately rate speech, background sound, and total effects. There are 21 ratings to make, 10 for the speech itself, 8 for the background, and 3 related to intelligibility, pleasantness, and overall acceptability. Lately, some vendors have begun to quote MOS values on a scale of 1 to 10, this being more natural than the 1 to 5 scale of MOS. The MOS is typically doubled in these vendors' product literature, so an MOS of 4.0 becomes 8.0, and so on.

A Sampling of Methods and Standards

By now it should be apparent that there are many ways to digitize voice. Not all of the methods are standardized, however, and even among those which are standardized, not all of them are on equal footing. When it comes to voice standardization, the main standards body is the International Telecommunications Union (ITU). The ITU-T

(International Telecommunications Union—Telecommunications Standardization Sector) governs all international telephony and has a lot of influence at the national level as well, since it makes little sense to use totally different methods domestically and worry about compatibility on international circuits.

ITU standards, which are rather ironically called "recommendations," have the force of law. They are called ITU recommendations because the ITU technically has no way to actually force member nations to adopt the standards once they have been voted on and passed. This is not to say that all ITU standard methods are universally followed either. For example, the North American T-carrier digital multiplexing hierarchy uses mu-law companding for voice, although the ITU recommendation specifies A-law companding for all international 64-kb/s PCM voice communications. Mu-law companding is an allowable option and is used domestically in the T-carrier system. The American National Standards Institute (ANSI) is charged with adapting ITU specifications for use within the United States.

Standards are used for interoperability, comparison reasons, and product stability. This section examines the voice digitization techniques mentioned in this chapter and shows which ones are standards. Naturally, the attractiveness of a VoIP voice digitization method is enhanced if the method is also a standard. When it comes to the ITU, all digital voice standards are issued by ITU-T Study Group 15 (SG 15) as part of the G-series of ITU-T recommendations. As an aside, MOS voice-quality procedures are described in ITU-T Recommendation P.80.

Which of the voice digitization methods have reached the level of ITU-T recommendation standards? These methods would be expected to be the most mature and stable, and of course, vendors should have a ready market for their wares in these categories. Some are part of H.324, the ITU-T recommendation for low-bit-rate (e.g., 33.6-kb/s) multimedia (voice and video as well as data) communications. One is used in wireless cellular personal communication systems (PCS). Table 2-1 lists them all.

The ITU-T is understandably reluctant to make any claims about the actual voice quality of any of these voice methods, since all of them depend on the implementation of the DSP chipsets and overall network quality more than on the standard documents on which the methods are based. And this is not the ITU-T's job anyway. However, in Recommendation G.728 Appendix II, SG 15 does state that 64-kb/s PCM voice has the best quality, followed by 32-kb/s ADPCM. G.728 Appendix II then states that 16-kb/s LC-CELP should have the same voice quality as 32-kb/s ADPCM, which has always yielded MOS numbers above 3.5.

TABLE 2-1

ITU-T G-Series Digitized Voice Standards

ITU-T Recommendation	Bit Rate	Coding Method	Comment
G.711	64 kb/s	PCM	Still the voice workhorse
G.723.1	5.3 kb/s	ACELP	Part of H.324
	6.3 kb/s	MP-MLQ	Part of H.324
G.726	40, 32, 24, or 16 kb/s	ADPCM	Most use 32-kb/s ADPCM
G.728	16 kb/s	LD-CELP	Low-delay version of CELP
G.729	8 kb/s	CS-ACELP	More refined ACELP for cellular PCS telephony

Some studies have put modern 32-kb/s ADPCM quality at about 3.75 and 8-kb/s CS-ACELP quality at about 3.85, giving a slight edge for the lower bit rate.

Many of these G-series recommendation methods are found in VoIP products. In fact, many products support more than one method. However, when it comes to VoIP, there are several standards groups that are working on aspects of VoIP standardization. Some are traditional standards bodies, while others are loose associations or vendor consortia whose members agree to implement VoIP in a consistent, but still often vendor-specific, manner. A formalized vendor consortium might become a regular forum whose members cooperate and move toward interoperability more systematically, although there is usually no way to "punish" or expel a member that violates a forum implementation agreement (IA).

Of course, a digital coding technique could still be totally vendor-specific (proprietary). Table 2-2 lists the voice digitization methods not already covered as ITU-T recommendations in descending bite-rate order.

Abate is a family of ADM techniques, and 32-kb/s Abate is used in the U.S. space shuttle program. RPE-LTP (regular pulse-excited linear transform prediction) at 13 kb/s is the basis for the GSM wireless systems and is seen in the United States as part of PCS-1900 wireless services. MRELP (modified residually excited linear prediction) is a Motorola method designed to be used with the international space station (ISS) program. Both SX9600 and SX7300 are Lucent methods used in Lucent's

TABLE 2-2

Miscellaneous Digitized Voice Methods

Voice Technique	Bit Rate	Coding Method	Comment
Abate	32 kb/s	ADM	Used on U.S. space shuttle
RPE-LTP	13 kb/s	RPE-LPC	Used in GSM and PCS-1900
MRELP	9.6 kb/s	CELP "upgrade"	Motorola proprietary
SX9600	9.6 kb/s	CELP "upgrade"	Lucent method used in VoIP
VSELP	8 kb/s	CELP "upgrade"	Motorola method used in IS-54
SX7300	7.3 kb/s	CELP "upgrade"	Lucent method used in VoIP
CELP	4.8 kb/s	CELP	AT&T Bell Labs product
STC	2.4-4.8 kb/s	LPC	Sinusoidal transform coder
QCELP	1-4 kb/s	CELP "upgrade"	Qualcomm CELP

VoIP gateways, although other coding methods can be used as well. VSELP (vector sum–excited linear prediction) is another Motorola creation, this time chosen for the Telecommunications Industry Association (TIA) North American Interim Standard IS-54 for digital cellular wireless telephony. The 4.8-kb/s CELP is a product of AT&T Bell Labs for the U.S. government. STC is another U.S. government project in low-bit-rate voice. QCELP is Qualcomm's CELP used in TIA IS-96, another digital cellular telephony standard.

The MOS, DRT, and DAM ratings of each of these methods are very good, typically above 3.5 for the MOS, in the low 90s for the DRT, and in the high 60s for the DAM. However, a lot of background noise will drive the MOS down to about 3.0 in many cases, and DAM ratings can nosedive due to the background-specific ratings in DAM. DRT and DAM scores sometimes vary quite a lot between men's and women's speech as well.

Note that several techniques offer variable rates, between 1 and 4 kb/s for QCELP, for example. But this just means that the circuit carrying the voice operates at either 1 or 4 kb/s or somewhere in between. A 4-kb/s voice circuit is still a circuit. Nothing in this chapter makes digital voice into a stream of IP packets. This is still a circuit-switched "all the bandwidth all the time" family of voice digitization methods—up to a point. Many of the methods described in this chapter can and have been used in conjunction with related techniques designed not only to *digitize* voice but also to *packetize* voice. This is the essence of VoIP: not low-bit-rate circuits, but low-bit-rate *packets* filled with voice bits that are indistinguishable from data packets. Digital voice in its simplest form runs at a constant bit rate, not a variable bit rate. Packet networks are designed for bursty, variable-bit-rate applications. Voice networks are designed for constant-bit-rate circuit applications.

Why has packetized voice in the form of VoIP taken so long to appear if digital voice under 64 kb/s has been around so long? Only because digitizing voice, as difficult as that was to perfect, was nothing compared with packetizing voice for the Internet. Thus most low-bit-rate digital voice just flows on low-bit-rate circuits, wireless or otherwise. The transition to VoIP packets has just begun.

Today, with the packet-switched Internet and Web everywhere, it is easy to lose sight of the circuit-switched networks that came before packet networks. Why not just take 8-kb/s voice and throw it on the same IP routers that the Internet uses? To see why packet-switched voice was so difficult to accomplish reliably (a point some would argue has not yet been reached), a chapter on the structure of the circuit-switched voice PSTN for those familiar with routers and packets is needed first.

CHAPTER 3

Telephony for IP People

Not so long ago, when people studied a global wide-area network (WAN) at a college or university, it was the public switched telephone network (PSTN) that was the main topic of investigation. Now, of course, attention has shifted to the global Internet Protocol (IP)-based Internet and its Web subset. College graduates are now much more familiar with the form and function of an IP router than a telephone central office (local exchange) switch. Even the term *switch* has come to mean "a really fast router" instead of a central office (CO) exchange device for connecting telephone circuits. This chapter is intended to bring people familiar with the Internet, Web, IP, packets, and routers up to speed with respect to the PSTN, circuits, and switches. After all, Voice over IP (VoIP) blends these two worlds into a working whole.

Many of the concepts presented in this chapter will be emphasized by comparison with the global Internet. It has been pointed out that the telephone network was not quite the nineteenth-century equivalent of the Internet, however. A strong argument has been made that the "Victorian Internet" was really the global telegraph network deployed between about 1845 and 1876, when the telephone came along. Certainly, early Internet use more closely resembles telegraph applications than telephony applications, given the text-based nature of both the original Internet and the telegraph. Many Internet concepts, such as data compression, encryption, and authentication, were first employed on the global telegraph network.

However, no one had a telegraph in their home. No one agitated for faster telegrams to the home. No one worried about being unemployable because they were not "Morse Code literate." No one needed access to a telegraph to finish school. The Internet is a good analogy for the global telephone network because the telephone was the first global network *for the people*, as the Internet, now free of its simple text basis, is today.

Thus it is entirely appropriate to assume that people are somewhat familiar with both the global telephone network and the global Internet. What is somewhat unique in this chapter is that the assumption is that people are now *more* familiar with the Internet architecture and hardware and software than they are with the PSTN architecture and hardware and software. After all, when was the last time the PSTN structure was taught in school to the same extent that basic Internet concepts are presented to students today? Probably never.

The Global Telephone Network

When country codes were added to the global PSTN in the early 1970s, many telephone companies included fliers in their customers' monthly bills that listed some of the codes to encourage international calling. The rates were listed, which were very expensive for the time, and callers still had to know the local telephone number format used in the destination country. But no human being was involved. Just pick up the telephone and dial.

Not surprisingly, a lot of interest in international calling was shown initially by telephone company employees themselves. Few had to worry about paying the bills for calls they placed at work, and Bell System employees had literally never seen a telephone bill since they had worked at the telephone company. Some 50-year veterans, joining right out of school, had never seen a telephone bill in their lives! Thus a lot of mail room workers, reading the fliers as they mailed them out, picked up the telephone and began calling other countries in their spare time, just for fun and to see if it really worked. It did. And perhaps there was some value in finding out by experiment that telephone numbers in England are not seven digits long, that in Italy they say "Pronto" when they answer the telephone, and that in Japan they say "Mish mish," even when rudely awakened in the middle of the night by some foolish person in the United States.

Today, just by pressing the right digits on a telephone, almost any telephone in the world can be made to ring without human intervention (a long-time exception is Cuba, calls to which have to be handled by a human operator). The PSTN is truly a global network today.

Although the topic here is the global PSTN, this chapter mainly explores the telephone network in the United States, for a number of reasons. First, national variations would make this chapter a whole book by itself. Second, the infrastructure in the United States is similar to the basic telephone structures used throughout the world. And finally, the infrastructure in the United States has been documented to a greater extent than almost any other country's telephone network. Moreover, in the United States, there was almost always a number of independent telephone companies that were separate from the "national" Bell System, an organization that is just emerging in many countries around the world today.

The PSTN was not always a global network. Before there was one global network, there were many isolated local networks. In some cases it was impossible to make a telephone call not only across the country but also across the street. To see how the many competing, isolated local telephone companies became one big global network, it is necessary to consider the basic building blocks of any telephone network at all, large or small.

Lines, Trunks, and Switches

All telephone networks consist of a few basic hardware elements, just like the global Internet. The basic hardware elements of the Internet are hosts (user devices or computers running TCP/IP), local-area networks (LANs, usually some form of Ethernet), access routers (needed to connect the LAN to the Internet service provider, or ISP), backbone routers (to shuttle IP packets between access routers and between ISPs), and WAN links connecting all the routers. The basic hardware elements of the PSTN correspond almost exactly to the hardware of the Internet. The PSTN has telephones (user devices), customer premises wiring (LAN), local exchange switches (access routers), interexchange carrier switches (backbone routers), and trunks (links) between the switches.

This is not to say that there are not significant differences between the Internet and the PSTN, or else this would be a very short chapter. For example, the Internet just has "links" between routers, whether used to connect access routers to the ISP or to connect backbone routers. The PSTN distinguishes between *access lines* (often called *local loops*) and *trunks* that multiplex voice channels between voice switches. A link that connects the customer premises to the local voice switch is an access line, and a link that connects voice switches to each other is a trunk, even though both may be unshielded twisted-pair copper wires. The network node of the Internet is called the *router*, and the network node of the PSTN is called the *switch*. And this is just for starters.

Figure 3-1 shows the basic structure of the PSTN. There are several details that should be discussed before showing exactly what happens on the voice network when someone makes a telephone call.

All telephone network customers (often called *subscribers*, a holdover from the days when customers could not own their own telephones) interact with the voice network using an *instrument*. The most common instrument is the telephone, as shown in the figure, but other common instruments are computer modems and fax machines. The telephone

Figure 3-1
Elements of the PSTN voice network.

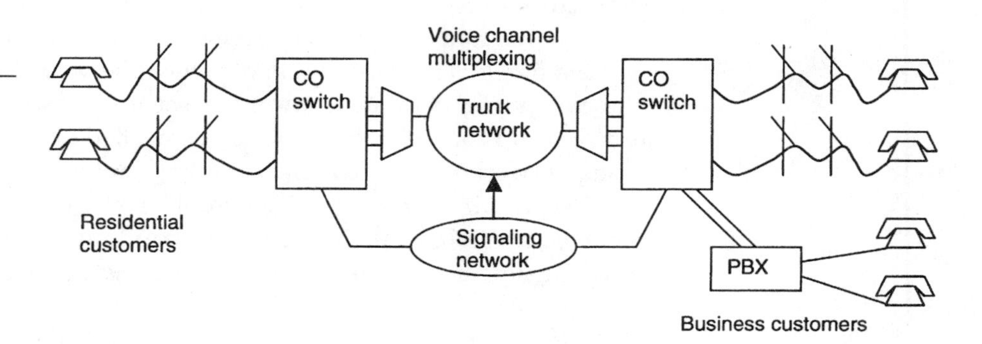

instrument is commonly connected to a central office switch, the network node of the PSTN, by means of a local loop or (more properly) an *access line*. The access line is typically a single pair of twisted-pair copper wires. However, many twisted-pair voice channels in a housing complex are multiplexed together onto fewer pairs of copper wire in an arrangement known variously as a *pairgain system* (since it can "gain pairs" for other uses), *digital loop carrier* (DLC), or *carrier serving area* (CSA) systems. These are not shown in the figure, since the main goal of these *distribution network* strategies is still just to bring a number of access lines to the central office (CO).

The central office switches' main job is to connect access lines to each other, based on the network address (telephone number) of the destination telephone instrument. The PSTN is a connection-oriented, circuit-switched network in which no information can flow from source to destination until a connection is established between a source and a destination. If the destination instrument is attached to the same central office switch, this job is easy. It is a more difficult task if the telephone number dialed is at the end of an access line on another switch, and in actual practice, the destination can be almost anywhere in the world. If the access lines that need to be connected are on different switches, the voice connection must be routed onto a trunking network, which is just a network that links telephone switches, not telephone instruments. The figure shows that many voice channels are routinely multiplexed when sent over the trunking network, although this is not strictly necessary. Trunks can be individual twisted pairs indistinguishable from access lines, although trunks are more likely to be carried on coaxial cable, fiberoptic cable, or microwave towers.

The figure also shows a *signaling network*. The signaling network is how the source and destination network address (telephone number) is sent to the destination switch and also how the voice channel to be used for the call on the trunking network is set up (shown by the arrow). In other words, all the call routing decisions needed to set up the call on the trunking network are done by the signaling protocol on the PSTN, which is called *signaling system 7* (SS7). On older switches, an SS7 interface may not be available. In this case, the trunking network itself can be used to transfer the signaling protocol "messages" from switch to switch.

This signaling network is vitally important in understanding the differences and similarities between the PSTN and the Internet. This is so because the signaling network is basically a connectionless, packet-switched network with routers as the network nodes, just like the Internet. The signaling network *must* be connectionless due to the simple fact that the messages that set up connections, known as *call setup messages*, cannot follow connections if their function is to set up connections (calls) in the first place. It is easy to make an "Internet" out of the PSTN: Just treat everything, from dialed digits to voice bits, like a call setup message and route them connectionlessly from source to destination on the signaling network, forgetting about the trunking network entirely. However, this is more properly a topic for the next chapter.

There is one other element to Figure 3-1. This is the business *private branch exchange* (PBX). The PSTN always distinguished between residential and business customers, going back to the days when Alexander Graham Bell charged twice as much per month for a business line as for a residential line ($40 versus $20). The thinking was that there were more people who could use the telephone in an office than at home, so the greater utility should cost more. In any case, businesses typically have a little voice switch all their own on the customer premises, as a "private branch" of the *local exchange* (the official name of the CO). The PBX is owned and operated by the customer, and the links to the PBX are properly called *trunks*, since they connect voice switches. The PBX trunks typically multiplex voice channels. Naturally, the links from the PBX to individual telephones are access lines. The PBX is the closest thing that the PBX has to an Internet access router.

One other common business arrangement is not shown in the figure, but only because it is just a variation on the residential access line architecture. Smaller businesses may not want the expense and complexity of configuration that a PBX requires. Thus a *centrex* service basically dedicates a piece of the CO switch for a particular business customer. Centrex service requires as many access lines as there are telephones in the office, and these voice channels also can be multiplexed.

The trunking network forms the backbone of the PSTN, in the same way that backbone routers form the global Internet. The trunking network consists of more voice switches and multiplexed links, but there are no access lines attached to the switches on the trunking network. This connectivity is strictly trunk-to-trunk. It would be prohibitively expensive to connect every CO switch with a direct point-to-point link, multiplexed or not, so the trunking network provides this connectivity. Some CO switches do have direct trunks to their closest neighbors, depending on actual calling patterns. However, it is more likely that calls are routed by the signaling network onto trunks leading to a *local tandem office*. The local tandem is most likely a centrally located CO switch with a second, trunk-only switch housed in "tandem" with the CO switch, which is where the name comes from. Thus the PSTN always seeks a direct trunk to the destination switch but can hand calls off the local tandem if the routing table cannot find an idle (free) trunk or if no direct trunk exists.

What if the local tandem cannot find a trunk that leads to the destination? As it turns out, a whole hierarchy of trunk switches was used in North America (the United States and Canada) when the Bell System was considered to be *the* telephone company. Even the independent telephone companies that were not part of the Bell System, such as GTE, had levels of trunk switches. If a call could not be completed at one level (such as the local tandem level), the call could be bumped up to a higher level of the switching and trunking hierarchy until it could find an available trunk voice channel to connect the call. Sometimes, such as on Mother's Day, Thanksgiving, and New Year's Day, no trunks were available at all from time to time.

In the Bell System, there were five levels of switches in the hierarchy. At the top, there were 10 class 1 switches in North America called *regional centers,* all mesh connected directly to the other nine and to each other with huge numbers of trunks. Below this level were the class 2 sectional center switches, then the class 3 primary center switches, class 4 toll center switches, and finally, class 5 local tandems or local wire centers (COs) that started it all. In the Bell System, the Bell Operating Companies (BOCs) operated the class 5 switches and tandems, and AT&T Long Lines ran the rest, which basically handled all long-distance calls.

Below the class 1 level, the actual numbers of each type of switch varied depending on who was doing the counting and when the counting took place. Figure 3-2 shows the structure of the switching and trunking hierarchy in the former Bell System (with their official symbols). Table 3-1 shows the approximate number of switches at each level around 1980 as well as the number of switches, if any, operated by independents at each level. Note the dominance of AT&T in the long-distance arena, since building nationwide trunking backbones was and is a very expensive undertaking.

The horizontal dashed line separates local calls carried entirely on the facilities of the local telephone companies from long-distance or toll calls. A toll call is just like a toll road: It is the same as a regular call or road,

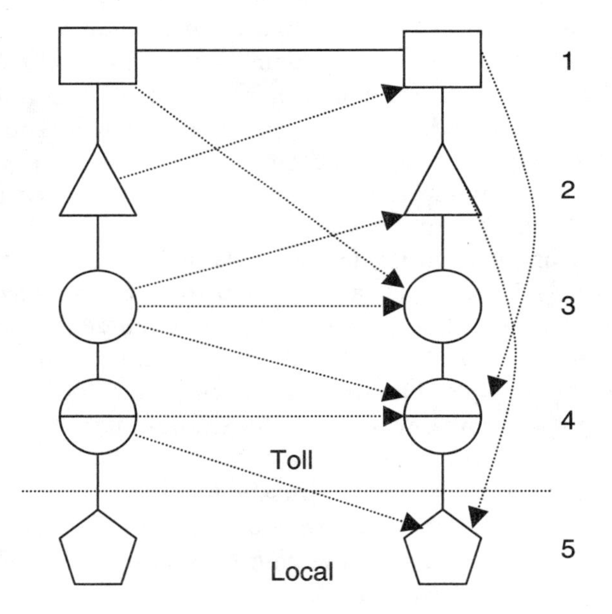

Figure 3-2 The North American trunking network around 1980. The numbers at right correspond to the classes in Table 3-1.

TABLE 3-1

Approximate Number of Bell System and Independent Switches circa 1980

	Class	Bell System	Independents
1	Regional center	10	0
2	Sectional center	50	0
3	Primary center	150	9
4	Toll center	550	475
5	Central center	9800	11,000

but it costs more to make it or use it. There are also local toll calls, but all long-distance calls are toll calls, more or less by definition. Long-distance calls in the Bell System were handed off the AT&T Long Lines at the nearest toll center. The dashed arrows show various shortcuts available by way of high-usage trunks (HUT) for routing calls through the network. For example, if a call originated at the left of the figure, the toll office always tried to find a trunk to the destination CO directly. If no trunk was available, the toll office would try to find another trunk to a toll office close to the destination. Finally, the toll office could hand the call off to a higher level of the hierarchy.

Sometimes it is the terminology itself that leads to confusion when Internet adepts look at the telephone network. Some very fast router architectures, and even LAN hubs, are called *switches*. However, this is not quite the same thing as is meant when a telephony specialist refers to central office *switches*. Every network node type developed by the telephone companies traditionally has been called a switch, so there are X.25 packet switches, frame relay switches, and ATM switches that function quite differently from voice switches. The unity of terminology is not entirely arbitrary. All network node switches, including voice switches, do have much in common.

On the other hand, when the Internet was being formed, the network nodes proper were first called *interface message processors* (IMPs) and later *gateways*. The term *router* is a relatively late development, and it was embraced in part to point out just how differently network nodes on the Internet treated packets from the way telephone carrier network nodes treated packets. Using a term like *router* pointed out just how radical in operation the Internet really was.

First and foremost, the Internet is a wide-area *connectionless* network. That is, no connections are defined at all between routers that establish fixed paths for IP packets to follow between the routers. There might be end-to-end connections between clients and servers (both are hosts), but this is done at the TCP layer, which is not active in IP routers. In fact, the Internet was the first major WAN architecture that was connectionless in operation between the network nodes. LANs are also connectionless, and there was a certain appeal to employing a connectionless WAN to link connectionless LANs. This is the real strength of the Internet: Packets need not follow fixed paths through the router network but rather make their way from source to destination router the best way they can (routers employ a "best effort" strategy). If a link that a previous packet from *A* to *B* is down, subsequent packets can follow another path. The Internet can "route around failures" of links or routers.

In a connection-oriented network, such as the PSTN, no bits can make their way from source to destination until there is a connection from source to destination that establishes a path through the network for the bits to follow. The bits could be digital voice, X.25 packets, frame-relay frames, or ATM cells. It makes no difference to the network: The connection determines the path followed by all bits between source and destination.

Connection-oriented networks, such as the PSTN, all have three distinct phases of operation. First, there is the *connection-establishment* phase, usually known as *call setup* when voice telephone calls are involved. In the PSTN, call setup is performed by a *signaling protocol* or set of rules used expressly for this purpose. Then there is the *data-transfer* phase, when the parties that are connected can talk. Finally, there is the *connection-release* phase, also known as *disconnect* or *call release*. This final phase also employs a signaling protocol to free up the network resources tied up in the connection so that the endpoints are free to establish other connections. Connectionless networks lack these first and third phases. Everything is a data transfer. If a router has a packet to send anywhere, anytime, it can just send it. The basic difference between connection-oriented and connectionless network operation is shown in Figure 3-3.

Obviously, it takes longer to transfer information over a connection-oriented network than over a connectionless network. Establishing connections takes time. And the user cannot do much of anything during this connection-establishment phase. Also, the signaling protocol adds another layer of complexity (and expense) to the network. Thus why is the PSTN connection-oriented when the Internet seems to do just fine without any connections between network nodes at all? What value do connections add to a voice network like the PSTN? As it turns out, quite

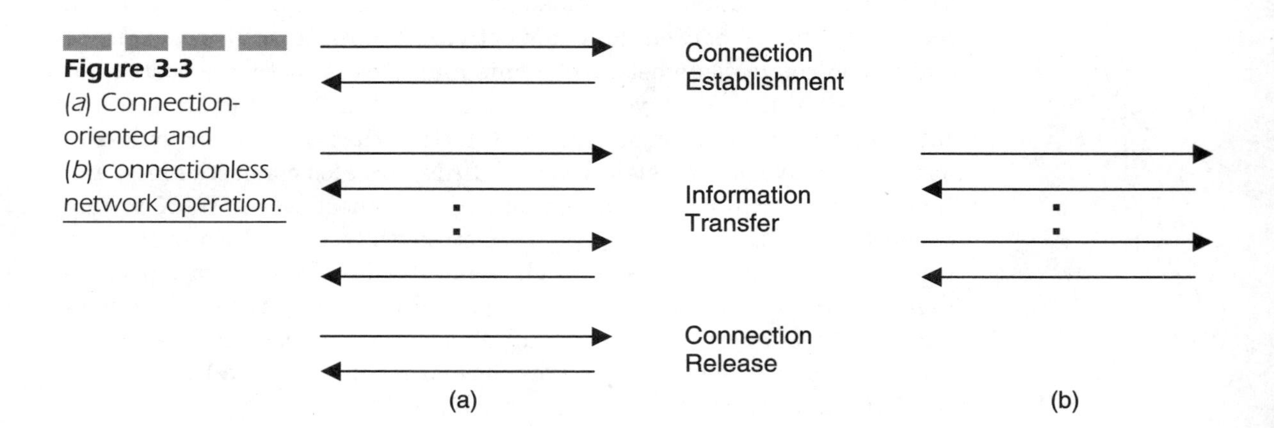

Figure 3-3 (*a*) Connection-oriented and (*b*) connectionless network operation.

a bit of value. So much value that the Internet is still struggling to provide the characteristics that voice PSTN users take for granted.

Circuit Switching with Guaranteed Bandwidth and Stable Delays

Circuit-switched networks provide "all the bandwidth all the time." Circuit-switched networks guarantee bandwidth from source to destination for the duration of the call. *Call* is just a word that is used to mean a connection on a voice network. It is a label for the circuit. For this reason, connections are often called *virtual circuits* in various data networks including the Internet. In the PSTN, however, circuits are not virtual. In the PSTN, circuits are real embodiments of dedicated bandwidths for a particular connection or call.

Circuits give the PSTN two essentials that a network optimized for voice applications needs to preserve quality. These are guaranteed bandwidth and stable delays. In addition, the delays must be low enough to satisfy voice users that they are exchanging information in real time. For voice purposes, the ITU-T has set this round-trip delay at 500 ms. This is a full ½ s, and many telephone users comfortable with voice networks with much smaller round-trip delays are less than pleased when they encounter a voice circuit with very high delays. The "symptom" of high-delay voice circuits, such as the circuits provided by satellites for international telephone calls, is that both speakers begin to talk at the same time. This can be quite annoying, and it is important to understand why this symptom occurs, since VoIP systems tend to have much higher delays than the PSTN.

When two people talk on a voice circuit, they are using a full-duplex circuit in a half-duplex manner. That is, even though the circuit can easily carry voice in both directions at the same time, most people will listen politely while the other party is speaking. But how does a listener know when it is his or her turn to speak? The cue is usually just a pause in the conversation. If the pause is long enough, the listener then knows that a response is expected. If the listener does not respond in a given length of time, the speaker figures, "Well, I guess he or she has nothing to say, so I'll keep talking," and the speaker begins to speak again.

As it turns out, the pause used to signal an end to a speaker's current "turn" in a conversation in most cultures is about ¼ s. The "timeout" is not fixed and varies quite a bit, naturally, and also changes from culture to culture, with Western cultures tending to be the most impatient in this regard. Why is this important for high-delay voice? Because on most satellite voice circuits, which rely on geosynchronous Earth orbit (GEO) satellites orbiting a tenth of the way to the moon (at about 22,000 miles), the round-trip delay is almost exactly ½ s. It takes ¼ s for the uplink portion and ¼ s for the downlink portion of the voice call. The propagation delay at the speed of light is about 240 ms, and the other 10 ms that make up the 250 ms in the ¼ s comes from the nodal processing delay in the satellite itself (older satellites had one-way delays of about ½ s). It is the coincidence of the ¼-s satellite delay and the ¼-s psychological listener timeout that causes all the problems.

To see why, consider Figure 3-4. The figure shows two people conversing over a high-delay voice circuit such as provided by a GEO satellite. In the figure, time passes from top to bottom, and the slant of the lines represents the delay between the time a speaker utters a word and the listener hears the word. On shorter, terrestrial circuits, the lines would be nearly horizontal, since even a coast-to-coast call on the PSTN in the United States seldom has an end-to-end delay above 30 ms. And the United States is one of the largest countries, telecommunications-wise, in the world. Only Brazil, Canada, China, and Russia are in the same category. In most countries, 10 ms or less is the typical delay on domes-

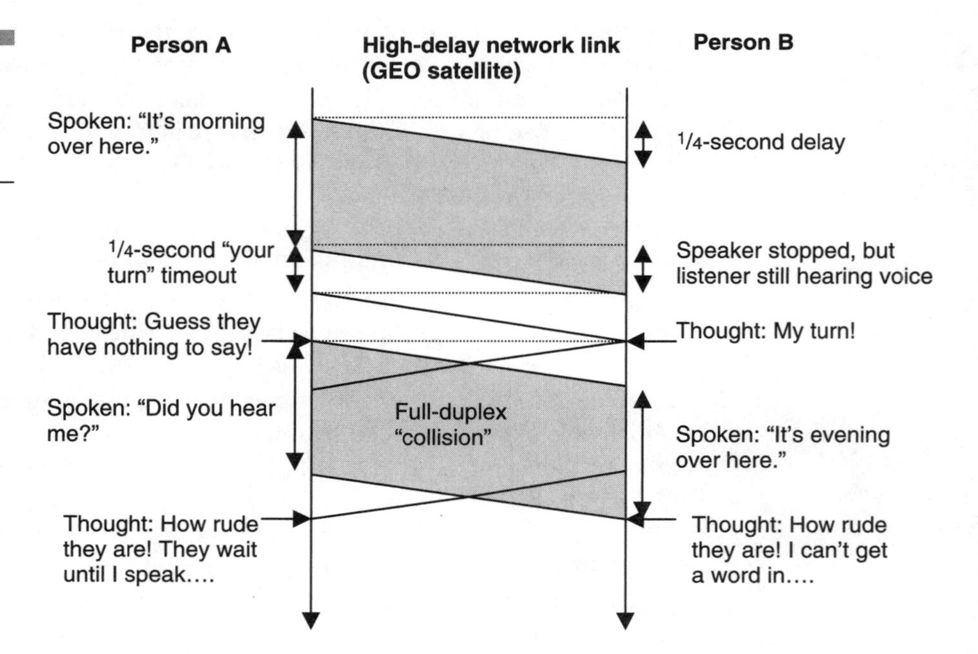

Figure 3-4 The trouble with high-delay voice circuits.

tic PSTN calls. The figure shows both what is spoken and what is just thought and not expressed in words by the two people.

The whole problem is that after person A finishes saying, "It's morning over here," it takes another ¼ s before person A hears the pause after the sentence. But before person B (the listener) can reply with, "It's evening over here," person A (the speaker) has thought, "I guess they have nothing to say in response. I'll keep talking." So person A, to avoid an awkward silence, speaks again, "Did you hear me?" Since the telephone is full-duplex, the words cross in midstream, and both hear the other while he or she is still talking. Naturally, person A thinks that the listener is very rude, since he or she waited until the speaker continued and then spoke himself or herself. And the feeling is mutual, since person B thinks that the speaker is so selfish as to speak when he or she should be listening.

No wonder that newer international voice circuits prefer undersea fiberoptic links with lower delays, and new satellite voice services use low-Earth-orbit (LEO) systems. Of course, for people who have never used a land-line telephone before, GEO satellite delays are not such a problem. They tend to accept the delay as a natural part of the system as a whole.

Thus low delays are essential for voice services in all developed nations with a legacy land-line telephone system. However, delays also must be *stable*. That is, the voice delay cannot vary appreciably during a sentence. If the delay does vary by more than a few milliseconds, the voice is distorted. If it assumed that the analog voice is digitized (not a bad assumption today) on the network, then what this means is that voice samples are arriving either farther and farther apart (called *dispersion*) or closer and closer together (called *clumping*). When the digital samples are converted back to analog sound, dispersion lowers the speaker's voice (people can sound like they have a sore throat), and clumping raises the speaker's voice (people can sound "squeaky").

Thus both low *and* stable delays are essential elements of a voice circuit. How does the PSTN supply low and stable delays? In two main ways. First, the PSTN voice switches must be able to move voice samples arriving on 64-kb/s input ports to the proper output ports very quickly. The ITU-T sets an upper bound on switching-element nodal processing delay at 450 μs, less than ½ ms. A *switching element* is defined as any PSTN component that does more than propagate 64-kb/s voice on a circuit, such as a voice switch (in actuality, a modern CO may contain several switching elements). This is much less than the typical propagation delay on a long voice circuit, and shorter circuits will have fewer switching elements anyway. On a typical coast-to-coast call within the United

States, the propagation delay of the bits on the circuit is generally about 20 ms, and there may be some 15 switching elements along the way. All together, however, they cannot add more than 15 × 450 μs, or 6.75 ms, to the total delay. And usually the delay through a modern switching element is much less. Thus variations in nodal processing delay do not significantly change voice quality.

The second way the PSTN furnishes low and stable voice delays is by mapping the voice connection (call) onto a fixed physical path through the network. This is why the PSTN is connection-oriented. If all the bits follow the same path, then sequential delivery is guaranteed, and the propagation delay is established at the beginning of the call and remains set for the duration of the call. This behavior can be contrasted with the Internet, where IP packets can be routed along different paths between a series of routers, causing variable nodal processing delays (due to varying router loads) and variable propagation delays (due to different physical paths being used). Sometimes the PSTN can reroute a call around a failed trunk, but this just reestablishes a new path in less than 60 ms or so. The people having the conversation notice a disruption, but once the new, stable path is established with the new delay, the conversation can continue as before. Internet delays can vary much more.

There is one other method that the PSTN often employs for monitoring delay. This is also a reason that the PSTN has retained fixed-bit-rate 64-kb/s PCM voice for so long. Delay variations can be detected in constant-bit-rate (CBR) voice such as a 64-kb/s DS-0 link very simply and efficiently just by counting bits. Suppose an input buffer on a voice switch receives not 64,000 bits in a second but 64,020. This means that the switch clock on the sending side of the link is running faster than expected, and adjustments should be made before the buffer overflows and voice bits are lost. Likewise, 63,950 bits could be counted, which indicates that the switch clock on the sending side of the link is running slower than expected, and adjustments should be made before the buffer *under*flows and voice bits are lost because there are no bits to send out.

This simple bit-counting clock-adjustment mechanism is one reason for the survival of 64-kb/s PCM voice when other methods exist and a large reason for the employment of CBR voice even at lower bit rates of 13 or 8 kb/s. A whole network synchronization system exists solely to deliver stable delays on the PSTN, a system that is totally lacking on the bursty, packet-switched Internet, where nothing forms a constant-bit-rate stream at all. Of course, if digitized voice were ever packetized, then the Internet would be an appropriate network for the delivery of VoIP packets. However, then other methods for providing the low and stable delays that voice demands must be found.

Analog Multiplexing

All the voice circuits in the PSTN typically originated and terminated at individual access lines. At the end of the access lines there was usually only one telephone handset. There was only one telephone because the telephone itself until the late 1960s could not belong to the customer. The telephone was telephone company property and leased on a monthly basis from the local telephone company. If anyone wanted a second telephone, the second telephone had to be paid for on a monthly basis also. And recurring costs always dominate over time compared with one-time, nonrecurring charges.

Not that there weren't telephones available for purchase. People cheated all the time and hooked up a second or even third telephone without obtaining the telephone through the local telephone company. Local telephone companies routinely checked for illegal telephones by sending ringing voltage down the access line and measuring the voltage drop. Any extra "ringers" would show up as an excess voltage drop, and telephone company representatives either literally showed up to disconnect the telephone or started collecting monthly fees. People quickly learned how to disconnect the ringing circuit to prevent the illegal telephones from ringing, but this compromised the telephone's usefulness for incoming calls.

In any case, only one telephone call at a time could be carried on one twisted-pair access line, this being the essence of circuit switching. However, once the voice circuit hit the central office, it made no sense to have a separate wire pair for every voice call that the central office was designed to carry. Thus the calls in progress were routinely multiplexed onto higher-capacity trunks. Note that until the 1960s, voice was strictly carried in an analog fashion. Since frequency-division multiplexing (FDM) was more efficient for analog voice conversations, the voice channels were combined into an FDM analog multiplexing hierarchy. Lower levels of the hierarchy used multiple pairs of ordinary twisted-pair copper wire, whereas higher levels used coaxial cable or microwave towers to provide the huge bandwidths that these trunking carrier systems required.

This section describes the analog multiplexing systems used in the Bell System. Other telephone systems could have their own methods, but most followed the Bell hierarchy simply because what the Bell System used became a standard way of doing things.

The analog multiplexing hierarchy began with N-carrier, a multipair system for local trunking needs. The N-1 (introduced in 1950) and N-2 equipment multiplexed 12 voice channels onto two pairs of twisted-pair wire identical to the wires used on the local loop. N-carrier links spanned anywhere from 5 to 250 miles, but most were used to link a central office to its nearby neighbors, to the local tandem, or to the nearest long-distance toll office. Later, N-3 (1964) and N-4 (1979) carrier equipment doubled the capacity to 24 voice channels while at the same time taking up less space and consuming less power, although everything was still analog.

Then there were two distinct short-haul microwave multiplexing systems. Microwave systems are line-of-sight radiofrequency transmission systems that require repeater towers placed along the path of the link. Short-haul microwave systems were used commonly to aggregate a lot of voice traffic between tandems or toll offices where running land-line wire would be very expensive, such as across major rivers. Despite the *short-haul* name, which applied to both the primary use and the fact that the repeater towers were often only 5 miles or so apart, these systems could span 10 repeater hops and have spacings up to 25 miles apart. Thus a short-haul microwave system could be anywhere from 5 to 250 miles long.

The 6-GHz systems included the TM-2, which carried 8400 voice channels, and the TM-2A, which carried 12,600 voice channels. This reflects the enormous bandwidths available on these microwave systems. The 11-GHz systems included the TL-A2, which carried 6000 voice channels, and the TN-1, which could have either 19,800 or 25,200 voice channels. By the early 1980s, deployment of both 11-GHz systems had ceased because the bandwidth consumed was far more than actually needed for the number of voice channels carried.

There also were land-line analog multiplexing methods based on coaxial cable, most notably the L-carrier system. The coaxial cable was similar to the cable used today in cable TV systems and Ethernet LANs but quite different electrically and in terms of frequencies used. L-carrier dates all the way back to 1941, but it was in 1946 that deployment began in earnest. In L-carrier, several coaxial cables could be grouped together as what were called *tubes* to multiplex even greater numbers of voice channels together. The early L-1 system carried only 1800 voice channels, but the L-5E system of 1978 carried 132,000 voice channels on 22 tubes. Ten tubes were for the voice channels in each direction, and two were for protection purposes. Thus each coaxial cable carried 13,200 voice channels. L-carrier links could be as short as 1 mile or as long as 150 miles.

Table 3-2 shows the L-carrier family in terms of date introduced into

TABLE 3-2

The Analog L-Carrier Family

L-Carrier Level	Year Placed in Service	Number of Voice Channels
L-1	1946 (1941)	1800
L-3	1953	9300
L-4	1967	32,400
L-5	1974	108,000
L-5E	1978	132,000

service and total number of voice channels in a single cable sheath. Each cable sheath could have many coaxial cables inside. Note the absence of an L-2 designation.

Undersea coaxial cables were used for international voice communications. This began with 24 voice channels on two coaxial cables from Key West, Florida, to Havanna, Cuba, in 1950. Transatlantic cables from the United States to Europe were installed every 6 years or so to keep up with the growing voice traffic between the United States and Europe. By 1976, analog coaxial cables could carry 4000 voice channels under the North Atlantic.

The next major category of analog carrier discussed here is the long-haul analog microwave. Again, the name is more intended to describe the purpose of the system, not the length of the spans, although few long-haul microwave repeaters were less than 20 miles apart, and 30 miles was normal. There were 4-GHz systems (the TD family) and 6-GHz systems (the TH family and AR6A) developed between 1950 and 1983. The TD systems carried from 2400 to 19,800 voice channels, depending on version. The TH-1 system carried 10,800 voice channels, the TH-2 system had 12,600 voice channels, the TH-3 system had 16,800 voice channels, and the AR6A system had 42,000 voice channels.

Finally, satellites often were used instead of undersea coaxial cable for international communications. Satellite systems basically point huge microwave dishes at the sky. The proof that this would work was provided by *Echo I* in 1961. *Echo I* was a big, shiny balloon inflated in orbit and used to bounce uplink and downlink signals between Earth stations. In 1962, *Telstar* carried all of the 12 voice channels in a low Earth orbit, but at least this satellite was an active system that received and retransmitted the voice channels through a system of *transponders*. *Telstar* moved in the sky also, limiting its use for long-telephone calls, unlike today's *geosynchronous* satellites, which are stationary over Earth's equator. Today, of course, international and private satellites carry 21,000 voice channels or more as well as television channels.

So far these descriptions have been of the *physical* multiplexing systems. However, for multiplexing purposes, the analog hierarchy is divided into conceptual levels that are observed regardless of the physical medium and multiplexing capacity on which the conceptual level is deployed. This is a big help in international trunking situations where different national carrier systems must interconnect.

The basic unit of analog voice multiplexing is the *group* of 12 voice channels. The ITU-T defines 10 groups as making up a *supergroup* of 120 voice channels, but in the United States, a *supergroup* has only 60 voice channels. Five supergroups make up a *mastergroup* of 600 voice channels. In the United States, the next level is the *jumbogroup* of 3600 voice channels (6 mastergroups), while internationally, the highest level is called the *supermastergroup* of 1800 voice channels (3 mastergroups). Thus no matter how the voice channels are carried physically, they are always organized into these logical groupings. And although the matchup between the U.S. and international versions is slightly different, the groupings are much the same. Table 3-3 shows the levels of the analog hierarchy.

TABLE 3-3

The Analog Group Hierarchy

Name	VCs	U.S. Equipment Level	ITU-T Equipment Level
Group	12	N-carrier	
U.S. supergroup	60		
ITU-T supergroup	120		
Mastergroup	600	L-1	G.341 (5 supergroups)
	1200	TD	
ITU-T supermastergroup	1800	TH/L-3	G.343 (15 supergroups)
	2400		G.344 (20 supergroups)
U.S. jumbogroup	3600	L-4	
	5400		G.332/G.345 (45 supergroups)
	6000	AR6A	
	7200		G.344/G.346 (60 supergroups)
	10,800	L-5	
	13,200	L-5E	
	21,600		G.333 (180 supergroups)

VC stands for the number of voice channels supported at each level. Note that in actual practice, the levels of the group hierarchy are "mapped" to the actual physical multiplexing carrier systems. In other words, there is no real multiplexing hardware that carries a single 60-voice-channel supergroup on coaxial cable, microwave, or anything else. Supergroups are called an *intermediate multiplex level* because of their conceptual existence, and hardware-based multiplexing levels are called *system levels*. Not very many levels even carry the same number of voice channels, making international connections that much more interesting.

Digital Multiplexing

Analog multiplexing in the PSTN is still around, if only because regulated telephone companies are only allowed to replace a certain amount of their equipment per year, and the useful lifetime in any case is generally set at anywhere from 30 to 40 years. Thus analog multiplexing systems installed in 1980 could be around for some time to come.

However, most analog multiplexing systems have been converted to *digital* multiplexing systems, and all new multiplexing systems installed in the PSTN are digital systems. Digital multiplexing has been around since 1962, with the introduction by AT&T of the highly successful T-1 multiplexing system. T-1 is the first level of what became a whole digital hierarchy that multiplexes voice channels not as analog voice passbands at 4 kHz using frequency-division multiplexing (FDM) but as digitized voice channels at 64 kb/s using time-division multiplexing (TDM). Both are still circuit-switching, "all the bandwidth all the time" systems, but the basic unit of analog multiplexing is the voice channel and the basic unit of digital multiplexing is 64-kb/s DS-0 serial bit streams. Digital multiplexing terminals (transmitters and receivers) cost less than their analog counterparts even in 1962 and had better performance characteristics (i.e., the voice sounded better to human listeners) than analog systems that amplified noise right along with the voice signal. As wire-pair installation costs increased and the cost of digital multiplexing equipment fell with the price of digital components, T-1 was cost-effective at shorter and shorter distances compared with analog trunks.

Technically, T-1 is a *physical* trunking system consisting of two pairs of unshielded twisted-pair copper wire, transmitter and receiver specifications, and repeaters usually spaced every 6000 ft in between. The organization of the bits sent on a physical T-1 system is defined as the *DS-1 frame structure*. A DS-1 carries twenty-four 64-kb/s DS-0s on the two pairs of copper wire. This gives a T-1 digital transfer rate of 1.536 Mb/s (24 × 64 kb/s), and there is 8 kb/s of overhead for a total line rate on a T-1 of 1.544 Mb/s, usually rounded off to 1.5 Mb/s.

Another way of looking at the T-1 line rate is to realize that 24 digital voice channels generate 8 bits and are sampled 8000 times per second (this is where the 64 kb/s comes from). The 24 channels therefore pack 192 bits (24 × 8 bits) in each T-1 frame. A 193rd bit known as the *framing bit* is added to each basic frame to allow receivers to synchronize on the incoming frame boundaries. Since there are 8000 samples per second in each digital voice channel, the T-1 link must operate at 1.536 Mb/s (192 bits × 8000 per second), or actually 1.544 Mb/s (193 × 8000 per second) just to keep up.

Thus each 193-bit T-1 frame is sent in 1/8000 of a second, or one in every 125 μs. This is below the threshold of human perception and much faster than the blink of an eye, which takes place in 1/10 s or 100 ms or 100,000 μs. Oddly enough, golfers actually may have the best perspective on T-1 frame speeds. The brief "click" sound of the driver club hitting the golf ball, a sound at about the limit of human perception, is about ½ ms long, or 500 μs. Thus, during that click, a T-1 transmitter has sent four frames out on a T-1 link.

A T-1 link can be used to carry digitized voice channels or serial bit streams from a computer in any combination. For instance, a T-1 multiplexer can carry 12 voice conversations from a corporate PBX to another PBX at another site within the same corporation and at the same time link 12 routers throughout the building to their counterparts at the other site at 64 kb/s.

Although voice channels transverse a T-1 link at 64 kb/s for the most part, serial data are often limited to 56 kb/s, or seven-eighths of the 64-kb/s DS-0 channel rate. This is so because sometimes the eighth bit of each DS-0 channel is used by the T-1 equipment for functions other than bit transfer. For voice applications, this intermittent "bit robbing" is not audible, but of course, computers rely on all the bits they send in a DS-0 channel to be delivered correctly. If bits are "robbed," the true content of the eighth bit is obliterated and cannot be restored. Now, not every eighth bit of every channel in every T-1 frame is "robbed," but since only the carrier knows for sure when the bit is used or "robbed," the customer has no choice but to avoid using the eighth bit in each DS-0 channel altogether.

There are actually four uses for this eighth "robbed" bit. However, one of these uses is only required when the DS-0 channel is used for voice trunks ("robbed bit signaling") and so does not apply when the DS-0 channel carries serial computer bits. The other three uses of the eighth bit are

1. To maintain *1s density* on the line
2. To signal *yellow alarm* to a transmitter
3. For internal carrier *maintenance* messages

Each of these uses should be explained in a few words. The T-1 line code, bipolar AMI, represents 0 bits by no voltage at all on the line. If too many 0 bits appear in a row, as may be the case when adjacent DS-0 channels carry computer bits, the receivers can lose *frame synchronization* ("Was that 30 bits of nothing, or only 29? Or maybe 31?"). The 1 bits, on the other hand, are represented by voltage pulses. So when a T-1 using pure AMI encounters 16 consecutive 0 bits to be sent on adjacent DS-0 channels, the equipment flips one of the 0 bits (the eighth bit in the first channel) to a 1 bit to maintain a minimal *1s density* on the line.

Yellow alarm occurs when a T-1 receiver loses the incoming signal entirely. Bits can still be sent, perhaps, but there is no sense in continuing operation if one pair is no longer functional. Thus yellow alarm sets every eighth bit in every channel upstream to a 1 bit in order to tell the sender to stop sending downstream (and potentially switch to a backup link if available). Thus computer bits that otherwise would arrive okay are obliterated by the equipment function. Finally, technicians can use the eighth bit in the channels to signal T-1 equipment to perform diagnostics, gather network management information, and the like. None of these methods is used to penalize data users, of course. They just began to be used when T-1 was exclusively a voice trunking method and continued to be used when T-1 began to be sold for data purposes in the early 1980s.

If a T-1 link allows data users to employ the full 64-kb/s DS-0 channel rate, known as *clear channel capability*, two things must be done by the T-1 service provider. First, instead of using straight bipolar AMI, the transmitter, receiver, and all repeaters and other equipment in between must use something called *binary 8 zero substitution* (B8ZS) line coding. Sometimes B8ZS is positioned as a distinct line code from AMI, but B8ZS is best thought of as a *feature* of AMI. This solves the 1s density problem. Second, another form of T-1 framing known as the *T-1 extended superframe* (ESF) is used to remove yellow alarm and maintenance messages from the channels themselves and place them in the 193rd framing bit pattern in 24 consecutive T-1 frames.

Oddly, the international equivalent of T-1, known as *E-1*, never suffered from this 56-kb/s limitation. The E-1 frame structure is different, so yellow alarm and maintenance messaging never became an issue. A special signaling channel in E-1 avoids robbed-bit signaling altogether. The E-1 line code is not bipolar AMI, and a special synchronization channel ("word") avoids 1s density concerns, so B8SZ is not required either. All E-1s support 64-kb/s transmission, and all circuits that begin in one country (as defined by the United Nations) and terminate in another must use E-1 instead of T-1. T-1 can only be used within the borders of the United States (Canada also uses T-1, but also internally).

E-1 multiplexes 30, not 24, digital voice channels into the basic E-1 frame. Frames are still grouped, but not for the same reason as in T-1. E-1 channels still run at 64 kb/s, but use a slightly different (and incompatible) form of voice digitization. The basic structure of the T-1 and E-1 frames are shown in Figure 3-5. For T-1,

193 bits sent 8000 times/s = .544 Mb/s

and for E-1

256 bits sent 8000 times/s = 2.048 Mb/s.

With either T-1 or E-1, the basic unit remains the 64-kb/s DS-0 circuit-switched voice channel. However, as the voice traffic loads offered to telephone networks increased, some method had to be found to multiplex more and more digital voice channels between the same pair of voice switches on the same physical facilities, the better to increase capacity on the trunk.

The Original Digital Hierarchy Over time, a whole digital hierarchy of TDM multiplexing levels was developed for T-1 and E-1. Japan, which

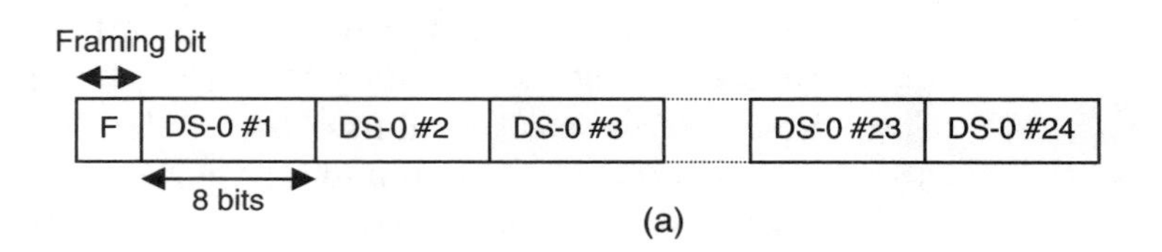

Figure 3-5 *(a)* T-1 and *(b)* E-1 frame structures.

used the basic T-1 structure, had its own variation, sometimes called *J-1*. The main difference between the newer digital and the older analog methods, other than the simple fact that they carried digital and analog voice channels, was that the digital hierarchy used TDM and the analog hierarchy used FDM as the multiplexing technique. This T-1/E-1/J-1 structure is being replaced by a newer digital hierarchy, but this section will examine the original. The next section explores the newer digital hierarchy used by the telephone companies.

It is sometimes claimed that analog signals cannot be TDM'd and that digital signals cannot be FDM'd. This is simply not true. There were Bell System analog multiplexers that TDM'd analog voice channels (called *time-slot interpolation equipment*), and the whole fiberoptic field of wavelength-division muliplexing (WDM) is based on the concept of FDM'ing digital channels. It is true enough, however, that before fiberoptics came along, analog TDM and digital FDM equipment was complex, bulky, expensive, inefficient, and downright silly to do without a good reason. Thus analog systems used FDM to multiplex, and digital systems used TDM to multiplex.

TDM divides the total bit rate into a series of *time slots* all of the same, short duration. Each time slot carries a fixed number of bits, generally 8, in a given *bit time*. For example, a T-1 frame sends 193 bits in 125 μs (1/8000 of a second). This gives a characteristic bit time of 125 μs/193 bits, or about 0.65 μs for each bit.

Each level of all the digital hierarchies multiplexes a certain number of voice channels into a frame structure with some overhead associated with the proper operation and administration of the link. This whole aspect of digital multiplexing overhead is known as *OAM* (operations, administration, and maintenance) to the ITU-T and usually is called *OAM&P* (operations, administration, maintenance, and provisioning) in the United States. (When asked about the *P* for provisioning, a Bell Labs engineer replied, "Provisioning determines when a customer can be billed for the link, so it's important enough to deserve its own letter in this country.")

Be that as it may, the overhead bits at each level cannot be used for carrying telephone calls or user data (serial bits between computers). The overhead bits are used by the telephone company to run the links properly. For this reason, users never really see them. However, the overhead required at each level of the original digital hierarchy became a cause for concern.

Table 3-4 shows the original digital hierarchy used in the United States (T-carrier), Europe (E-carrier), and Japan (J-carrier). At each level, the figure shows the line rate employed, the number of voice chan-

TABLE 3-4

The Original Digital Hierarchy: PDH

Multiplexing Level	Carrier Designation	Line Rate, Mb/s	Voice Channels	Overhead	Overhead Percentage
Level 1	T-1/J-1	1.544	24	8 kb/s	0.518
	E-1	2.048	30	16 kb/s	0.781
Level 2	T-2/J-2	6.312	96	168 kb/s	2.661
	E-2	8.448	120	768 kb/s	9.090
Level 3	J-3	32.064	480	1.344 Mb/s	4.191
	E-3	34.368	480	3.648 Mb/s	10.519
	T-3	44.736	672	1.728 Mb/s	3.862
Level 4	J-4	97.728	1440	5.568 Mb/s	5.697
	E-4	139.264	1920	16.384 Mb/s	11.764
	T-4	274.176	4032	16.128 Mb/s	5.882
Level 5	J-5	397.20	5760	28.560 Mb/s	7.190
	E-5	565.148	7680	73.628 Mb/s	13.028

nels supported, the amount of overhead (obtained by subtracting the number of 64-kb/s voice channels from the line rate), and the percentage of the line rate that is devoted to overhead.

In the United States, a fifth level was never defined, and the fourth level was seldom deployed at all. Instead, multiplexing equipment vendors used inventive methods to combine various numbers of T-3s onto fiberoptic systems, all in a vendor-specific fashion. Note that within each carrier system the overhead percentage increases with line rate, mostly due to efforts to deal with ever-decreasing bit times (as the line rate increases, serial bit times decrease proportionally) on the link. This need to *synchronize* bits was only partially successful most of the time (i.e., bits tended to disappear or appear in transit). Thus the whole system came to be called the *pleisiochronous* (a Greek word meaning “sort of synchronous”) *digital hierarchy*, or *PDH*. Much of the overhead was devoted to inserting or deleting bits on the link to maintain overall synchronization.

It should be noted that each level of the hierarchy is inherently channelized. In other words, a T-3 is expected to carry 672 DS-0s, each limited to 64 kb/s, whether used for voice or not. Of course, today, data users need more than 64-kb/s links between routers and other computers. Thus almost all the levels are also available (“provisioned”) in *unchan-*

nelized forms. This means that all the line rates derived from the voice channels are now available for a single source (i.e., computer serial port) to use. Thus an unchannelized T-1 represents not 24 DS-0s at 64 kb/s but a single bit stream at 1.536 Mb/s (the overhead still cannot be used).

The PDH is still around and will be for a long time, again partially due to limitations arising from regulation but also due to simple economics. Replacing all of anything is expensive and time-consuming. However, there exists today a method for better digital multiplexing, a method based on fiberoptic systems and better synchronization techniques. In the United States, this new hierarchy is called the *synchronous optical network* (SONET), and the ITU-T calls it the *synchronous digital hierarchy* (SDH) to acknowledge the minor differences between SONET (which usually replaces T-carrier) and SDH (which usually replaces E-carrier).

Enter SONET/SDH The PDH, both T-carrier and E-carrier (J-carrier will now be treated as a variation of T-carrier), suffered from a number of liabilities. The fact that the multiplexing was pleisiochronous meant not only that overhead increased as the number of voice channels grew but also that the 8-bit voice samples could not be kept together above the T-1 and E-1 levels. Thus, above the first level of the hierarchies, multiplexing was done bit by bit. The only way to extract a voice channel from a T-3, for example, was to completely recover (demultiplex) all 672 voice channels just to get at the one needed. Also, there was no standard way to carry the higher levels of the hierarchies on fiberoptic links, which led to many vendor-specific implementations that made it next to impossible to hand off trunks between different telephone companies above the T-1 level in the United States.

Fiberoptic links had been around in the United States since the late 1960s, but AT&T deployed the first commercial fiberoptic links in Atlanta in 1980. By 1984, the need for a new fiberoptic-based hierarchy was apparent, and in the United States, a system was developed known as the *synchronous optical network* (SONET). The term *synchronous* referred to the fact that SONET multiplexing always kept 8-bit voice samples intact at any level of the multiplexing hierarchy. This made accessing individual voice channels that much easier and cost-effective. This also allowed the overhead percentage to be fixed and not grow in proportion to the number of voice channels multiplexed.

In the United States, the lowest SONET level was intended to align nicely with T-3, which ran at about 45 Mb/s and carried 672 voice channels. Since T-3 did not exist outside the United States (although some T-3s were used in the Philippines, installed by the U.S. military), the

ITU-T adjusted the SONET levels for its own purposes and renamed the final result the *synchronous digital hierarchy* (SDH).

SONET/SDH defines both frame formats and physical transmission systems, just as T-carrier and E-carrier do. In SONET, for example, the frame formats used are designated *synchronous transport signals* (STSs) and use an *N* suffix to indicate the level of the hierarchy. Although *N* technically can take on any value from 1 through 256, in actual practice, only a handful of the values have been implemented in SONET equipment. The physical network of fiberoptic cable, transmitters, and receivers on which the STS frames flow are designated as *optical carriers* (OCs), also with an *N* suffix for the level. Thus an STS-3 frame is used on an OC-3 link, in the same way that a DS-1 frame was used on a T-1 link in the T-carrier hierarchy. However, just as people spoke loosely about T-carrier and said "T-1" when they really meant "DS-1," SONET people usually say "OC-3" and rarely, if at all, mention STS-3 frames.

The ITU-T uses the term *synchronous transport module* (STM) instead. There are optical STM levels (STMO) and electrical STM levels (STME) for the frame structures, both with an *N* suffix. Usually, just STM-N is used, and whether the issue is electrical or optical must be determined by context. Oddly, the *N* in the SDH is interpreted differently from the *N* in SONET. Basically, STM "counts in threes" when compared with SONET. Thus an OC-3 does not run at the same speed as an STM-3, as one might expect, but an STM-1. An STM-3, of course, would run at the same speed as an OC-12. SDH has been as selective as SONET when it comes to supported speeds. Table 3-5 shows the speeds of the supported levels of both SONET and SDH, along with the number of voice channels each can carry when used as a PSTN trunking mechanism in SONET (voice channel numbers are slightly different in SDH).

TABLE 3-5

The SONET/SDH Hierarchy

Line Speed	SONET Level	SDH Level	Number of Voice Channels (SONET)
51.84 Mb/s	OC-1	"STM-0"	672 (same as T-3)
155.52 Mb/s	OC-3	STM-1	2016
622.08 Mb/s	OC-12	STM-4	8064
2.488 Gb/s	OC-48	STM-16	32,256
4.977 Gb/s	OC-96	Not used	64,512
9.953 Gb/s	OC-192	STM-64	129,024

Note that all levels are simple multiples of 51.84 Mb/s, the basic SONET line rate.

All forms of both SONET and SDH are also available in unchannelized formats, just as in the PDH. The unchannelized format is more appropriate for serial bit transfer between computers or routers. When unchannelized, SONET and SDH frames carry not a huge number of voice channels but a single stream of IP packets, frame-relay frames, ATM cells, or even other things. Somewhat perversely, unchannelized levels of SONET and SDH are indicated by adding a lowercase *c* to the multiplexing level. The *c* stands for "concatenation." So an OC-3c represents a single bit stream that can take IP packets to and from a router serial port at 155.52 Mb/s. Technically, the *c* should be attached to the STS level (i.e., STS-3c), since it is the *frame* that has changed (it has been concatenated), not the optical link. But the OC-3c form is used most often, and sometimes even with a capital *C*, as in OC-3C (this came about supposedly because equipment vendors could not print lowercase on their SONET line cards).

Loops and Trunks: Two-Wire and Four-Wire

One other point about loops and trunks should be made before moving on to a discussion of the use of the PSTN rather than its physical components. While it is true that there is no physical difference between the media used for local loops (access lines) and trunks (multiplexed interswitch voice circuits), this does not mean that there are no differences in the way that loops and trunks operate. This can be seen in the fact that it has already been pointed out that local loops generally carry only one conversation on a single twisted pair, whereas a trunk can multiplex many conversations on the same type of twisted pair (although usually on two pairs).

In fact, this is the crucial difference in operation between twisted pairs as loops and twisted pairs as trunks in the PSTN. A local loop is typically just two wires: a single twisted pair. Electrically, the local loop is dc powered (so simple batteries are used in the CO to power the access line), operates full-duplex (both parties can speak and be understood at the same time), and has no amplifiers in most cases. In contrast, a trunk in its simplest form consists of four wires: two twisted pairs. This was before coaxial cable, microwave towers, and fiberoptic cable were used as trunking methods. Electrically, a twisted pair trunk is ac powered (ac travels farther than dc without expensive amplifiers), operates half-duplex (one pair is used for transmitting and one pair is used for receiving), and amplifiers are used (trunks are much longer than loops, on average). The reason for two-wire local loops is simple: Two-wire loops cost less to buy and install, and there are many more local loops than there are trunks.

Ironically, within the telephone instrument itself, voice is treated in the four-wire manner, except for the ac powering. If ac power is required by the telephone instrument, such as a telephone, fax machine, or modem, it is usually to do more than just allow the device to interface with the telephone network The ac power is used mainly for storing dialed numbers, displaying digits, and so forth.

In order to convert between two-wire and four-wire operation, an electrical device known as a *hybrid* is required. The hybrid is a simple enough electrical device that isolates the two-wire and four-wire circuits from each other, performs a function known as *impedance balancing*, and so on. For the purposes of this discussion, the important thing that hybrids introduce on a voice circuit is *echo*. Echo can cause telephone users to be distracted by hearing their own speech delayed as they speak. In other words, echo lowers voice quality dramatically.

The hybrids in the telephone handsets are not the problem. These hybrids can be used to provide *sidetone*. Sidetone is useful voice feedback that prevents users from shouting on the telephone because the telephone handset cuts off half a person's hearing. Since sidetone is not appreciably delayed from the speaker's own words, sidetone is not interpreted psychologically as an echo at all.

However, the hybrids at each end of the local loops where the trunks between the switches interface with the two-wire loops are a significant source of echo. Generally, whenever there is an impedance mismatch in an electrical system, echoes (signal reflections) are the result. In the PSTN, these two sources of echo on trunks are called *near-end echo* and *far-end echo*. The end-to-end structure of loops and trunks, with echo sources, are shown in Figure 3-6.

Even when trunks are provisioned on coaxial cable, microwave towers, or fiberoptic cable, the voice channels they carry are emulating *four-wire equivalent circuits*, and a hybrid function must be provided. In practice, when digital voice switches are involved, the hybrid function is supplied on the line card interfacing the two-wire local loops with the digital switch. This makes sense, because the digital switch cannot handle analog voice on the local loop anyway. It is common to characterize these digital line card functions by the acronym BORSCHT. BORSCHT stands for the line card functions of *battery* (the dc power for the local loop), *over-*

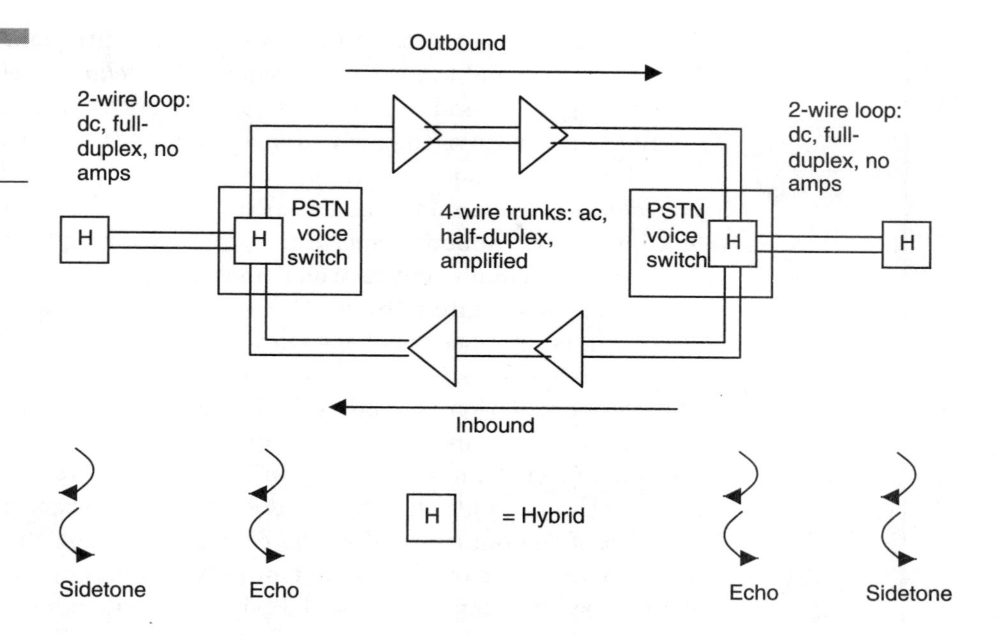

Figure 3-6 Two-wire loops, four-wire trunks, and echo.

voltage protection (so that things like lightning strikes on the local loops will not fry delicate digital chipsets in the CO), *ringing* (so that the user's telephone rings), *supervision* (now called *signaling*, and represents all on-hook, off-hook, and other functions), *coding* (from analog to digital), *hybrid* (balancing the loop electrically), and *testing* (the line card has a test-set interface). Naturally, any device that terminates a local loop must perform the same functions, even it if is an IP router or VoIP server.

Hybrids introduce echo. But what should be done about echo? One option is nothing. If the echo delay is short enough, then the echo is indistinguishable from the sidetone, which the user requires anyway. Generally, the echo from the near-end hybrid at the end of the local loop is not delayed enough to cause a problem. Studies have shown that echo must be delayed during its round-trip by about 50 ms compared with sidetone to be perceived as a problem.

The main problem with echo is with the far-end hybrid's echo. This echo must be canceled if people are to use the voice service productively. This echo "leaks" into the part of the half-duplex trunk that leads back to the speaker. The echo gets amplified right along with the far-end voice and winds up at the speaker's ear. This *speaker echo* is what must be dealt with if the propagation delay and the nodal processing delays (switch delays) on the circuit are above some threshold value (typically set at 45 ms round trip). Thus in some cases *echo cancellation* is needed on longer circuits and in some cases on shorter circuits, depending on the number and speed of the switching elements between the hybrids.

Echo cancellation is today provided by using special electronic *echo cancelers* on the trunks. The details of operation of the echo cancelers is not important. What is important is that echo cancelers are common on international circuits but rare on domestic circuits outside large countries like the United States, Brazil, Canada, China, and Russia (occasionally, Australia is added to this list). Only in these countries do domestic calls have round-trip delays that routinely require the use of echo cancelers.

Echo cancellation is another PSTN function that must be supplied by VoIP systems. Oddly, when echo cancelers are present on a voice circuit, full-duplex modem operation is impossible. The reasons this is so are fairly technical, but what it boils down to is that echo cancelers will cancel part of the data signal, so the echo cancelers on the PSTN trunking network are more of an impediment to full-duplex operation than a help. Thus a system for turning off PSTN echo cancelers—if present—on trunk was devised for analog modem operation. This is the purpose of the 2100-Hz "whistle" that modems generate when they first complete a circuit, as anyone who has mistakenly dialed a fax machine has discovered. All PSTN echo cancelers will disable themselves when they sense this special tone (in the range 2010-2240 Hz) for at least 400 ms (most modems generate this for about a full second). The echo cancelers enable themselves at the end of the modem connection (actually, when the signal vanishes for more than 100 ms), although occasionally they get stuck "off" and annoy the next voice callers to use the trunk. Figure 3-7 shows the function of the echo cancelers on the trunks. Note that an echo can-

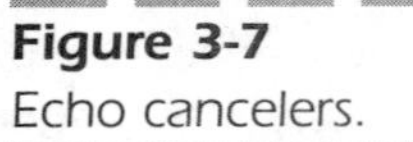

Figure 3-7
Echo cancelers.

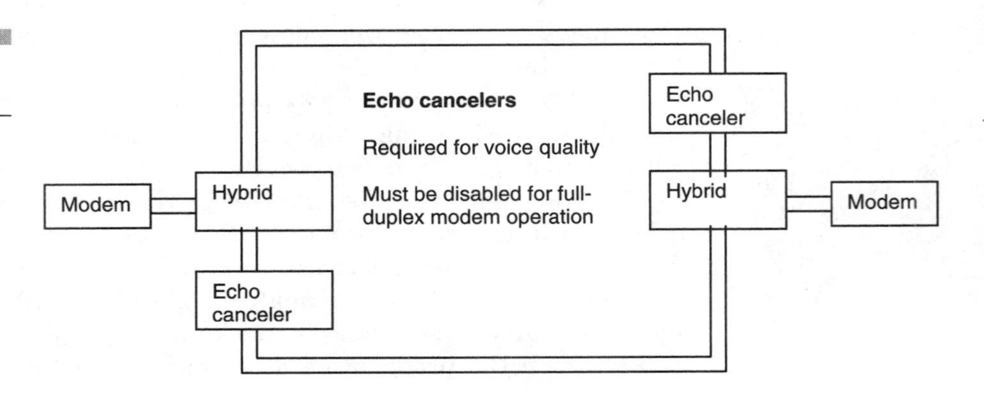

celer is needed on each four-wire trunk, since the trunks are half-duplex and each has a speaker.

Why spend so much time on voice two-wire and four-wire distinctions? Because the vast majority of digital links used to connect routers on the Internet are already four-wire equivalent circuits. If an analog local loop were ever hooked up directly to an Internet router for VoIP purposes, then all the functions that a voice performs to make analog two-wire voice ready for digital four-wire circuits must be done in the router, including echo cancellation. The router becomes the voice switch, and the network becomes the trunking network. Even if the analog local loops were to terminate on a special VoIP telephony server, then *this* server device becomes the equivalent of the PSTN switch. If either router or server were to map fixed-bit-rate digitized voice onto a T-1 link (for instance), this would be no different from supporting four-wire equivalent circuits on trunks between PSTN switches.

Of course, there is little point in taking fixed-bit-rate voice directly onto a packet-switched network. This is properly a topic for the next chapter. For now, it is enough to note that this can be done, but it makes little sense from a networking and economic perspective.

So far this chapter has discussed the physical components of the PSTN, especially the trunks. This is so because the unchannelized versions of these trunks are commonly used as links between the routers on the Internet. Thus many VoIP networks essentially replace the trunking network with a packet-switched network like the Internet, using unchannelized versions of exactly this same trunking technology as backbone links. However, the time has come to look at just what happens on the PSTN when a person (user) makes a telephone call.

Making Telephone Connections

Some telephone calls are known as local calls, and some are known as long-distance calls, although technically there are just telephone calls that cost more than other calls. Generally, the longer the distance between telephones, the higher is the toll for the call, just as in driving, when driving farther on a toll road requires a higher toll to be paid by the driver. Thus there are local toll calls as well as long-distance toll calls.

Even a local call requires the use of trunks and other switches, except when both telephones are served by the same CO. This section assumes that the telephone number dialed is within the local calling area of the originator, not a long-distance call requiring the use of another telephone company. This call may be a local toll call, but that is not important to an understanding of this scenario. What is important here is that whatever the PSTN does to complete the call also must be done by a network providing VoIP services if the users are allowed to dial telephone numbers to complete calls.

The general flow of the telephone call completion process is shown in Figure 3-8. It is assumed that there are no special features involved such as speed dialing, call forwarding, call waiting, or the like. This is just a person picking up a telephone handset, dialing a number, and seeing if someone at the destination answers the telephone or not.

The figure is somewhat condensed but accurate. The call starts with the originator (calling party) lifting the telephone handset and going *off-hook* in preparation for making a call. The local switch constantly monitors (supervises) the local loops waiting for just this moment, which is an opportunity to make revenue for the PSTN. To users, it may seem that the dial tone is present as soon as this off-hook condition is detected, which may take place in one of several ways. However, dial tone is not instantaneous. Before generating dial tone, the local switch must first reserve a *digit register*, which is essentially a series of memory locations capable of storing a single dialed digit. Then the accounting (billing) equipment is engaged, basically by making an entry that says "get ready to bill for a call." Only after both operations are completed is the dial

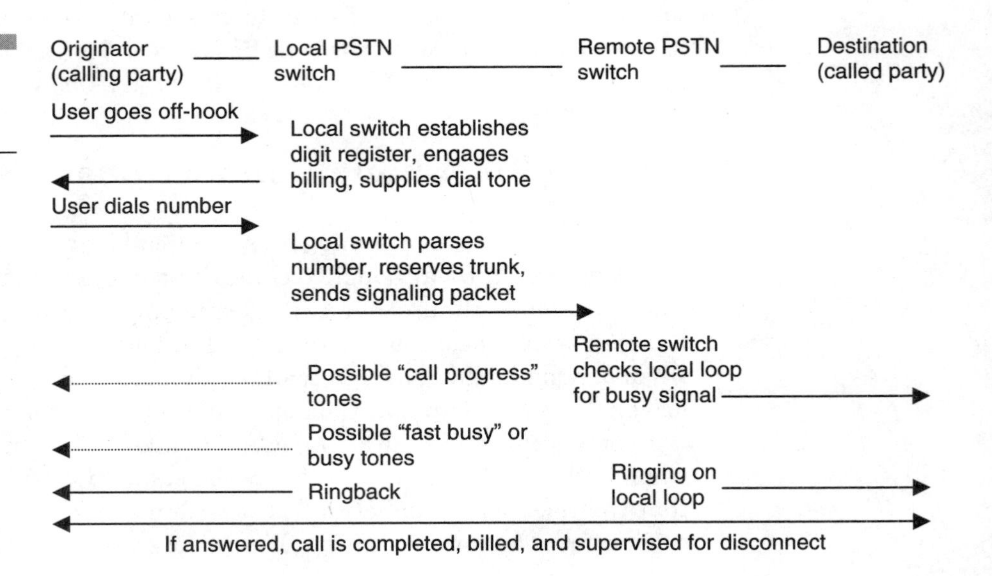

Figure 3-8 Making a PSTN telephone call.

tone placed on the line. In the United States and many other countries, the switches are fast enough to perform these operations in a fraction of a second, so the dial tone is usually present by the time the earpiece reaches the ear. In some countries, however, dial tone can be delayed for several seconds (extreme delays of 45 s or more have been reported). And dial tone must be denied if no digit register is available. This can happen under very congested (lots of calls) switch conditions, since digit registers are switch resources that must be configured ahead of time, just like router memory.

Once the caller has received dial tone, the caller can dial the digits that represent the network address of the destination of the called party. More details of this dialed number are explored in the next section. All that is needed to understand the figure is the assumption that the call is local but requires the services of a remote PSTN switch.

The next step is for the local switch to parse the dialed number and perform the required action. Sometimes, the number requires special handling, such as 611 dialed repair service calls and 800 numbers, which also require special billing treatment. If the dialed number is just another subscriber, then the local switch reserves a trunk based on a routing table (quite similar in form and function to an IP routing table) and sends a signaling packet onto the signaling network containing the originating number (network address), destination number, and other information.

At this point the caller may receive *call progress tones* that formerly represented analog signaling tones on a trunk but now are just used to let the caller know that things are happening even though the telephone at the other end is not ringing yet. If there is no trunk available to the destination switch, which should happen only under extremely congested conditions, then the caller will receive a "fast busy" tone, technically called *reorder*. Most callers know that this fast busy means to hang up and try the call again later. Of course, the remote switch must check the destination local loop to see if a call is in progress on the destination loop. If there is already a conversation taking place, then a busy signal is given to the caller. Sometimes a congested switch may give a "false busy" just to get callers to try a call again later, presumably when there are fewer calls to handle.

If the destination loop is free, ringing voltage is put on the loop. Note that the caller also hears ringing, but this is generated by the *local* switch and is not necessarily synchronized with the remote ringing cycle. This *ringback* is initiated by a signaling message from the remote switch. This is why callers sometimes hear a telephone answered on the third ring (for example) and say, "What took you so long?" The puzzled recipient often replies, "I answered the telephone the first time in rang." This feature prevents people from arranging codes such as "If the telephone rings twice at three o-clock, I have arrived safely" and not paying for the call.

Naturally, once the telephone is answered, the call proceeds, and billing commences. Notice that the PSTN has performed an enormous amount of work even *before* collecting a penny from the completed call. Thus, in some telephone systems, billing is allowed even if the called telephone is not answered or, in some cases, if the called telephone is busy. Even in the United States, it is usual for the telephone company to play a recording such as, "The party you called is not answering..." and offer to perform some other action. After all, the trunk is in use and switch resources are consumed, without any revenue coming in.

The switches at both ends monitor the call looking for a disconnect (on-hook) condition. This generates a signaling message that frees up the trunk. Billing records are also terminated. There is a lot more that could be said about the whole call-processing procedure, but this is enough to show that VoIP must have some way of doing these same types of things over an IP network such as the Internet.

Telephone Numbers: The PSTN Network Address

All access lines on the PSTN are represented by a network address called the *telephone number*. Note that this address basically identifies a port on a PSTN switch and not an individual telephone. This is one reason why telephones can be freely moved around from place to place. The telephone number is not like an IP address that must be configured in a host or router and changed if the device moves around on a network.

All countries have their individual numbering plans that specify how telephone numbers are structured within the country. This section describes the North American numbering plan used in the United States and Canada for illustration purposes. Before about 1960, all telephone numbers in North America had seven digits. The first three digits were the *exchange code*, which identified the CO switch itself on the PSTN. The last four digits were the subscriber extension, which identified the port on the switch to which the local loop attached. If someone wanted to make a "long distance" call, the caller dialed the "0" on the telephone, gave the operator the desired city and seven-digit telephone number, and often hung up to wait for the call to be completed. It could take up to 10 minutes or so to switch a coast-to-coast call manually like this, proceeding trunk by trunk and operator to operator across the country. Naturally, this process did not generate any revenue until the call was ready to go, but trunks were tied up for many minutes.

Therefore, in the early 1960s, the AT&T Bell System introduced *area codes*, technically called *numbering plan assignments* (NPAs). The original 10-digit NPA system had to have a digit from 2 to 9 in the first position (designated *N*), a 0 or a 1 in the second position, and could have any digit from 0 to 9 in the third position (designated *X*). Thus the original NPA plan had an *N*0/1*X* structure. This made it easy to tell when a caller dialed a "long distance" call, since exchange code had an *NNX* structure that meant a 0 or 1 never appeared in a valid local call. This allowed long-distance calls to be routed immediately onto special long-distance trunks, enabling quicker call completion and thus more calls handled per unit time. The last four digits had an *XXXX* structure, although some numbers were routinely reserved for special uses (99*XX* numbers were usually used internally). All calls that began with a 0 or 1 were special in the sense that they were neither local nor long distance.

In some parts of the United States, mostly those with major metropolitan areas, the *NNX* structure limited the number of central offices too severely. Rather than add many new areas codes, *NXX* was allowed in these areas, starting in 1971. This allowed a 0 or 1 in the second position of the dialed number for local calls, so there was now no easy way to identify long-distance calls immediately. Thus, in these areas, users were forced to dial a 1 before making a long-distance call, adding an eleventh digit to the number. It could be argued that the switch could just wait until the caller dialed 7 or 10 digits, but the digit register was assigned *before* the digits were dialed. Since most calls were local calls, the default number of digits captured was 7. Also, waiting for all dialed digits slowed call processing, users often paused during dialing, and so on.

Today, of course, the pressure of second lines for Internet access, mobile phone services, and so on have put so much pressure on the area code structure that in 1995 the North American numbering plan allowed NPAs in the form of *NXX*. In some areas, callers have no choice but to dial the full 10 digits all the time, even when calling a neighbor next door. This evolution of the North American numbering plan is shown in Table 3-6.

Pressure continues to expand the address space of the PSTN. However, since most people still dial telephone numbers, adding three or four new digits is not a preferred solution. However, there are only so many telephone numbers to go around. And increasing the number of

TABLE 3-6 The North American Numbering Plan

Date	Area Code	Central Office Code	Access Line Code
1960s	*N0/1X*	*NNX*	*XXXX*
Before 1995	*N0/1X*	*NXX*	*XXXX*
1995	*NXX*	*NXX*	*XXXX*

Note: $N = 2$ through 9, $X = 0$ through 9.

area codes is only a temporary solution. California, for example, which had 13 area codes in 1995, has 26 as of 1999 and will have 40 in 2001.

Electronic Switching and Routing

The original PSTN "switches" were human beings sitting in front of electromechanical switchboards. The two wires of the local loop were carried by the *tip* of a metal plug with a *ring* around it, and the two wires are called *tip and ring* to this day. Operators were slowly replaced by electromechanical switches after a crippling operator strike during World War I drove home the utter dependence of the PSTN on human beings. The step-by-step switch (also called the *Strowger switch* after its inventor) had been around since 1890 but was very expensive to deploy. However, by the early 1920s, cities like Chicago had central office switches that allowed callers to dial local calls directly without human intervention. Operators were still required when the caller did not know the local number and for long-distance calls. Operators, believe it or not, also spent a lot of time talking to people who were just lonely.

These "stepper" switches are still around, mostly in rural areas. However, a new generation of switches known as *crossbar* switches replaced stepper switches in metropolitan areas. The new switches performed three main tasks: signaling, control, and switching. The signaling function reserved trunks and sent the dialed number to the remote switch. Signaling also monitored the status of the lines and trunks and passed status information to the control function. The control function processed the received dialed digits from a local loop or incoming trunk and performed the appropriate actions in conjunction with the signaling element. The switching *matrix* actually linked the switch ports together.

The crossbar matrix was an analog device in the sense that the voice traveled through the switch in analog form. However, there was more to a switch than just the matrix. The control function represented very rudimentary forms of digital computing, aided by the fact that telephone numbers were just digits. By 1960 or so in the United States, the presence of area codes and the fact that people could afford to make long-distance calls meant that the PSTN could generate much more revenue for the telephone companies if more calls could be completed in a shorter amount of time.

The Bell System installed its first electronic switching system (the No. 1 ESS) in 1965. The matrix remained analog, and so did the signaling. The control function, however, was provided by a computer. This class of switches was called *stored program control (SPC) switches*. It would be another 10 years before the switching matrix was digitized as well, enabling digital DS-0s to pass through a voice switch as a stream of 0s and 1s for the first time.

Once the switch had a real computer, many features associated with the PSTN became feasible for the first time. The No.1 ESS introduced speed dialing (abbreviated dialing), call forwarding, call waiting, and three-way conference calling. SPC switches also simplified many of the administrative and maintenance tasks. Telephone numbers could be changed more easily, for example, and the computers provided record keeping, traffic statistics, call tracing, and other features.

The ESS family grew to include the No. 2 ESS, the No. 3 ESS, and the No. 1A ESS, all with an analog matrix based on the *reed relay*. Outside the Bell System, the GTE EAX (electronic automated exchange) also used a control computer. However, the No. 4 ESS, introduced in 1976, had a digital switching matrix as well as control, thus making the whole central office switch into a big computer for the first time. Today, the Nortel DMS, the GTE No. 5 EAX, and the AT&T No. 5 ESS are all digital computers used as PSTN switches. Thus IP routers and modern PSTN switches have much in common.

Signaling: The Connectionless PSTN

The preceding section described the digitization of the PSTN central office control and switching matrix. This process was more or less completed by 1976, although even today there are plenty of old analog switches around.

But what about signaling, the third switch function? Before a computer can be said to be able to do everything a PSTN switch does, signaling must be digital as well. As it turns out, digital signaling followed a development parallel to the development of the digital switch. This was so because it made sense to control and supervise a digital switch with a digital signaling protocol.

Another benefit of digital signaling was that it allowed the telephone companies to move signaling off the analog trunks that were then used to carry the voice call. This was called *in-band signaling*, and call setup signals could share the physical trunk because they were usually done at different times. When a caller heard beeps and tones ("call progress tones") over the telephone once they had completed dialing but before the telephone at the other end rang, this was the called number being sent through the switching network and setting up the call on the trunks "behind" it.

In-band analog signaling had a lot of limitations. It was slow, taking anywhere from 10 for 15 s to set up a long-distance call. There were usually only 16 tones available, so the signaling repertoire was limited. And people could fool the PSTN into thinking that the signaling was internal when it was generated by "phone phreaks" seeking to make free long-distance or pay phone calls. By the 1970s, it was clear that the telephone companies would have to do something about in-band signaling.

The quest for *out-of-band signaling* secure from hackers did not immediately lead to digital signaling systems. The first efforts used a separate trunk just for carrying *E&M signals* (so-called for their association with the ear and mouth). Then E&M signaling was mapped channel by channel onto a digital T-carrier trunk using robbed-bit signaling.

The quest for faster and more secure out-of-band signaling, however, culminated with the development of a family of common channel signaling (CCS) methods. And the most advanced CCS now in use is known as *signaling system number 7* (SS7).

SS7 basically established a totally separate, connectionless, packet-switched network for PSTN signaling. Thus, when a caller dials a number, the SS7 network takes the dialed number, the caller's number, and other information and places the information into a packet, similar to an IP packet. The packet gets routed through the SS7 network by a series of routers and sets up the call as it goes. There are even databases that are consulted to see if anything special should be done with the call, such as call forwarding.

More details on the SS7 network will be given in the next chapter. The architecture of a CCS network using SS7 is shown in Figure 3-9.

The *service switching points* (SSPs) are where the PSTN switches pass and get information from the SS7 packet network. Think of these as the clients. The *signal transfer points* (STPs) are where the packets are routed from switch to switch or region to region (e.g., local company to

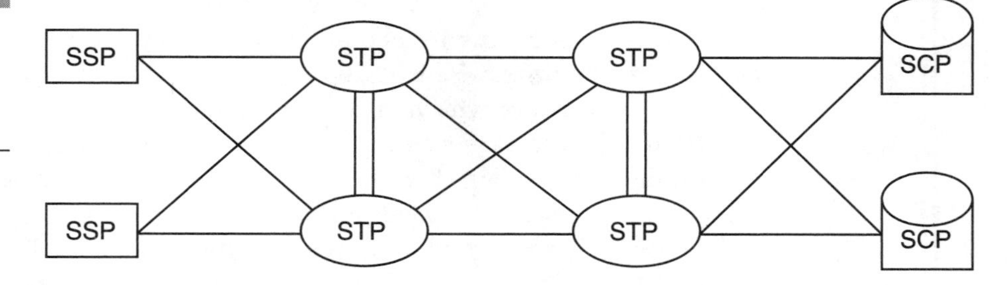

Figure 3-9 Common channel signaling (CCS) with SS7.

long-distance company). Think of the STPs as routers on the SS7 network. Finally, the *service control points* (SCPs) are databases of information needed to complete the call correctly. Think of the SCPs as a server of the Internet. The links between the clients, servers, and routers have a structure all their own, also discussed in the next chapter.

If the SS7 network of clients, servers, and routers were considered alone, without the PSTN voice switches, then the SS7 network would look a lot like the Internet and Web.

The Bell System

Alexander Graham Bell invented the telephone during an effort to multiplex telegraph signals on the same wire using frequency separation. What happened instead is that he and his assistant Watson inadvertently expanded the frequency on the telegraph wire so that it was capable of carrying voice. In February of 1976, Bell filed a patent application for this "improvement on telegraphy" hours before a competitor. By 1877 it became obvious that the telephone was distinct enough from the telegraph to be a separate invention, and the patent on the electromagnetic telephone was granted on January 30, 1877.

Strangely enough, Bell and his financial backers, Hubbard and Sanders, knew that what they had was of value but were not really anxious to found a new industry based on an unproven market for the new technology. Thus they offered the telephone to Western Union, the leading telegraph company and, after all, a communications company for $100,000 (admittedly a fortune in those days). The offer was rejected, so the Bell Telephone Company was organized in Massachusetts on July 9, 1877. The company immediately had money problems and was reorganized as American Bell in 1880. Initially, the company just leased telephones to users who had to install their own lines. Naturally, most of these telephones were installed on point-to-point lines, often leased telegraph lines. A pair of telephones cost $40 a year for businesses but only $20 when used for "social purposes."

Subscribers immediately realized the potential for any-to-any connectivity, and the "central office" was born when a burglar alarm company run by Bostonian E. T. Holmes began using young boys to connect voice calls from those with telephones using the burglar alarm lines. By August 1877, Holmes had 700 customers linked by this "exchange." As other exchanges sprang up in New York, Albany, Boston, Philadelphia, and Washington, some means of using "long lines" to connect these exchanges was sought.

Such a backbone network would be very expensive to build and beyond the means of American Bell. Thus, in 1885, the American Telephone & Telegraph company was incorporated in New York as a wholly owned subsidiary of American Bell. The whole idea of AT&T was to build long-distance networks to link local exchanges, leaving local services to American Bell. The ambitious plans of American Bell and AT&T meant that a steady supply of high-quality wire, electrical components, and telephones was needed to fuel the explosive growth of this early Bell System. In 1888, the Western Electric Company of Chicago was purchased to supply the material and allow the Bell companies total control.

By the late 1890s, independent telephone companies had sprung up and tried to get a piece of the action. American Bell had to keep rates low to compete in areas where there was a choice of telephone service providers, but AT&T was making money left and right, since the "long lines" network was the only real way to connect local exchanges. Everyone had to hook up to AT&T, or they were limited to local calling only. Sometimes AT&T denied the small independents interconnection with the long-distance network, usually for obscure reasons, making it hard for the local companies to grow. In many cases, the local Bell company lowered rates in competitive areas so that the cash-poor local independent had no choice but to join the growing Bell System or face bankruptcy. All contributed to cash inflow at AT&T and cash problems at American Bell.

In 1900, in an extraordinary step, the subsidiary AT&T actually bought the parent American Bell, and the Bell System took on the overall structure that it retained until 1984. Corporate direction was set by the parent AT&T, with AT&T Long Lines as the long-distance company and the Bell Operating Companies (BOCs) as the providers of local service, often organized along state lines (New York Telephone, New Jersey Bell, and so on). The other pieces were Western Electric and later Bell Telephone Laboratories as the research and development (R&D) branch.

Bell Labs began as the Engineering Department of American Bell in 1881. Bell Labs emerged in 1925 and contributed such advances in telecommunications as the sound motion picture, coaxial cable, the transistor, information theory, the laser, and the communications satellite. In the computer field, Bell Labs developed UNIX and C language, software familiar to anyone with an Internet background.

The Independents

Not all telephone companies were founded by or absorbed by the Bell System. There was always a class of independents, so-called because of their independence from the Bell System (and the Bell telephone patents in some cases). The independents started in mostly rural and less densely populated areas as the Bell System concentrated on the Northeast. For instance, the company that later became Sprint started at Brown Telephone Company in Kansas City in 1899. Many of the independents later became part of GTE, which eventually became large enough to have its own long-distance network independent in large part from AT&T Long Lines.

AT&T: The Nation's Telephone Company

In 1913, as a result of threatened U.S. government action, AT&T agreed not to buy up competitors and link independents to the AT&T long-distance network or engage in other practices that hurt competition. In return, AT&T BOCs enjoyed a monopoly on local telephone service in most areas and agreed to be regulated by the appropriate state and federal authorities. This effectively made AT&T "the nation's telephone company," as close as the United States would come to the way telephone networks were run in most other countries, as a branch or bureau of the government.

Over the years, AT&T successfully fought off a series of efforts by government bodies such as the Federal Communications Commission (FCC) and the U.S. Department of Justice to "break up the Bell System." Most efforts focused on ownership of Western Electric (WECO), since no other supplier was approved to furnish equipment to the Bell System. Since money paid to WECO remained in the Bell System, WECO was accused of "gold plating" even simple telephones so that they would be more expensive than necessary (on the other hand, WECO telephones were virtually indestructible). Any costs incurred by the local BOCs could be passed along to customers, as long as the state authorities approved of the rate increases. Figure 3-10 shows the structure of AT&T around 1980.

Between 1913 and 1984, AT&T fiercely protected its turf, sometimes a bit too zealously for many people's liking. The telephone itself, and even the wiring inside the home, belonged to the local telephone company. Anything non-Bell that attached to the telephone or access line was prohibited. So when a simple plastic cone designed to provide speaker privacy on business telephones (the Hush-a-Phone) was marketed, AT&T sued and won, although many thought that the whole thing was getting somewhat out of hand. A few years later, when AT&T tried to claim that only WECO telephones could be attached to the network without causing harm, the court decision went against AT&T (the U.S. government had used its own telephones on Bell lines for years). This paved the way for non-WECO customer premises equipment. It also launched a whole new, competitive industry for telephones, fax machines, and later computer modems.

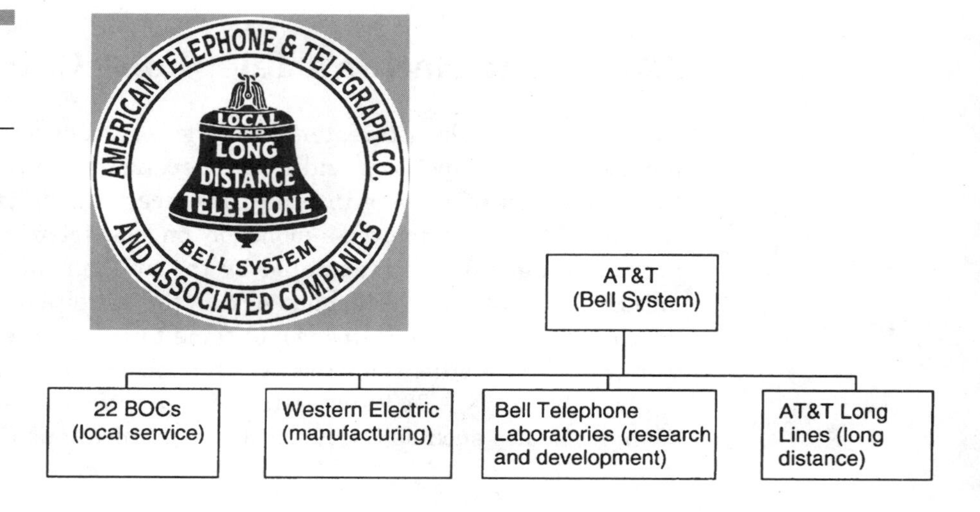

Figure 3-10 The AT&T Bell System.

One of the biggest boosts to the computer industry and later the Internet happened because of this regulated environment. To protect competitive industries, Bell Labs was prohibited from marketing any of its inventions to outside industry, patented or not. So transistors could be used by the Bell System, but not sold on the open market. Anyone who wanted to make transistors could send away for the patent filing, and for the cost of shipping and handling could make and sell transistors without owing a penny to Bell Labs. Naturally, the computer industry, which at the time used many transistors, benefited greatly from this.

Later on, when Bell Labs created UNIX, anyone could obtain the complete UNIX operating system, royalty-free, for the sum of one dollar. Many colleges and universities, seeking both to use a single operating system for diverse computing platforms and to minimize expenses, sent in their dollars. And in 1955, AT&T, although owning computing pioneer Bell Labs, was explicitly forbidden from making and selling computers in competition with IBM, although computers could be made and used within the Bell System. This agreement was called the Final Judgment, and settled yet another attempt by the U.S. government to spin off Western Electric from AT&T.

Things got so bad at Bell Labs that when the laser was invented in the late 1960s, no one was inclined to bother to patent it. After all, no money could be made from the discovery. Fortunately, some genius decided to patent the laser anyway. Perhaps some day Bell Labs would be able to profit from the innovation. That day was actually closer than most imagined at the time.

The Breakup of the Bell System

By 1971, the Bell System was an astonishing business phenomenon, although quietly so. AT&T stock was the most widely held stock around; its solid returns supposedly made it the stock of preference for "widows and orphans" who could not risk market downswings. In terms of number of employees, revenues, profits, and assets, AT&T was almost as big as the top five companies making up the 1971 Fortune 500 *combined*. For example, AT&T in 1971 had *one million* employees, at a time when the entire population of the country was slightly more than 200 million. Out of adult males, therefore, almost 2 percent of the whole population worked for "The Telephone Company."

Of course, as a regulated service industry and not a manufacturer, AT&T never appeared as a Fortune 500 company. This bothered some AT&T executives, who wondered what an unfettered and competitive AT&T might be able to accomplish if AT&T were free to enter competitive markets. At this same time, AT&T growth was relatively stagnant by the 1970s, since everyone who wanted a telephone basically had one.

While AT&T pondered its future, another entity was considering the future course of AT&T. This entity was the U.S. Department of Justice, which was in the midst of yet another effort to split Western Electric off from AT&T and allow electronics companies and others to compete for the AT&T supply business. By the early 1980s, the stage for telephone deregulation had been set by the wildly popular breakup of the airline industry monopolies on air routes.

What the Airlines Wrought

The airline industry had been as firmly regulated, and therefore as monopolistic, as the telephone industry. For instance, one airline was granted a monopoly on the flights from New York's LaGuardia airport to Chicago's O'Hare (then Orchard Field), and that was that. No one else could fly that route. This practice was designed to protect the early airline industry in the 1930s from wasteful competition and failed businesses. But by the late 1970s, it was obvious to all concerned that airline regulation resulted in higher ticket prices, no incentive to improve service, and various other symptoms of protected markets.

So with great fanfare, the airline industry was deregulated. The initial results were mixed, but the general public mostly embraced the lower ticket prices caused by the new competition, although in many cases these proved to be temporary. Some airlines, and large ones at that, went bankrupt, and others were merged and acquired out of existence. But the experiment was largely viewed as a success, especially by the airlines that prospered in the new environment.

None of this was lost on the parties involved in the telecommunications industry, both regulators and AT&T. Everyone agreed that whatever the woes of competition, it had to be better than the constant warfare of regulation. So AT&T surprised everyone in 1983 by proposing that it be broken up, but not along the lines that the government assumed. AT&T wanted to keep Western Electric and Bell Labs, but in return for a freer hand in the long-distance arena, such as setting rates, AT&T offered to spin off the local BOCs.

After some discussion, the divestiture by AT&T of the local monopolistic BOCs was approved as the Modification of Final Judgment (MFJ) of 1955. The BOCs were organized into seven larger Regional Bell Operating Companies (RBOCs). AT&T Long Lines basically became AT&T, a long-distance company free to compete (with some restrictions) in a deregulated long-distance industry with MCI, Sprint, and a host of others. AT&T retained Western Electric and Bell Labs.

The RBOCs were firmly under state regulation and retained existing monopolies on local telephone services. The RBOCs could carry calls on their own facilities (lines, trunks, and switches) if both ends of the call were within the same local access and transport area (LATA). These were intraLATA calls. All other calls were interLATA in nature, and callers could choose to use any one of a number of long-distance companies to carry part of the call, as long as the long-distance company had a switch (called a *point of presence*, or POP) within the originating LATA. All LATAs were artificial constructs applied to the RBOCs to make sure that there were enough long-distance calls to support the new interexchange carriers (IXCs) now challenging AT&T. Independents that were never part of the Bell System were unaffected by the LATA rules. These independents could join an RBOC LATA (many small independents did), establish their own LATAs, or ignore the LATA structure altogether.

So the local exchange carriers (LECs) had a monopoly on local service, but interLATA calls had to be handled from POP to POP by the competitive IXCs. These interLATA calls were listed on the bill sent to customers by the LEC, which collected the entire amount on the bill. The IXCs paid for this service, which avoided the need for the IXCs to bill the customers directly. But how did the IXCs recover their rightful portion of the LEC bill? And how did the LECs charge the IXCs for the use of their facilities from POP to customer? The answer was through a system of *access charges* that are still an item of discussion with regard to Internet access and the ISPs.

LATAs and Access Charges

A local call is carried over the facilities owned and operated by the local carrier for just this purpose. A local call may be a toll call, but that just means it costs more. The MFJ defined local calls precisely: both endpoints must be within the same LATA. Whenever a LATA boundary is crossed, this is a long-distance call, even if it is only a few miles from end-to-end across a state line (very few LATAs crossed state boundaries). These were handled from IXC POP to IXC POP on interexchange carrier facilities.

So all long-distance calls included two local calls. There was a local call on each end, and an IXC call in the middle. The local calls went to the POP on the originator side, and from the POP to the destination in the other LATA. But if the IXC in the middle gained all the revenue from the long-distance call (even though the call was billed by the originating LEC in most cases), how could the LECs at each end be compensated for the use of their lines, trunks, and switches? After all, these were in use for the duration of the call and could not be used for local calls or to make money for the LEC in any way.

The answer was to install a system of *access charges* as part of the MFJ. The IXCs pay the LECs for the use of their facilities to complete local calls. Now, the trunk to an IXC POP might be a T-1 carrying 24 voice calls. That is not subject to access charges. The LEC voice switch ports that link to the T-1 are, however. So the T-1 is still a leased line, but each call that is carried to and from the POP consumes LEC resources and is subject to a per-call access charge system.

All entities that can cross LATA boundaries for services must pay access charges, except for one major category—the ISPs. Data services were explicitly exempted from paying access charges according to the MFJ. This was done to make sure that the data services segment, just emerging in 1984, would not be crushed by costs that it could not afford. This also reinforced the basic split between *basic transport services* provided by the protected LECs and *enhanced services* provided by the ISPs. This principle meant that the monopolistic LECs could only transport bits through their network, not store, process, or convert them in any fashion. Changes to bits were enhanced services, and reserved for competitive entities such as the ISPs.

Recent regulatory rulings have made it clear that there is no reason why ISPs should not pay access charges (no one denies that Internet traffic does cross LATA and state boundaries); it is just a matter of what policy best serves the public interest. For now, the reactions of regulators at the federal and state level to proposed ISP access charges have ranged from "please don't charge the ISPs for now" to "LECs and ISPs must work this out on their own" to "don't make me come over there!"

If anything seems clear, it is that sooner or later the ISPs will have to pay some form of what most of the ISP see as an "Internet tax" to the LECs in the form of access charges. The only question is how much.

Postdivestiture PSTN

Since 1984, the architecture of the PSTN is taken on a structure of LEC, LATA, POP, and IXC. Instead of a hierarchy of switches, SS7 has allowed for a system of IXC backbone switches linked in a flatter structure.

There is often still a hierarchy of access switches and backbone switches, but nothing as elaborate as the "class" switching structure that prevailed before 1984. The architecture of the postdivestiture PSTN is shown in Figure 3-11. All of the blocks are voice switches, distinguished primarily by their size and function (tandems switch trunks, and so on). There can be direct trunks between COs (dotted line) and trunks can run from an IXC POP to toll office switches, access tandems, or even central office switches themselves. This is just a representative figure, and therefore not too much should be read into the lack or presence of a trunk here or there (toll offices do not really link to only one CO).

In Figure 3-11, IXC A can carry calls from LATA no. 1 and LATA no. 2, but not LATA no. 3, since there is no IXC A POP in LATA no. 3. IXC A can complete calls to LATA no. 3, however, but only by passing the call off to IXC B, which has a POP in LATA no. 3. Naturally, IXC A pays IXC B for

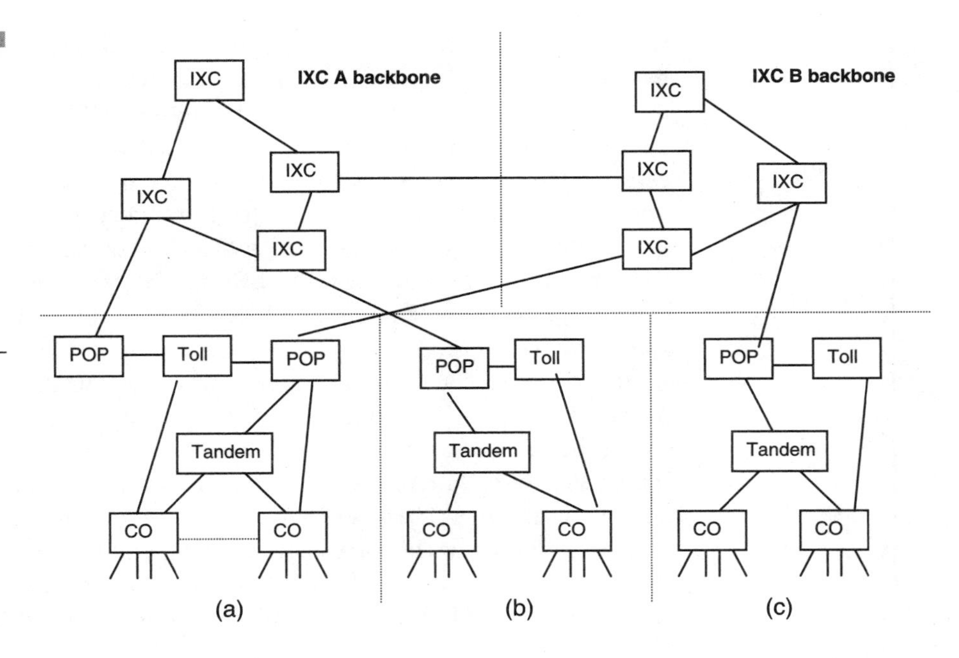

Figure 3-11 The structure of the postdivestiture PSTN: *(a)* LATA no. 1, RBOC; *(b)* LATA no. 2, independent; *(c)* LATA no. 3, RBOC.

this capability, and it all depends on whether IXC A and IXC B have trunks between them (as they do in this example). Note that only callers in LATA no. 1 have a choice of long-distance service providers, since the other two LATAs have only one POP from one IXC.

There is now no longer a clear distinction between local and long-distance calls (the distinction was never as clear-cut as people imagined anyway). There are instead intraLATA local calls (the lowest rates apply), intraLATA toll calls (still carried on LEC facilities, but costing more), and interLATA toll calls (which might be intrastate or interstate, but are carried by an IXC). Sometimes, intraLATA toll calls are called "local long distance," but thankfully this term is not common. The artificial nature of the LATA construct is reinforced by the possibility of interstate but intraLATA toll calls and of special "corridors" between some LATA where LECs can use their own facilities. However, these are special cases that do not invalidate the general scheme of things.

TA96

By the early 1990s, the breakup of the Bell System and the rise of competitive long-distance services were seen as unqualified successes by most people. Rates were down, people could choose their own long-distance service providers on a per-call basis (as long as there was a POP reachable in the LATA), and a free-market atmosphere was well established. Attention then turned to the local telephone service market.

LECs claimed that they should be able to offer long distance on their own facilities, without sharing the wealth with the IXCs. LATAs were established to make sure that there were plenty of long-distance calls, and there certainly were. LECs also felt that they were in a competitive environment anyway, since many states had certified local competitors in the form of competitive LECs (CLECs) who often purchase "unbundled" local loops from the incumbent LEC (ILEC) and ran them to their own voice switches.

So with much fanfare, the Telecommunications Act of 1996 (TA96) was passed by the U.S. Congress. TA96 proved to have such a Byzantine structure of clauses and rules that no one is quite sure what difference, if any, TA96 has had on local service competition. TA96 allowed ILECs to offer long distance if and only if their local territories were open to competition, whatever that actually meant. Much of the debate focused on the resale price of unbundled loops offered to the CLECs. What amount should be charged monthly to a CLEC by an ILEC that had installed the wire in 1940? Wasn't it effectively paid for many times over from ILEC services? But what about the incremental costs of maintaining, upgrading, and otherwise making sure the loop was available?

There have been no easy answers to the questions TA96 poses. Depending on the source, CLEC services are either more plentiful than they ever have been before, or the ILECs are still in the habit of placing roadblocks in the path of any serious competitor. In fact, both statements may be true.

ILECs, CLECs, and ISPs

Today, the structure of the PSTN at the local level finds three types of service providers offering connectivity services for voice and data. These are the incumbent local exchange carriers (ILECs) which formerly had a monopoly of local services, whether RBOCs or not. Then there are the state-certified competitive local exchange carriers (CLECs) which compete either by building their own facilities or by leasing "unbundled" facilities for resale from the ILECs.

Naturally, there are the ISPs that provide Internet access as well. Sometimes the ISP is a subsidiary of the ILEC, but state regulation means that these ISPs must be customers of the ILEC, just like everyone else. So BellSouth.net and BellAtlantic.net are technically separate entities from their parent organizations. This arrangement retains the split between basic transport and enhanced services.

In the near future, ISPs of all kinds are expected to have to pay access charges to the ILECs just like the IXCs. Most ISPs just interpret this as a type of "Internet tax" and chalk it up to the cost of doing business. The ISPs handle the local access needs of the Internet for data. But perhaps the ISPs might handle the local access needs for *voice* as well. This is what VoIP is all about.

However, there is more that must be done to make voice fit for IP transport than just the digitization of analog voice. Voice must be packetized as well. This raises new challenges for voice that have only recently been addressed, and not always to everyone's satisfaction. But this is a topic for another chapter of its own.

CHAPTER 4

Packet Voice Fundamentals

It is one thing to make analog voice into digital voice, but it is quite another to take digital voice and make packets out of the voice bit stream. This is so because on the public switched telephone network (PSTN), digital voice is still circuit-switched voice. Packet voice must be inserted, carried, and delivered over a packet-switched network. This circuit-switched to packet-switched adaptation may sound easy enough to perform, but in practice, it is quite difficult to accomplish.

This chapter will explore all the issues involved in preparing digital voice for packet networks such as Internet Protocol (IP) networks. This is the essence of voice over IP (VoIP), of course, although much of this chapter could be applied easily to voice over a frame-relay network or voice over an asynchronous transfer mode (ATM) network (at least with regard to some ways of transporting voice over an ATM network).

However, before exploring the specifics of packet voice, this is the place to investigate more fully the differences between circuit-switching and packet-switching networks. Then the difficulties of packet voice delivery with VoIP can be appreciated.

Circuit Switching and Packet Switching

The PSTN developed as a circuit-switched network to better serve the needs of voice communications. Data networks like the Internet developed as packet-switched networks to better serve the needs of data communications. Some of the differences between the needs of voice communications and the needs of data communications have been explored already, but this is the place to be more precise about the differences between voice and data networking needs. These needs were so distinct that until VoIP came along, it was so hard to support circuit voice on a packet data network that separate physical networks (separate network nodes, separate links) were always used to carry voice and data. The major differences between telephone users and computers using networks are listed in Table 4-1.

Table 4-1 lists only the most obvious differences between two people making a telephone call and two computers transferring a file (for instance). Digital voice needs at most 64 kb/s, a very modest amount of bandwidth today. Giving voice more than 64 kb/s will not make the voice network function any better. However, data transfers will always benefit

TABLE 4-1

Telephone Users and Computers Using a Network

Network Feature	Telephone Users (Circuits)	Computers (Data Packets)
Bit rate	Fixed and low (≤64 kb/s)	Wide variation (to Gb/s)
Bursts	Nonexistent	High (100/1000:1)
Error tolerance	Own error control ("Say again?")	Must be error-free
Resending information	Cannot be done (real time)	Can do it fast enough
Delay	Must be low and stable	Can be higher, can vary
Connection	Connection-oriented	Can be connectionless

from more bandwidth, and some applications can consume 1 Gb/s (1000 Mb/s) with ease. Variations in bit rate are commonly called *bursts* and are nonexistent in circuit voice networks but common in data networks, where the bit-rate requirement can range from very low to very high, sometimes at different times when running the same application (e.g., a Web browser). The data burst ratio can be anywhere from 100:1 to 1000:1. This means that over time, the average rate of bits sent between client and server can be as low as 1 Mb/s, but the application may require short periods where the rate of bits needs to be as high as 100 Mb/s or more or the application will hang, freeze, or fail.

Error tolerance on a voice network is high because humans do their own error detection and correction, using short phrases such as "Okay" and "I see" (positive acknowledgments) or "What?" or "Can you repeat that?" (negative acknowledgments). This relieves the voice network of much of the burden of error control. The same is not true of a file transfer, though. If the file is not completely error-free at the conclusion of the transfer, it will not work correctly. Error control is difficult in voice networks because voice cannot be resent if there are errors. Resends can be used as an error-control strategy if the data network is fast enough or the application can tolerate the wait (this file transfer example certainly can).

The network delay for voice must be both low and stable. Most data networks like low delays, but these delays do not always have to be stable. Who cares if the beginning of the file arrives faster than the end of the file? Nothing is done until the whole file is present anyway. In order to best provide low and stable delays, voice networks traditionally have been connection-oriented to provide a stable path. Data networks, with only minimal delay concerns, can afford to be connectionless (no stable paths for bits to follow between network nodes), although many data networks have been as connection-oriented as voice networks.

What all these differences boiled down to was that it was easier to build separate voice and data networks than to attempt to have one network do it all. The separate voice and data networks were a necessity, not an ideal. If links and network nodes could be shared more easily between voice and data (not to mention video), then many attractive benefits such as efficiency through shared use, economies of scale, and so on potentially kick in. There were schemes like ATM designed to merge voice and data, but only by building a third architecture, the ATM network, which was optimal neither for voice nor for data (so ATM is equally indifferent to the specific needs of both).

It would be better for migration and backward compatibility (not to mention effort and expense) to put packets on circuit switches or circuits on packet switches rather than to put both on new ATM switches. The problems were which approach to try and how to accomplish it.

In truth, every time a user employs a modem to link a home PC to the Internet, he or she is using a circuit-switched network for packet use. The sending of packets on voice circuits has lead to many frustrations because of the inability of the fixed-bandwidth PSTN to deal gracefully with data bursts and instead to parcel out low voice bandwidths to data applications. As a result, data sessions are slow and lead to long holding times that congest the PSTN designed for 3-minute voice calls. And VoIP is simply a way of "packetizing voice" and using packet switching instead of the PSTN to deliver circuit-switched voice. Most observers agree that advances in digital voice make it more attractive to put voice on packet networks than data on circuit networks. But VoIP has struggled to supply the same performance levels that PSTN voice users take for granted.

Why should this merging of voice and data be so difficult? It is one thing to point out the differences between voice and data users and create separate networks for each, it is quite another to understand the architectural differences embodied in networks like the PSTN and Internet and see the basic issues that must be dealt with before packetizing voice on a large scale.

Switching with Circuits

A PSTN voice connection is known as a *call*. Even when a connection is set up for use of a modem to access the Internet, this is still just a voice call to the PSTN. The modem sounds different than human voice, but the analog signals make their way through the central office switches and trunks and loops just like a voice call to a friend. Thus a call is just a connection, and a connection is just a series of voice circuits set up to handle the call. One part of the voice circuit is the local loop, another part of the voice circuit is through the voice switch, a third part is a voice channel (think of a channel as a piece of a multiplexed circuit) on a trunk, and so on. The connection is a kind of "label" for the pieces of the voice circuit that make up the path that the voice follows through the PSTN.

The *path* is the map of the connection through the network. The distinguishing characteristics of circuit-switched networks are that they are connection-oriented (paths must exist before information flows) and the circuits reserve "all the bandwidth all the time." In other words, a voice call, even with both ends of the connection on hold, still busily generates and sends 64 kb/s of digital silence back and forth through the PSTN. The connection is set up by using a *signaling protocol* that in most cases on the PSTN flows on an entirely separate network employed for that purpose.

Thus circuit-switching networks such as the PSTN have the following three characteristics in common:

1. Signaling protocols set up the connection.
2. All bits follow the same path through the network.
3. The path consumes all the bandwidth all the time.

Although the term used here is *path*, it is good to keep in mind that a *path* is made up of a *circuit* labeled by a *connection* known as a *call* in the PSTN. Although the terms *path* and *circuit* and *connection* are often used in data networking, it is always good to remember that *call* is a term used almost exclusively in voice to mean a connection.

The way that these three features relate in a voice network is shown in Figure 4-1. In the figure, the end nodes of the network are telephones, and the light lines represent the local loops. The network nodes are telephone switches, and the heavy lines are multiplexed trunks carrying many voice channels. The dashed arrows represent the path chosen for this particular call. How was the dashed path chosen? Through the signaling nodes at the bottom of the figure. The links between voice switches and the signaling nodes themselves are shown as dashed lines. Another dashed line separates the voice network, where the calls are, from the signaling network. The signaling network is physically separate from the voice network. Note that the end nodes have no direct access to the signaling network, only the switches.

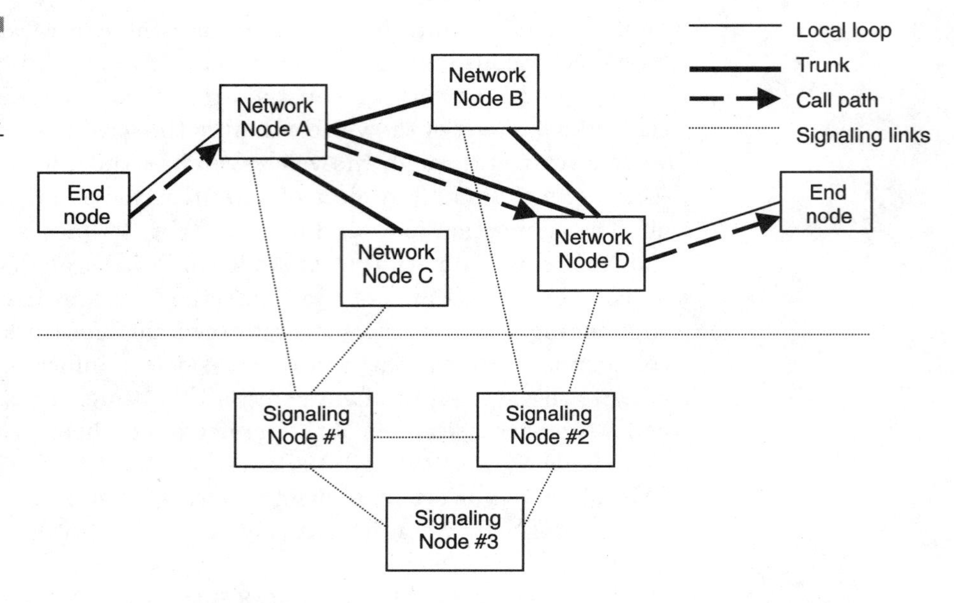

Figure 4-1 The characteristics of circuit switching.

Thus the voice call is set up with a signaling protocol. This establishes a path for the call to follow through the network. Now the route selected for the call could not go through network node C for the simple reason that the destination is not reachable through node C. The signaling network could route the call through network node B to network node D, but why? There is a direct path from node A to node D, which is why the path was used in the figure. Naturally, if all trunk channels are in use between node A and node D, it would be possible to route the call through node B. However, this decision can only be made at call setup time by the signaling nodes. Calls in progress are not rerouted, and calls that are *blocked* are simply dropped (the "fast busy" is returned to the user).

The signaling nodes play a crucial part in the call setup process. In Figure 4-1, signaling node 1 is attached only to network nodes A and C, while signaling node 2 is attached only to network nodes B and D. Yet the signaling network must make routing decisions based on *global* knowledge of the network. Information about trunk use and other voice network parameters must be shared by the signaling nodes. On the Internet, routers (which are essentially the functional equivalent of the signaling nodes in the voice network) exchange information through the use of a *routing protocol*. The same thing happens in the PSTN, although the routing protocols used are appropriate for a circuit-switched network. The signaling network in the PSTN is a router-based client-server connectionless packet-switched network built on the same principles of operation as the Internet. In Figure 4-1, switching node 3 can handle communications between signaling nodes 1 and 2 if the direct link between them fails. Thus setup requests are routed around failures and are not blocked, as are the telephone calls.

Once the call is set up and accepted (the other end answers), the same path is followed for all voice bits. There is always 64 kb/s flowing in each direction, even during periods of absolute silence (e.g., hold). This is not true of the signaling network. Signaling messages are bursty and do not use the same amount of bandwidth to set up a call as to release a call.

Thus the characteristics of the signaling network used to set up voice calls are as follows:

1. The signaling network is connectionless (it must be; there are no connections yet).
2. Messages need not follow the same path through the signaling network (rerouting is possible).
3. The path does not consume any fixed amount of bandwidth.

The signaling network that controls voice calls in the PSTN is not a circuit-switched network. It is a packet-switched network, and a connectionless one at that.

Switching with Packets

When computers first began to send files and other forms of digital information back and forth between each other, direct point-to-point circuits were used. However, for the reasons outlined earlier, this use of "all the bandwidth all the time" methods was extremely wasteful.

Packet networks came about to address the separate needs of data, which included variable bit rates to accommodate bursts, a relative indifference to delay variations, and a heightened concern for error-free information transfer. The breakthrough that packet switching represented was all predicated on the concept of the packet.

A *packet* is usually defined as the unit of data that leaves one site and traverses the network intact until it reaches the destination site. This is the whole point of the network: to get the user information from one place to another. The packet always contains some *network address* information to allow the network nodes to route the packet properly to the destination.

Bursty data packets also demanded a new way of multiplexing. Instead of the voice-oriented time-division multiplexing (TDM) methods used in the PSTN, data-oriented packet networks used what was variously called *statistical TDM* (*stat muxing* for short) or *asynchronous TDM* (later made popular in ATM). Since the data were so bursty, it made no sense to divide the total capacity of a link up into fixed-bit-rate channels as did the PSTN. It made more sense to *unchannelize* the total capacity of the link and allow each packet to use the full capacity available but not all the time. So "all the bandwidth all the time" circuit-switching multiplexing became "all the bandwidth part of the time" packet-switching multiplexing. This concept is illustrated in Figure 4-2.

With statistical TDM, invented for bursty data applications, the link between the multiplexers is unchannelized. The computers are not limited to 64-kb/s channels (in this example) with TDM but can use the full 128 kb/s available, so long as both do not attempt to use the link at exactly the same time. When this happens, as when both senders "burst" into the network simultaneously, the multiplexer has no choice but to *buffer* one packet while sending the other. When the burst activity dies down, the multiplexer can send the buffered packet. The only other choice is to discard the incoming packet.

Networks that employ statistical TDM are often said to provide *bandwidth on demand* to applications. It would be better to use the more technical term *flexible bandwidth allocation* because users get the idea that bandwidth can be created out of nothing when the term *bandwidth on demand* is used ("I have 1.5 Mb/s, but if I demand 2 Mb/s, I'll get it"). Buffering allows the bandwidth on the unchannelized link to be dynamically allocated to the currently bursting applications. An idle application consumes no bandwidth on a packet network.

A *buffer* is just a memory area reserved for networking purposes. Because memory is a fixed resource in a network device, buffers are also a fixed resource in a network device. When there are no buffers available, as when many sources of bursty traffic all bursting at once have

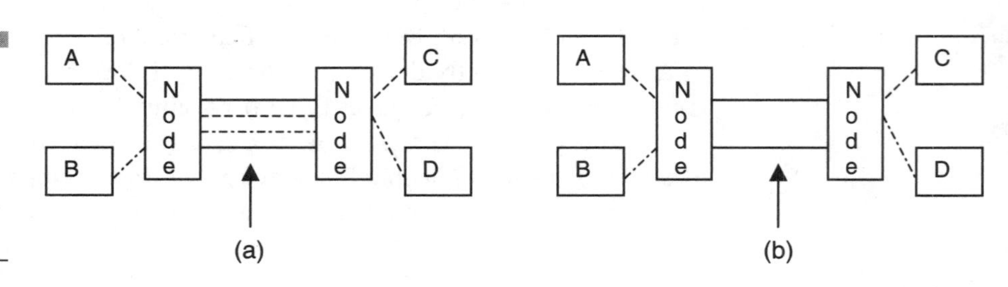

Figure 4-2
(a) TDM: 64 kb/s for A-C and B-D. *(b)* Statistical TDM: 128 kb/s for A-C and B-D.

filled all available buffer space, then incoming packets are discarded. The number of packets a device can *queue* for output is strictly dependent on the device's buffer space. In the past, when memory was expensive, buffers had to be kept small for economic reasons. Today, buffers can be quite large, but this practice of larger buffers introduces a problem into packet networks that becomes more pronounced as buffers grow larger and larger, namely *variable delay*.

Voice quality is heavily dependent on low and stable delays. Packet networks are generally unsuited for voice delivery because the buffers required to deal with bursts in a statistical multiplexing environment always lead to higher and more variable delays. Suppose a packet containing delay-sensitive traffic arrives at a network node. It does not have to be voice. The packet could contain financial information ("Sell right now!") that is much more important than regular e-mail. However, if there are already six packets in the output buffer waiting to go out, the packet is delayed by at least the time it takes the link to send the six packets. The next network nodes may have ten packets, or four, or seventeen. The delay varies according to network load.

In a circuit-switched network, there is no buffering to speak of, and the packet flies right through. This is the tradeoff between circuits and packets. The more efficient use of unchannelized packet network links is balanced by the need for extensive buffers and processor-intensive buffer management schemes, which lead in turn to higher and more variable delays.

Now the network node could easily place the delay-sensitive packet at the head of the output buffer queue, all set to send as soon as the current packet finishes. This is what packet network quality of service (QoS) is all about. However, although this method will *minimize* delay variations, priorities alone will not guarantee stable delays. The reason is a phenomenon known as *serial delay*.

Serial delay results when a delay-sensitive packet is stuck in a buffer behind a packet that has *already begun to be sent*. All links used in a *wide-area network* (WAN) operate in a serial, one-bit-after-another fashion. Once the first bit representing a data unit has been sent out on a serial port, there is no easy way to interrupt the data unit and send priority data. (Some efforts to supply a "forward abort" for this purpose have been made, but they proved to be more trouble than they were worth and are rarely used today.) Thus the delay-sensitive traffic must sit until the outgoing data unit has passed through the serial port. The delay variation introduced by this serial delay depends on the maximum allowable length of the data unit itself. Longer data units result in less link overhead and better network throughput but maximize serial delay. Extremely short data units, such as ATM cells, minimize serial delay but have a lot of link overhead and can lower network throughput compared with other methods.

There are only two ways to deal with serial delay in a packet network. First, the data unit size can be limited. In IP, this is known as the *maximum transmission unit (MTU) size*. Second, the speed of the unchannelized link being shared can be raised to the highest speed affordable or available. In many packet networks, both approaches are used to lower and stabilize delays.

All packets contain a network address to tell the network nodes to which output port to queue the packet. The network address information in a packet comes in two basic forms. In connection-oriented packet networks, the network address is a single number, the *connection identifier*. Only one number is needed because a connection, by definition, connects two network endpoints. Thus connection 22 leads from point *A* to point *B*, and the network nodes (and end nodes) need no other information than this. All packets that arrive at point *B* on connection 22 came from point *A*, by definition, and all packets sent to point *B* from point *A* are sent on connection 22 for the same reason.

In connectionless packet networks, the network address is the full, globally unique source and destination addresses of the end nodes that send and receive the message. Thus a connectionless packet would have something like "TO: B, FROM: A" in the network address field of each and every packet. Without connections, receivers need to know where the packet came from.

The first packet networks were as connection-oriented as the PSTN itself. There were a number of reasons for this. First, many packet networks were constructed by the telephone companies, companies that had always dealt with connections. Second, connections mapped packet flows onto stable paths. Third, connections provide an easy and well-known metric to bill for network services (connection time). Finally, and most important, networks spend a lot of time and effort delivering packets. In a connectionless network, packets are just dumped into the network with a delivery address, much like dropping a letter into a mailbox. If the destination node is not able to receive the packet (the computer may be turned off, for instance), then a lot of work has been done for nothing. Telephone companies usually did not charge for calls that were not accepted, and users did not like to pay for packets that were not delivered. Thus connections coordinated the whole exchange. Accepting a connection said to the sender, "I'm here and ready to receive what you send me."

Thus a connection-oriented packet network requires that the senders and potential receivers establish a connection before information transfer takes place. There are two ways that connections can be set up on a packet network. First, connections can be mapped out ahead of time, by hand, during what is known as *service provision time*. In other words, when a company contracts with a network service provider to link sites A, B, C, and D, the customer hands over a map of the connections needed to the network service provider. The map basically tells the network service provider which connections to establish between the sites. Not all customers need or want "full mesh" connectivity with a connection between each site. These connections are called *permanent virtual circuits* (PVCs) in the world of packet networks. A perfectly functional network might look like the one in Figure 4-3.

The figure shows that there is no PVC defined between node A and C. This does not mean that nodes A and C cannot communicate, just that they must do so though node B. Perhaps a direct link between nodes A and C is not required given the level of traffic between the two sites. Perhaps the organization cannot afford another link. There are many possible reasons, all of them valid.

However, what if a virtual circuit between nodes A and C is needed? In a PVC environment, this must be configured by hand, usually within 24 hours, but never right away. The alternative is to use demand connections, or *service virtual connections* (SVCs). SVC support requires a signaling protocol to be used both by the end nodes and by the network nodes (network nodes are not shown in the figure). The signaling protocol involved with SVCs will perform the call setup, monitor the information-transfer phase, and handle the release of the demand connections.

In a packet network, the call setup messages of the signaling protocol are sent on the same physical network as the data to follow once the connection is made. Compared with the PSTN and SS7, all network nodes in a packet network are both SS7 nodes and central offices.

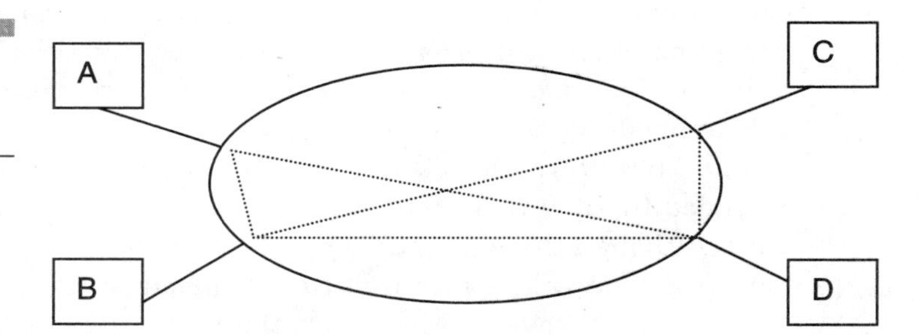

Figure 4-3 A packet network with PVCs.

X.25 Packet Switching

The first international standard for packet switching was the ITU-T's Recommendation X.25. This effort grew out of the popularity in the 1960s for computer *time-sharing* systems. Just as packet networks are designed for bursty applications, time-sharing computers are designed for the same purpose: applications that only occasionally need computing resources. Time-sharing computers and their applications did not need dedicated circuits to function but could share a network as long as the messages sent back and forth from terminal to mainframe (the time-sharing equivalent of client and server) were broken up into packets. The 1960s saw the birth of Tymshare (later Tymnet) and GEIS (General Electric Information Services) even before the Internet (originally ARPAnet) came along in late 1969.

All these early packet networks, which later came to include GTE's Telenet in the United States, Datapac in Canada, and Transpac in France, had their own rules and methods of forming and sending packets. They had much in common, such as the fact that they all used layers, but they had many differences as well. These differences made interconnecting packet networks difficult, if not downright impossible.

Therefore, in 1972, the ITU-T (then the CCITT) put together a study group to create an international standard for packet-switching networks. Working at what in those days of networking was breakneck speed, they came up with X.25 in just 4 years, in 1976. The grand title of Recommendation X.25 was *Interface between Data Terminal Equipment and Data Circuit Terminating Equipment for Terminals Operating in the Packet Mode on Public Data Networks*. Thankfully, everyone just called it X.25, and most of the abstract terms quickly received acronyms and common examples to help people relate to them. *Data terminal equipment* (DTE) was basically the terminal itself, although it would be a PC today. *Data circuit terminating equipment* (DCE) was essentially an external modem that attached the terminal to the packet network. Naturally, the DTE had to function in packet mode, not circuit mode. The packet network itself was a *public data network* (PDN). Not that there could not be *private* packet networks (and there were many), but X.25 only applied to public packet networks.

The 1976 version of X.25 had many problems and was quickly amended in 1980 to a more stable form. Further emendations came in 1984. Although there are still some 1980-version X.25 networks around, most X.25 networks today are firmly founded on the 1984 specifications. Note that X.25 says nothing about the DTE or the DCE, only about the interface between them and the fact that the DCE attaches to a public data network on one side and the DTE on the other.

X.25 was a layered protocol based on the famous Open Systems Interconnection Reference Model (OSI-RM), itself being put together around the same time by the International Organization for Standardization, often called ISO after its documentation (*ISO* is just the ancient Greek word for "equal"). The OSI-RM was a seven-layered architecture that does all the Internet protocol suite (TCP/IP) does in four layers. X.25 covered only the bottom three layers of the OSI-RM: the bits, the frame, and the packet. X.25 says nothing about the messages inside the packets. The rough relationships between layers, which can never be precise, are shown in Figure 4-4.

At the risk of confusing an already complex relationship between protocol stacks, a few words about the intent of each OSI-RM layer are in order. A lot of this is needed because over the years the purpose and function of the OSI-RM layers have been mangled somewhat as terminology has been used more loosely and familiarity has fostered a false sense of understanding in some cases.

All layers of the OSI-RM have both names and numbers. Officially, all layers have *protocol data units* (PDUs) except Layer 1, which just has a stream of bits. Some PDUs also have official names. Thus a *packet* is not only a *network layer PDU* but also just as properly a *Layer 3 PDU (L3 PDU)*. The layers of the OSI-RM are as follows:

Physical layer. Defines how bits are sent on the medium. Often involves just a connector type (e.g., V.35), but all modem standards are at this layer as well. In X.25, the physical interface was supposed to be X.21, but V.35 was used frequently, especially in the United States.

Figure 4-4 OSI-RM layers, TCP/IP layers, and X.25 layers.

OSI-RM	TCP/IP	X.25
Application	Application Services	
Presentation		
Session		
Transport	TCP (Transport)	
Network	IP	X.25 PLP
Data Link	Network Access	LAPB
Physical		X.21, V.35

Data link layer. Defines a frame as a *first-order bit structure*. When bits are sent, they represent a frame. When bits arrive on a link, they are assembled into a frame. Most frames have *delimiters* that tell the receiver where the frame starts and ends. There are also *link access procedures* defined for the flow of frames on links between *adjacent* systems (i.e., those nodes directly connected). In X.25, a frame structure and protocol known as *Link Access Procedure Balanced* (LAPB) is used.

Network layer. Defines the *packet* as the content of the frame. Packets have the network address (connectionless packets) or connection identifier (connection-oriented) needed to *route* the packet through the network nodes. Frames get built and processed link by link, but packets flow essentially intact from end to end through the network. In X.25, the *Packet Layer Protocol* (PLP) is used.

Transport layer. This is the content of the packet, often called a *message*, but really just an L4 PDU. TCP/IP calls this a *segment*, which is a very good term. The content of a packet is typically a segment of a Web page, or large file, or anything else that cannot fit inside a single packet (which has an upper limit on size).

Session layer. A *session* is just the "history" of a connection between client and server. Thus, if the link fails during the transfer of a large file, both ends can start up where they left off when they reconnect. The Internet lacks true OSI-RM sessions, as anyone who has downloaded large files for an hour or so knows all too well.

Presentation layer. Masks any differences in client and server *internal representation schemes* from the applications. Thus transfers from a Motorola-based Apple computer to an Intel-based Windows computer are fine, thanks to the presentation layer. So are transfers from 16-bit computers to 32-bit computers, and all without the applications knowing or caring.

Application layer. Has no "real" applications in the user sense at all. Contains basic operations for common network tasks such as checking e-mail, asking for passwords for remote logins, and so on. However, there is still plenty of code to write at the user interface to make the application level a useful network package.

In practice, the application, presentation, and session layers are often combined into some form of "application services." Thus, for example, in TCP/IP, a *message* at the application services layer is chopped up into *segments* that are sent as *packets* inside *frames*.

The whole idea behind packet switching in general and X.25 in particular was to reduce the cost of networking by combining traffic streams on a single, unchannelized link to a network using virtual circuits instead of relying on time-division multiplexed channels over a circuit-switched network.

As might be expected, as soon as packet networks for data came along, efforts were made to put voice on the packet network. Why not use virtual circuits to save money on voice networks as well as data networks? Throughout the 1970s and 1980s, none of these efforts were really successful. All the characteristics of packet-switching networks described so far conspired to defeat early efforts to packetize voice. These can be summarized briefly as follows:

Voice is not bursty. The standard 64-kb/s PCM voice required the full 64 kb/s all the time. Since many packet networks had only 64 kb/s available for the whole link, a single "packet voice" conversation brought the data-transfer functions on the link grinding to a halt for the duration of the call.

Voice has no packet structure. Packet networks need packets, of course. However, digital voice in its simplest form is just an unstructured stream of bits unsuited for packet transport. Voice either had to have its own packet structure created just for its own needs or use an existing packet structure such as X.25 or IP.

Voice cannot be buffered for long. Voice is delay-sensitive. The network delays must be low for voice to work. Early packet networks had very slow processors for network nodes by voice standards. Packets made their way from input to output in whole seconds, not milliseconds, as in a circuit voice switch. It was not only an issue of echo cancellation but of the viability of two-way conversation in the first place. Thus data could be transferred leisurely and buffered everywhere, but not voice.

Voice cannot tolerate long serial delays. Not only does the network delay need to be low, the network delay must be stable for voice. Delay variations (jitter) distort voice in a variety of ways.

Voice is connection-oriented. Many packet networks can operate in both connectionless modes and connection-oriented modes. Connectionless modes are faster for data but introduce delay variations (since there are no stable paths for bits to follow). Connection-oriented modes slow data down, and unless the connection setup process is fast enough, connection-oriented voice on a packet network is too slow. Note that connections for voice to follow are not even an option for packet networks such as the Internet, which has no connections at all between network nodes (the routers).

Assuming that these limitations could be overcome, as they have been in the 1990s, it is now feasible to reconsider putting digital voice on a connectionless packet network. And the telephone companies even have a connectionless packet network readily available. This is the SS7 signaling network.

SS7 as a Connectionless Packet-Switched Network

The SS7 signaling network of the PSTN is a connectionless packet network. SS7 has to be connectionless; one of its main jobs is to set up connections in the form of telephone calls. In fact, the leap of genius that led to the creation of the connectionless Internet came with the realization that *all connection-oriented networks that supported SVCs had to be connectionless networks as well*. This was true if the network was circuit-switched or packet-switched. It was true if the network was for voice or for data. In the PSTN, the connectionless SS7 network is physically separate from the connection-oriented voice network to minimize hacking and fraud. In connection-oriented data networks like X.25, the call setup signaling messages were sent over the same links as the data.

If the Internet had not been invented already, someone could have basically looked at X.25 and said, "Hmm. All X.25 switches have to handle call setup routing tables for the signaling messages *and* build connection maps for the data. The nodes handle source and destination addresses and connection identifiers. What if we just made the data all look like call setups? Then all that would be needed is the routing tables and addresses. Wouldn't that be a whole lot simpler?"

Or if SS7 had already existed, perhaps the reasoning could go, "Hmm. Here is a connectionless network that sets up connections with signaling packets. And then we can send packets on the voice circuits. And the signaling and voice networks are totally separate. Why don't we just put the data on *another* separate network, one that looks just like an SS7 for data?"

Neither of these lines of reasoning actually happened, of course. SS7 is much more than just a connectionless packet-switched network. But

SS7 is a good analogy compared with the Internet, even more so now that VoIP calls that extend to the PSTN must interact with SS7 to set up calls. Thus a closer look at SS7 is needed to understand how VoIP calls can be made to and from the PSTN.

SS7 is a signaling system that uses out-of-band signaling to send call control information between PSTN switching offices. SS7 includes databases in the network for doing things like 800/888 number translation and providing other services (such as caller ID) that can be a source of added revenue for telephone companies. Although designed for call setup-related functions, non-circuit-switching applications, such as connection-oriented or connectionless X.25 packet delivery, can be performed by the SS7 network.

SS7 replaces the older in-band signaling systems that were vulnerable to fraud and slow to set up calls (anywhere from 10 to 15 s). SS7 is an example of a common channel signaling (CCS) network and is in fact the international standard for CCS systems. Not only is the signaling moved out of band with SS7, but the signaling messages are actually moved off the voice network entirely. Call setup time dropped to about 3 to 5 s with SS7, more trunks were freed up to carry calls, and costs were minimized because less equipment was needed for SS7 than for other signaling methods. Also, CCS networks like SS7 can offer advanced services to users such as caller ID, credit card validation, and so on.

A CCS network like SS7 has a definite architecture, as shown in Figure 4-5. The figure shows the portion of SS7 that might be used by a local exchange carrier (LEC).

More details on the signaling network are provided by the figure. The *service switching points* (SSPs) are the clients of the network and usually are implemented as software in central office switches. The signaling messages originate at the SSPs (and usually end up at an SSP as well).

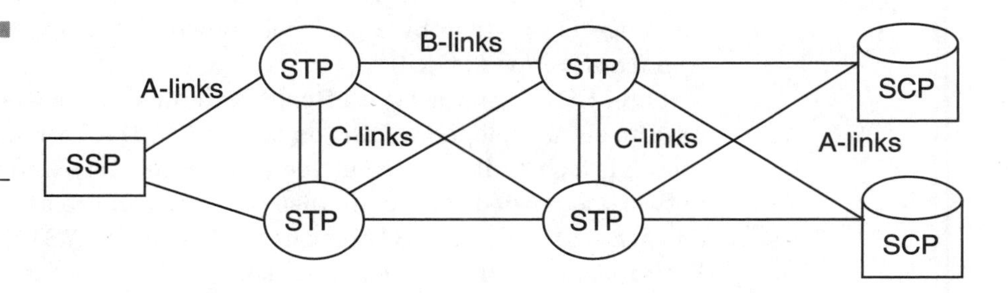

Figure 4-5 Common channel signaling components and links.

The *signal transfer points* (STPs) are the routers of the network and shuttle the signaling messages around based on routing tables. Finally, the *signaling control points* (SCPs) are the databases that store the information needed to provide the customer services, from 800/888 number translation, to credit card validation, to call forwarding.

Each of the components is connected by a special link type. The link architecture is designed to provide redundancy to the signaling network. After all, if the signaling network is down, no calls at all can be completed, with a huge impact on revenue. Access links (A-links) connect an SSP or SCP to two STPs. Bridge links (B-links) interconnect the STPs in different regions. Cross links (C-links) connect "mate" STPs and are used when A-links or B-links fail.

There are even three other types of links used for a variety of purposes. These are not shown in the figure but would be used when calls are routed through an IXC, for example. Diagonal links (D-links) connect an LEC SS7 network to an IXC SS7 network. Extended links (E-links) connect an SSP to a second set of four STPs to balance high traffic loads. Finally, fully extended links (F-links) connect two SSP switches or SCP databases directly for load balancing and further reliability.

An SS7 network uses links at 56 or 64 kb/s, but can be as fast as 8.448 Mb/s if the load calls for it. The signaling messages can be up to 272 octets in length. The routing of call setups is done based on the address of the destination office, and the routing is purely dynamic, the same strategy used for the Internet.

SS7 is a layered architecture. The bottom three layers of the OSI-RM (a frame and packet structure) are provided by the Message Transfer Part (MTP) protocol. These are called MTP level 1, MTP level 2 (the frame), and MTP level 3 (the packet), although MTP level 3 does not do all that the OSI-RM Network Layer or IP does. The rest of the network layer in SS7, if needed, is provided by a protocol called the *Signaling Connection Control Part* (SCCP). Some signaling messages do not need SCCP. When the SCP database is queried by an SSP, the database query language is provided by a protocol known as the *Transaction Capability Application Part* (TCAP).

Thus SS7 forms a type of connectionless packet-switched network used to set up telephone calls on the PSTN. If a VoIP call has one end on the PSTN, some interface between the IP network and the SS7 signaling network is needed. There are also scenarios where a call consists of VoIP on each end with the PSTN in the middle or the PSTN on each end with VoIP in the middle. In each case, some VoIP to or from the SS7 gateway device is needed. These cases are shown in Figure 4-6.

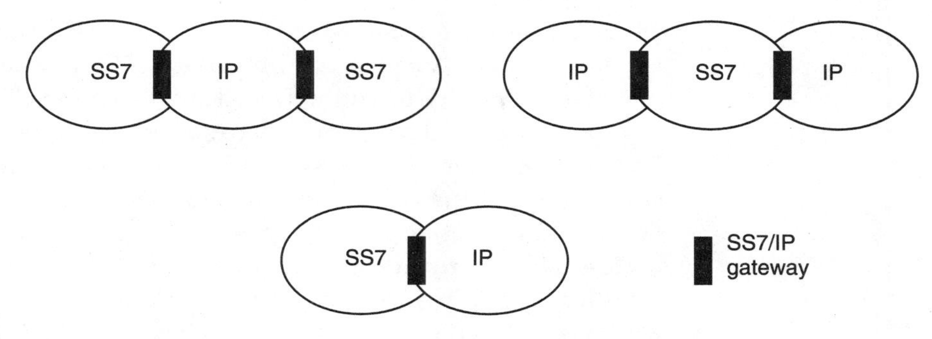

Figure 4-6
IP to SS7 gateways for VoIP services.

The SS7/IP gateway is easy enough to implement in the IP to SS7 direction. SS7 messages have a standard structure that must be observed, and IP packets can trigger the proper signaling messages as an *intelligent peripheral* attached to an SSP. For example, an IP router can be an intelligent peripheral with a direct link to the SSP module of a voice switch as long as the router can generate the proper signaling messages and packets. This is sometimes called the *tail end hop off* (TEHO) of a VoIP call.

Things are a little more complex in the SS7 to IP direction or where two SS7 clouds are linked only by an IP network such as the Internet. There is no standard way (yet) to convert SS7 messages and packets to an IP packet stream. The IP network can be used to *transport* the SS7 messages, of course, but the whole idea is to set up the VoIP call over the IP network. Work continues on standard methods to interface SS7 to IP.

The Internet as a Connectionless Packet-Switched Network

The Internet began in 1962 when Paul Baran at the Rand Corporation first published a paper outlining the idea of packet switching. The plan was to create a more survivable network than centralized circuit-switched networks by breaking messages into packets, routing each one independently through a network, and reassembling the message at the destination. This is pretty much exactly what the Internet does today.

However, it was not until 1969 that the ARPAnet was formed to try these ideas out. Sponsored by the U.S. Department of Defense's Advanced Research Project Agency (ARPA), the network linked four sites. These were the Stanford Research Institute, the University of California at Los Angeles, the University of California at Santa Barbara, and the University of Utah. The idea was to use the network to allow government-contracted scientists to share information over the network. The network nodes were assembled by Bolt, Beranek, and Newman (BBN), and in 1972, BBN's Robert Kahn introduced ARPAnet to the public at the International Conference on Computer Communications.

The demonstration was an instant hit, and the Internetworking Working Group was formed to explore more network interconnection methods, set standards, and develop protocols. Vinton Cerf of Stanford chaired the group for the first 4 years. In 1974, Cerf and Kahn published the first work on a protocol intended to link packet-switching networks, which was called TCP/IP. IP was the first major WAN protocol that was connectionless between the ARPAnet network nodes (first called *gateways*, and only later called *routers*).

It is important to realize that in those days the Internet was not a monolithic IP network from end to end. The ARPAnet was all about *inter*networking. In other words, portions of the total system could still implement whatever protocols they liked. Only when these constituent networks were linked to form an "internetwork" were the standard protocols required. For example, in 1979, Duke University and the University of North Carolina started something called Usenet, a system of newsgroups that sent news between computers using Bell Labs' UNIX-to-UNIX Copy Protocol (UUCP). In 1980 and 1981, City University of New York and Yale University began Bitnet (originally the "Because It's There Network" but later known as the "Because It's Time Network") with network software bundled with IBM computers. BBN also began the Computer Science Network (CSNet) for researchers without access to the ARPAnet. The key is that all these individual networks could be linked to the "internetwork" with the proper gateway, and many of them were.

By 1986, the system had become so popular that TCP/IP became the standard protocol whenever the ARPAnet "internetwork" was used to link networks; many university and research networks began to link to the ARPAnet, and the National Science Foundation (NSF) introduced its own backbone network, the NSFNet. The NSFNet linked five supercomputer centers, and the U.S. military began to dissociate itself from the ARPAnet it had created. ARPAnet was officially decommissioned in 1989, an event that can be used to conveniently mark the transition to the Internet as it is structured today.

The basic structure of the Internet consists of clients (Web browsers in many cases), servers (Web sites in many cases), access routers at individual sites, and backbone routers to tie it all together. Now the clients and servers are almost universally Internet hosts, devices that run the Internet Protocol stack (TCP/IP).

It sometimes seems that more has happened on the Internet since 1990 than in the 21 years before that. Much of the change has been fueled by the rise of the multimedia Web, of course. The Web debuted in 1992, and Mosaic appeared in 1993. By 1994, the White House and the U.S. House and Senate had Web sites, and the President, Vice President, and First Lady all got e-mail addresses (which the Secret Service screened). Radio stations appeared on the Internet in 1994 with RT.FM from Las Vegas. Even the Rolling Stones performed, complete with video.

Packetizing Voice

Early efforts to packetize voice were confounded by packet networks that were designed for bursty, variable-bit-rate data applications. Data applications, such as e-mail and file transfers, were indifferent to high delays and variable delays. It was not merely a matter of taking voice samples and stuffing them into a packet and launching the packet across the packet network.

In order to packetize voice effectively, something must be done about each of the following:

Packetization delay. Packetization delay occurs when voice samples, generated every 1/8000 of a second (125 μs), must be accumulated until they reach a number that makes it reasonable to place the voice samples inside a packet. This adds delay to the voice "circuit" in direct proportion to the size of the packet. For example, an IP packet has a 20-byte header, so it makes little sense to take fewer than 20 voice samples, each 8 bits, and place them inside a packet of their own. In actual practice, many more than 20 voice samples are placed in a packet. The default size of an IP packet in many implementations is about 570 bytes (the maximum size of an IP packet is 64,000 bytes, but no IP packets ever approach this size in reality). This packetization delay would add $125\ \mu s \times 570 = 71{,}250\ \mu s$, or 71 ms, to the voice "circuit." The higher end-to-end delay may require the extra expense of echo cancelers even if the VoIP call is over a short distance.

Serial delay (jitter). Serial delay occurs when the delay-sensitive voice packet is queued behind a varying number of data packets (or just less delay-sensitive traffic). This occurs at every IP router along the way, so the jitter (delay variation) resulting from this repeated procedure can distort the voice in a number of ways.

The "high" voice bit rate. The most commonly used standard digital voice bit rate is 64 kb/s. However, this also happens to be the basic speed of links used between IP routers. Thus voice packets that contain 64-kb/s PCM voice can force every other form of packet off the link. And this does not take into account the IP packet header overhead, which must be accommodated—but not on 64-kb/s links.

Constant-bit-rate voice. The 64-kb/s PCM voice standard is also constant bit rate, generating 64 kb/s each and every second, even during silent periods. However, packets are best suited (and designed for) bursty applications. Perhaps removing silence could make voice bursty, since 50 percent of most conversations is silence due to listening (actually, the periods of active voice are closer to 40 percent of the total bit rate in both directions). But then the result is total silence. Callers are used to some background or "comfort noise" during a call. Without comfort noise, listeners think that the line has failed while they are speaking. The natural consequence is annoying "You still there?" queries intended to provoke a voice response from the listener (which counteracts the enforced bursty effect). This also results in a problem with detecting resumed bursts of voice, a problem known as *voice activation detection* (VAD).

Resends resulting from errors. IP networks are connectionless "best effort" networks and perform no error recovery at the IP layer. IP routers just toss packets with errors. However, since IP was designed as a data protocol, the TCP layer running above IP will attempt to resend any segments not received at the destination. TCP seems like a natural choice for VoIP transport, mainly because TCP is connection-oriented and so is voice. However, segments and packets containing voice samples cannot be resent (voice is "real time"). Lost voice samples just cause gaps in the conversation, but humans perform their own error recovery ("Say again?"). The alternative protocol, UDP, is attractive because UDP does not resend, but UDP is connectionless. This makes it much harder to map VoIP connections to the TCP/IP networks.

Fortunately, there is something that can be done about each of these issues in VoIP. Some have even been standardized, although there are variations. A close look at each one is definitely called for.

Small Packets

The only real way to deal with the effects of packetization delays is to use small packets for voice samples. There is no standard size for a VoIP packet. However, it is tempting to use the default size of the IP packet for this purpose, since this is quite small to begin with.

Priorities

The use of priorities for VoIP packets can cut down on the effects of serialization delay and the jitter that results. There is no accepted standard way to employ these priorities, and most major router vendors have championed their own way of prioritizing IP traffic. Many proposals have been made, and some have even been tried, but none has emerged as the clear winner. Ironically, the IP packet header itself has a type of service (TOS) field that can be used to ensure VoIP packet priority. However, this field has never been used consistently by routers, and many router vendors simply ignore the TOS field or use a vendor-specific implementation that is worthless unless each and every router in the network looks at the TOS field in the same way.

Priorities can guarantee that the VoIP packets go to the head of the queue in an output buffer. But this does not address the situation that arises when a variable-length data unit has just begun to be sent out of the serial port. Jitter still results when a VoIP packet encounters longer or shorter packets in progress on the serial output ports on a series of routers. Something else is needed, such as a jitter buffer.

Jitter Buffers

The effects of jitter or delay variations on VoIP can be counteracted by the use of jitter buffers at the receiver. Jitter buffers are memory areas used to store voice packets arriving with variable delays so that it appears that each voice sample has arrived in the same amount of time.

The steady output of the voice samples from the jitter buffer is called *playout*. The playout is steady and constant, and as long as the jitter buffer receives an ample supply of voice packets, the system appears to have a fixed delay.

As an example, consider two VoIP packets containing voice samples that arrive with different delays through the router network. The first packet might arrive in 100 ms, whereas the second packet arrives with a delay of 90 ms. The packets still arrive in sequence (although this is not always a given), but the network delay has fallen between the time it has taken to generate and send the second packet full of voice samples. However, the jitter buffer holds *both* packets so that it would appear that each packet made its way through the IP network with a delay of 120 ms (for example). The concept of a jitter buffer is shown in Figure 4-7.

The delay chosen for the jitter buffer is critical. If the delay is set too low, then the whole scheme will not work. Packets that arrive "late," i.e., after the jitter buffer delay (anything delayed on the network more than 120 ms in the example given), must be discarded. This discarding can cause a noticeable gap in the conversation, since a sizable portion of a second of conversation may be discarded. If the delay is set too high, then the jitter buffer may overflow with similar loss of conversation, or the voice is delayed needlessly.

Another consideration is how the jitter buffer knows exactly how long it took the voice packets to make their way across the network. Several mechanisms are used in practice, from simple timestamps to more elaborate delay-determination methods. The simplest method just plays out a voice sample every 125 μs from the jitter buffer. As long as fresh packets arrive quickly enough and the buffer space is adequate, the buffer will not be empty. If the buffer does empty, the jitter buffer can simply repeat the last packet's worth of voice samples.

Jitter buffers are the most practical way of dealing with the delay

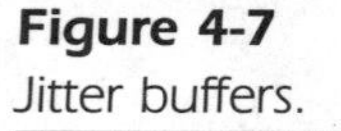

Figure 4-7
Jitter buffers.

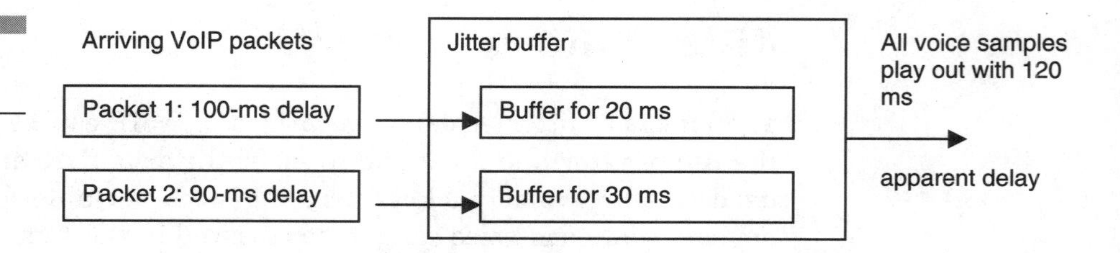

variations introduced by packetized voice. They are simple and easy to implement. Jitter buffers are used not only in VoIP but also in voice over frame relay (VoFR) and even in ATM voice applications (where the process is known as *network conditioning*).

Voice Compression

The simplest way to deal with the fact that digital voice is typically generated as 64-kb/s PCM voice is to apply some form of *voice compression* to the PCM voice. Although the term *voice compression* is used commonly for this process, it is somewhat misleading. It is not the voice itself that is being compressed, but the 64-kb/s PCM words themselves. Toll-quality voice can be achieved today at 13 kb/s or even 8 kb/s. These lower voice rates can be generated in one of two ways. First, completely new chipsets can be developed to digitize analog voice directly to these bit rates. Second, the 64-kb/s PCM voice itself can act as input to a second stage of voice digitization. The output of this voice-compression step would be the lower-bit-rate voice. This second method does not require a completely new chipset from analog to digital and can just be added on to existing 64-kb/s PCM voice chipsets.

Note that voice compression itself does not automatically generate bursty voice. Voice at 8 kb/s is still constant-bit-rate voice. In order to make the compressed voice bursty, it is necessary to remove the periods of silence in the conversation.

Silence Suppression

Silence suppression removes the periods of silence that occur naturally within a voice conversation. The main cause of silence is when one of the parties to the conversation is listening, but shorter periods of silence occur between sentences, phrases, words, and even within longer words. All together, silence accounts for nearly 60 percent of the bits sent during a two-way 64-kb/s PCM voice conversation.

The biggest problem with silence suppression is detecting when the speaker begins to talk again after a period of silence. Why should this be a problem? Only because there is always some background noise at the speaker. This ambient background noise must not be packetized and sent, only the speaker's voice. The trick is to reliably detect when the speaker's voice level has risen far enough above the background noise to determine that the speaker has actually begun to talk again. This rise in "noise" level could be caused by a random variation in the background noise level (a dog barking, for example). This is the issue of *voice activation detection* (VAD) at the speaker.

If the VAD level is set too low, then listeners will suddenly hear the dog barks, trucks backfiring, and all sorts of extraneous and disconcerting sounds. If the VAD level is set too high, then the speaker's resumed speech will be *clipped* because the initial sounds will not be digitized, packetized, and sent. This is due to the fact that the voice digitization process must be done in real time. By the time a VAD process set too high realizes that the speaker has begun to talk again, it is too late to "capture" the initial vocalizations.

Coupled with voice compression, silence suppression makes it possible to packetize voice and send it over a 64-kb/s link along with other packet traffic. The VoIP packet stream is almost indistinguishable from a file transfer or an e-mail transfer on the link.

Comfort Noise

Comfort noise is used to compensate for the loss of sound at the listener when silence suppression is in use. Total silence leads the listener to think the line has failed. Some ambient background noise is generated at the receiver to maintain the illusion of a constant stream of noise across the network. In actuality, of course, the comfort noise is not sent across the network at all.

The process of adding comfort noise to a packet voice system is called *comfort noise generation* (CNG). There are many ways to perform CNG. The easiest is just to put a chip in the receiver to generate some random "white noise" hum at a low volume. The problem is that listeners quickly become aware of the artificial nature of this comfort noise unless the variation is quite sophisticated.

More elaborate CNG methods actually will sample and retain the short pauses that are not eliminated by silence suppression at the beginning of a conversation. Later, during periods when silence suppression is in use, the CNG can vary and play out the actual background sound. This method results in quite natural comfort noise. (Audio engineers typically record a few minutes of "open mike" silent room sound that can be used to fill gaps caused by editing for the same purpose.)

RTP

IP is a network-layer protocol. But between IP and the application services layer in the Internet protocol suite is the transport layer (frequently called the *TCP layer*). The main protocols used here are connection-oriented Transmission Control Protocol (TCP) and connectionless User Diagram Protocol (UDP). Connection support makes TCP a natural choice for VoIP, but TCP will always resend missing segments discarded by the IP network due to errors. This is useless for real-time VoIP and just causes the VoIP conversation to fail. UDP seems better suited, since UDP does not resend, but the connectionless nature of UDP makes it more difficult to map VoIP connections onto the IP network.

Fortunately, there are more protocols available for use inside IP packets than TCP and UDP. In fact, there are 256 protocols that can be placed inside IP packets (it is an 8-bit field in the IP header). About 200 have been defined, and among these is one known as the *Real-time Transport Protocol* (RTP). RTP is quite similar to TCP, with one major exception. RTP does not resend when errors result in lost packets. This makes RTP suitable for all sorts of real-time applications, including VoIP.

The issues raised so far with regard to packetizing voice are summarized in Table 4-2. Just having an answer to a VoIP issue, however, does not make the solution a standard. Standard methods of implementing each of the answers must be found to ensure interoperability among VoIP products.

TABLE 4-2

VoIP Issues and Answers

VoIP Issue	Answer
Packetizing delay	Small packets for VoIP
Serial delay	Priorities and jitter buffers
"High" bit-rate voice (64 kb/s)	Voice compression
Constant-bit-rate voice (PCM)	Silence suppression, voice activation detection (VAD), and comfort noise generation (CNG)
Resends due to errors	Real-time Transport Protocol (RTP)

ISDN: What It Did Not Do

Most people are somewhat familiar with the concept of an integrated services digital network (ISDN). Today, ISDN is frequently used for high-speed Internet access from homes at 128 kb/s. However, ISDN was designed initially for end-to-end voice digitization and as a platform for a host on new network services. What ISDN was *not* designed for, and did not do, though, was packetize voice.

ISDN voice is just 64-kb/s PCM voice on a circuit-switched network. This is worth pointing out because ISDN channels can function in *packet mode* as well as in circuit mode. However, an ISDN channel is used in packet mode to attach to an X.25 network, not an IP network. While it is conceivable to packetize voice inside X.25 packets, using all of methods outlined in the preceding section, this is obviously not VoIP.

VoIP has no relationship with ISDN packet-mode channels. VoIP, if carried for any portion of the journey over ISDN, still travels over the ISDN channel in straight circuit mode. ISDN is an international standard for digital voice but not for packetized voice.

G.729B: How to Packetize Voice

Ironically, there *is* an international standard that can be used to packetize voice. However, it is not ISDN. The standard is ITU-T Recommendation G.729B, which has the imposing title, *A Silence Compression Scheme for G.729 Optimized for Terminals Conforming to Recommendation V.70*. V.70 just mandates the use of G.729 coding techniques. G.729B is Annex B to G.729 itself, which describes the coding of speech at 8 kb/s using conjugate structure algebraic-code-excited linear-prediction (CS-ACELP). Think of G.729 itself as providing the voice compression. G.729B provides the silence suppression (called *discontinuous transmission*, or DTX, in G.729B), VAD, and CNG needed for packetized voice applications.

Note that G.729B says nothing about packet size, priorities, jitter buffers, or the use of real-time protocols. Some of these issues will remain beyond the scope of the ITU-T, since they involve customer premises equipment. However, G.729B is a start toward the full standardization of all packet voice techniques. The major sections of G.729B describe the standard way to implement VAD, DTX, and CNG.

CHAPTER 5

Packets for Telephony People

Anyone taking a tour of a telephone company central office will begin to get an idea of how seriously the modern world takes its telecommunications. From the often-paranoid physical security, to the banks of 48-V batteries, to the neatly arranged cable runs, to the diesel generator on the roof, central offices are physical manifestations of the solid design philosophy that has been a hallmark of the phone system from the beginning.

As with anything that a culture values enormously, there are a large number of written and unwritten rules about the provision and use of the telephone network. The arcane world of telecommunications regulation is a topic best left for later, but perhaps the first guiding principle of the telephone network (both written in law and assumed by all) is that the service it offers should never be interrupted. Telecommunications service providers go to great lengths to make their physical lines safe from harm and provide redundant backup lines and procedures for the dynamic rerouting that occurs if a cable should get cut. Usually, this happens without the knowledge of the people using the telephones.

The second unspoken law of telephone systems is that nothing should ever disturb a call that is taking place, even if it means not handling calls on behalf of people who would like to use the phone. While it is thus possible to encounter a "fast busy" signal on Mother's Day (informing callers that the trunk lines between them and their moms are busy), it is virtually impossible for a caller to be "kicked off" the telephone by other people eager to speak with their own mothers.

Finally, a third law says that the services of the telephone network should be available to all, or at least to those willing and able to pay the "reasonable" amount that basic monthly service costs. This last rule is the precept behind the "universal service" requirement that influences telephony economics in the United States (although not always throughout the world, where the universal service concept is not always a given). Presumably, the same law dictates that operating a telephone should be so simple that many children can place a call at 3 or 4 years of age.

As one might imagine, these ideas also have dictated how the telephone network equipment itself operates: how a call is placed, how individual calls can be made to coexist on a single, integrated network, how the switches and other telephony equipment inform each other of incoming calls, and so on. These operations and all the other functions of your phone system embody the underlying solidity of the United States and, to a lesser extent, the global telephone network. Shortly after the telephone was invented, people realized that they could not do without it, and the people in charge have taken pains to design all its components accordingly ever since.

What has all this to do with the Internet Protocol (IP)? Only that just

30 years after it was created, the world has fallen in love with the Internet. However, even among those who use the Internet daily, there are few people who would argue that it is as important to the everyday life of the world as the global public switched telephone network (PSTN). The fact is that the PSTN is not betrayed by the regular brownouts, slowdowns, and outright failures that, while not usually disabling the entire network, are regularly experienced by all Internet users. Some of this stark disparity is simply a result of the money spent on the PSTN and the comparative dearth of it historically spent on the Internet. The last few years have seen an explosion in the number of Internet users, and while the Internet service providers (ISPs) have been adding capacity and robustness at an incredible rate, their rate of improvement lags far behind that required to bring a PSTN-like degree of reliability to Internet service. As the rate of growth of the Internet begins to ease (as it must before too long, or everyone on Earth will be working for a router manufacturer), and as people begin to depend on it in their daily lives and business dealings, improvements can be expected in Internet infrastructure quality that will begin to make it as "bulletproof" as the telephone network. However, does this mean that in the presence of a solid, reliable, dependable, ubiquitous Internet, people can simply plug in their telephones and have a chat between any two locations? Is it simply the issues of reliability and capacity that keep the Internet from being a second, and perhaps later the primary, telephone network?

Unfortunately, making a "PSTN" out of the Internet will not be this simple. High reliability and enormous capacity are two necessary attributes of any global phone system, but they are not the only ones. The Internet was never designed to live up to the standard set by the PSTN, with the individual message (or phone call) being important enough to spend large amounts of money and resources on to protect. Rather than place, track, trace, and do not forget bill for each individual message as the phone system does with calls, the needs of the Internet's earliest users led to a design approach in which the individual message was not terribly important. After all, if there were problems in the network, a message could be routed around them or be sent again later. While such an assumption may result in a network design that works splendidly for electronic mail or file transfers, it is vastly incompatible with information that must always be delivered with as little delay as possible. And, of course, real-time conversations such as phone calls are a primary example of this type of communication. Thus there are two distinct ways to handle information on a network, and they form the underlying principles of how these two types of communications protocols work.

The development of these two approaches to data communications, which are usually called the *connection-oriented* and *connectionless* approaches, can most easily be traced when seen through the eyes of their developers. The concept of connections has been introduced already. This is the place to see how the Internet can handle connection-oriented telephone calls when there are no connections at all between routers. First, it may be best to briefly go over some consequences of networking with connections and physical circuits.

Connections and Circuits

As shown in Chapter 3, circuit-switched networks like the PSTN focus a large amount of effort on the setting up and breaking down of a call's physical connection. After the caller dials the requested telephone number, the network switches proceed in a methodical, step-by-step fashion to find a path through the network(s) to the destination, reserve a certain amount of bandwidth all along the call's path, ring the destination phone, and make the connection when that ring is answered. At this point, the two telephones are for all intents and purposes connected by a point-to-point circuit; anything uttered into one handset is immediately repeated out the other handset. Of course, since the existence of a call between two phones uses resources (such as bandwidth on the lines and memory and ports on the switches) as long as it is up, the call must be terminated in some kind of orderly fashion so that those resources may be returned to the network to be used by subsequent callers.

There are a lot of great things about designing a network that operates this way, regardless of the payload being carried. Once a call has been set up, both ends are dealing with a link of known quality to a known destination, with at least some expectation that when they put information into one end, it will come out of the other end and be delivered to the party they expect. Functioning in this manner is termed *connection-oriented operation* because the "thing" that all nodes—end points as well as switches—are working together to maintain is the connection itself. The existence of the connection means that very little else must take place to move information beyond submitting it to the network in a predefined manner. The connection takes care of the rest.

Another nice thing about connection-oriented networks is that the switches, which along with the network links actually comprise the network, can operate very simply once the connection is established. Of course, the signaling that is necessary to make the switches set up the connection takes some amount of time, but once complete, the switches need only forward the information from one port to another according to the information in their circuit tables. This simplicity means that connection-oriented switches can be very fast and therefore add little delay to the information they forward. And when it comes to networking, fast is good—regardless of the payload.

Since the switches in the PSTN seem to form an actual electronic circuit from caller to receiver, they are referred to as *circuit switches*, and the PSTN is referred to as a *circuit-switched network*. This is significant because it places very few restrictions on the format of the information that the ends of the circuit may send to each other, besides limiting the *bandwidth*, or range of frequencies, that may be used to that which the circuit is capable of handling. In normal human speech, the range can be limited to between 300 and 3300 Hz, and most people accept this as the attributes of a "phone circuit" without question. However, once the connection is made, the people on the ends of the phone lines have guaranteed access to a 3000-Hz circuit from end to end, with the ability to use it in whatever manner they please.

Circuit switching has the wonderful attribute of this guarantee, which leaves both users secure in the knowledge of exactly what they are getting for the duration of their call. The down side of such an arrangement is that the users probably do not need all of it. This, in turn, has two consequences: that users pay for network time they do not need and that the network has no flexibility to use the wasted bandwidth, since it has been "promised" to the users on the ends of the circuit. This is just the concept of "all the bandwidth all the time" multiplexing seen from a different angle.

Virtual Circuits

There are ways, however, to provide a transport that acts—from the users' point of view at least—like a circuit while remaining capable of redistributing unused bandwidth to other users. Such a network is capable of allowing a number of users to share the same physical circuits simultaneously, thus allowing the provider of the network to maximize use of the circuits. This generally means the providers can charge less for access to such a network than they would for access to a true circuit-switched environment. Networks such as these are referred to as *virtual circuit networks*, and while they remain connection-oriented, the ability to provide shared access to circuits (good for the providers) while acting like dedicated circuit-switching networks (good for the users) makes them the backbone of most public data networks.

Recall that at the physical layer, the ability to share a single physical circuit between many users is called *multiplexing*, and there are a number of techniques that can be used to accomplish this. The method used by virtually all digital data networks is *time-division multiplexing* (TDM), which divides the physical circuit bandwidth by assigning a certain amount of time to each of a set of users, during which the circuit is theirs to use as he or she wishes. When a user's time is up, the next user gets his or her turn, and so on, until all have had their chance—whereupon the first user again has a chance, and the round-robin process continues. The device responsible for implementing the time division is called a *multiplexer* or *MUX*, although devices such as switches usually integrate the multiplexing task into their other protocol tasks, and therefore, no device is dedicated to multiplexing as its sole function.

Figure 5-1 shows that in TDM there is a separate conversation occurring between each pair of devices labeled A, B, and C over the single circuit between the two multiplexers. The multiplexers assign each device a *time slot*, or portion of available bandwidth, and then handle the combining and separation of traffic over the shared line, as well as the transmission between themselves and the user devices at the edges of the network.

Of course, with data traffic usually being bursty in nature, it often occurs that A, B, and C may have differing needs for the shared circuit. For example, at any particular time A may be extremely busy and have a large amount of data to send to its counterpart, while B has much less and C has very little. If the circuit is divided according to standard TDM, C's time slots will be largely wasted by the circuit, while A will not be able to transmit as fast as it would like. To address such a situation, a variant of TDM called *statistical TDM*, or *statmuxing*, is available. Statmuxing allows the multiplexers to assign bandwidth dynamically as

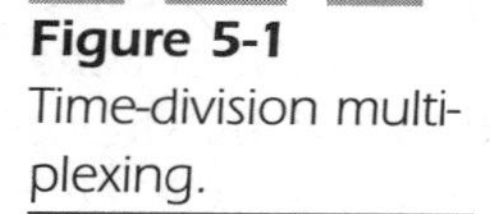

Figure 5-1
Time-division multiplexing.

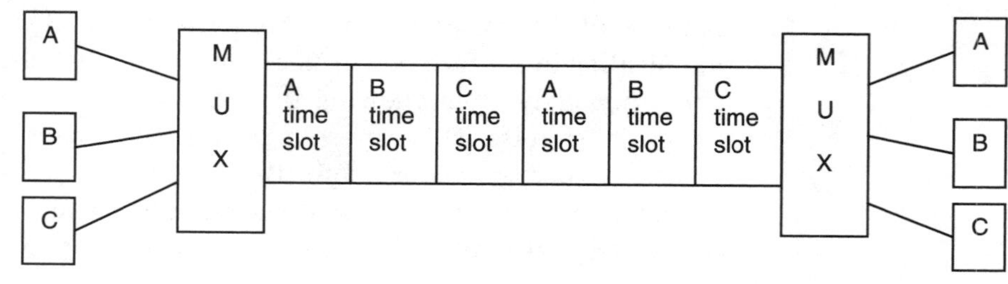

needed to each of the devices that use the shared circuit. This is shown in Figure 5-2.

The mechanics of statistical multiplexing differ from those of standard multiplexing in one important respect. Since time slots are assigned dynamically instead of in a fixed, round-robin manner, the data in each slot must be marked by the transmitting multiplexer to identify the virtual circuit, and thus the end device, to which it belongs. This enables the receiving multiplexer to deliver the data contained in each slot to the correct recipient. This is accomplished by having the transmitting mux attach some type of identifier to the data at the beginning of each time slot, which can be read by the receiving mux and associated with a particular end device.

Most virtual circuit networks are considerably more complex than the one described above, particularly in that there is more than one physical link over which the data travel between end devices. In fact, to take advantage of the economy of scale, as well as to make their service available to large numbers of potential customers, data network providers often construct enormous systems of physical lines and switches, which can be used to connect any two user devices, regardless of their location. Figure 5-3 shows the concept of virtual circuits.

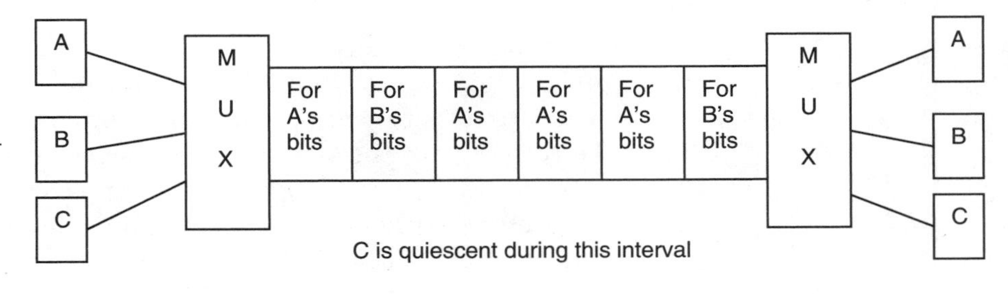

Figure 5-2 Statistical time-division multiplexing.

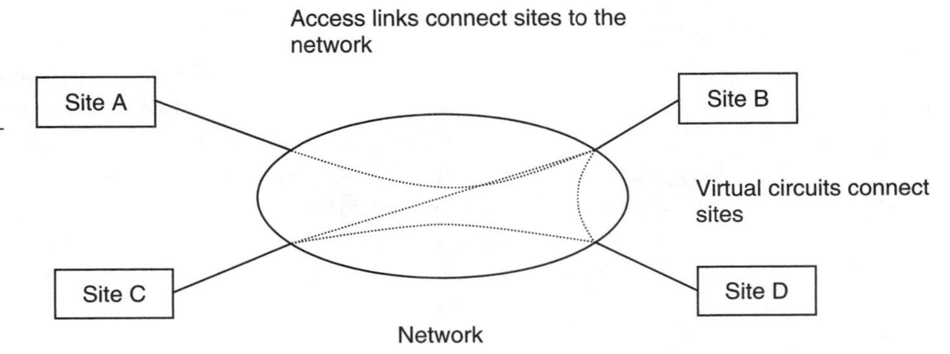

Figure 5-3 A virtual circuit network.

Establishing Virtual Circuits

If a virtual circuit network is to be statistically multiplexed, the multiplexers (and any switches between them) must be able to learn which time slots "belong" to which virtual connection. In Figure 5-3, a mux at site A may identify those time slots to be sent to the mux at site B by inserting a field with the virtual circuit identifier 47. Of course, the mux at site B must be told that the data in all slots marked 47 have arrived from the mux at site A. This process is analogous to the call establishment and breakdown that occurs in a circuit-switched network.

Establishing a virtual circuit can be accomplished either manually by the network engineers or dynamically by the communications protocols themselves. Manual, or static, configuration of virtual circuits is most useful when the circuit is intended to be permanent, or at least exist for a considerable period of time measured in months or years—indeed, circuits provisioned this way are usually referred to as *permanent virtual circuits* (PVCs). Note that in such a system, no communication between the muxes is necessary to set up a virtual circuit, since each is configured by the network administrator and thereafter associates each virtual circuit number with a particular end device until configured otherwise.

Dynamic virtual circuits, or switched virtual circuits (SVCs), however, must provide a means by which muxes can inform each other about which virtual circuits belong to which end devices, as well as a way to choose the correct path through the network between any two of them. This means that some type of signaling circuit assignment must be used that all the muxes understand, and some type of routing between switches must be present for path selection. In such a network, the network engineers need only make sure that this signaling protocol operates correctly, leaving the muxes to set up and break down the virtual circuits themselves.

There are many types of data communications systems that operate as either (or both) switched or permanent virtual circuit networks. Among networks designed for the transport of digital data, there are the X.25 protocol, frame relay, asynchronous transfer mode (ATM), and others. Before taking a more detailed look at X.25 virtual circuits as the archetypal connection-oriented communications protocol, a look at alternatives to connection-oriented networking is in order.

Connectionless Networking

What other choices might there be for a philosophy around which to design a shared network? Isn't a connection of some type, either physical or virtual, always necessary to get information to the correct place? In fact, many data networks use no defined circuits to identify the end points of information transfer, leaving that job to a unique identifier or address that is assigned to each potential recipient. Such a network is commonly referred to as a *connectionless network*, for obvious reasons.

In a connectionless network, it is the job of the sending device (using the term *caller* here makes no sense, because no call is being established) to attach to each message the identifier or address of the intended recipient. It then forwards that information to a *router* (the connectionless world's name for a packet switch) whose job it is to somehow help the information reach its intended destination. Note the *somehow*—sometimes the router may be able to deliver the information to the destination itself, and sometimes it may pass it to another router. In this manner, the information moves hop by hop through the network until it reaches a router that can in fact give it to the eventual recipient.

In order to keep a device from beginning to send information and never stopping (perhaps it is programmed to transmit a never-ending stream of data, like weather information or stock market prices), connectionless networks require their participants to send only a limited amount of data in a single message, often called a *datagram*. There is much debate about the continued use of the term *datagram* for "connectionless packet." Initially, the Internet founders wanted a term that would underscore the fundamental differences between connectionless routing and connection-oriented switching. The reasoning basically went, "Our network is so different from traditional packet switching that we need a different name for the data unit." The problem was that an IP packet is still a packet. Today, in all Internet documentation, under the entry for *datagram* definition, it says, "See packet." However, use of the term *datagram* is common enough to justify its use here.

In any case, limiting the length of datagrams is a good idea because it makes multiplexing easier by creating more spaces between the messages that make up the contents of the datagrams (which can be multimegabyte files) and therefore more chances for different devices to transmit their own datagrams. Conveniently, it also defines a unit or package to which the sender may affix the addressing information. Think of a datagram as a variable-length time slot used for statistical multiplexing.

However, if there is no circuit showing the way to the destination, how does a router know what to do with a packet that is given to it? The answer lies in the essential difference between a connectionless router and a connection-oriented switch. While a router lacks the circuit table that allows a switch to quickly forward data according to virtual circuit identifiers, it does have a routing table that lists not preexisting circuits but *network addresses*. These are essentially the ZIP codes of all the network segments available to the router and the addresses of other routers that are connected to those segments. When a router receives a packet, it therefore must look at the packet's destination address, find in its router table the corresponding "ZIP code," and move the packet one step closer to its destination by sending it to the router who is connected to that destination network.

Note that no setup, call establishment, or circuit overhead is necessary to begin this process. The sender must know only one thing before addressing and sending a packet: the address of the device with which he or she wishes to communicate. The packet (or datagram) makes its own progress toward its destination, simply by having its desired "ZIP code" studied by each router it encounters. At some point, the router that is directly connected to the network bearing the destination "ZIP code" will be given the packet, which it may then deliver directly to the recipient. Table 5-1 summarizes the differences between virtual circuit (connection-oriented) and datagram (connectionless) packet networks.

TABLE 5-1

Attributes of Virtual Circuit and Datagram Networks

	Virtual Circuit	Datagram
Design philosophy	Connection-oriented	Connectionless
Addressing entity	Circuit identifiers	Node and network addresses
Unit of information	Packet, frame, or cell	Packet, frame, or cell
Network entities	Switches	Routers
Bandwidth use	As allowed by network	Depends on number and size of submitted frames
Examples	Frame relay, X.25, ATM, most other WANs	Internet Protocol (IP), Ethernet, most other LANs

Switching versus Routing

The issue of what type of network to implement—connection-oriented or connectionless—is very closely related to the nature of the devices used to forward the packets through the network. Of course, both types of networks attempt to get the packets to their intended destination, but how that destination is identified and how the devices act on each packet differ considerably. While hardware manufacturers tend to confuse the issue by claiming that their latest machine combines all the best attributes of both approaches, it is best to attempt to define here the difference between a router and a switch and to differentiate between routing and switching as forwarding techniques.

A connection-oriented switch has a relatively simple job to do. The switch must

1. Accept an incoming packet
2. Examine the connection ID field (which identifies the virtual circuit that the packet is following)
3. Relate that ID to a directive in its circuit table
4. Replace the ID with the one indicated in the table
5. Transmit the redirected packet on the indicated circuit

This process is at the heart of all virtual circuit network technologies. Figure 5-4 shows how a packet is sent through a switch. All arriving packets are processed by the *routing table* (even switches have routing

Figure 5-4
A switch.

Input Port	Conn.ID	Output Port	Conn.ID
0	34	1	18
0	47	0	12
1	77	1	12
1	47	1	23

Connection ID: 47

Connection ID: 12

tables). The routing table is organized by input port (this simple switch has only two ports). The switch just replaces the "label" and queues the packet to the correct output port. Thus traffic on connection 34 goes to output port 1 and traffic on connection 47 goes to output port 0, even though both arrive on input port 0. Note that connection identifiers can repeat in the table, just not on the same ports. This is called *local significance*. Thanks to local significance, huge networks can use a relatively small number of connection identifiers.

What this operational mode means for the switch is important. Having received an incoming packet, the switch can do only one thing: relay the packet one step farther along its journey. Very little intelligence is required to do this: A simple lookup of the ID and the ability to follow the directions in the circuit table will suffice. The primary advantage of this simplicity is its speed, because the less work that must be performed to process the packet, the less time it must spend in the switch. The main disadvantage of such an approach is its lack of adaptability to changing network conditions. If something unfortunate happens to any of the links that define a circuit, that circuit is effectively unusable until the problem is corrected. This is not to say that it is impossible to reassign the circuit to other links; in fact, many switched networks have some way of doing this. However, this rapid reassignment is a complex process, one that usually requires some degree of manual intervention, and it adds a lot of management and control traffic to a network.

There is a subtle difference between this process and that used by routers in connectionless networks such as the Internet. A router, being a somewhat more complicated beast by nature, trades simplicity for adaptability. A packet arrives at a router carrying not a connection ID but rather a globally (not locally) unique destination address—a single identifier that distinguishes the intended destination node from all others in the world (well, at least the world known to that router). It is the router's job to

1. Read the destination address at the front of the packet.
2. Determine the network that this address resides on.
3. Find, in its own routing table, the address of a router that is closer to the destination network than itself.
4. Transmit the packet to that router.

While on the face of it this seems simple, the fact that destinations, not circuits, are being represented by the packet addresses has enormous implications for the amount of processing that must be done to decide how to forward a packet. Since an Internet router may receive, on any port at any time, a packet destined for any destination network on Earth, it must be prepared to compare the packet's destination address with every destination in its own routing table. The Internet routing table, in early 1999, listed about 50,000 routes. This comparison thus involves a lot of work, especially since high-speed routers are required to do this at rates in the tens of millions of packets per second. A switch's job is easier because the number of circuits that can be represented on a single port is typically much, much smaller.

An Internet router's job is to list reachable destinations, a huge task. In contrast, a PSTN switch only has to keep track of connections currently in progress. This makes all the difference.

In reality, there is rarely a purely connection-oriented or connectionless network application. Even the circuit-switched PSTN has elements of connectionless networking in some of its control functions, as well as in its connections to other networks. Observing the operation of both a connection-oriented and a connectionless network is instructive in discovering the advantages and disadvantages of both kinds of networks. And since eventually this will wind up at a technology (voice over IP) that is a challenger to the PSTN, a closer look at the X.25 service for a brief example of connection-oriented data networking is in order. Following this, the Internet and its associated protocols will provide an example of the ins and outs of the connectionless network paradigm.

The X.25 Network Service

Some of the aspects of X.25 were discussed in Chapters 3 and 4. Now is the time to add some detail in the operation of the X.25 protocol, the better to understand how the IP functions. The X.25 protocol is a specification for connection-oriented access to a public switched packet data network (PSPDN). As illustrated in Figure 5-5, it is comprised of three parts, or layers, each of which maps more or less to one of the bottom three layers of the Open Systems Interconnect (OSI) reference model (OSI-RM). It is important to recognize that the X.25 specification does not address the responsibilities of the end stations or user equipment, leaving the top four layers of the model to be implemented by users in whatever way they wish.

The physical layer is not of much concern here; it is enough to note that X.21 and X.21*bis* describe the electrical and mechanical specifica-

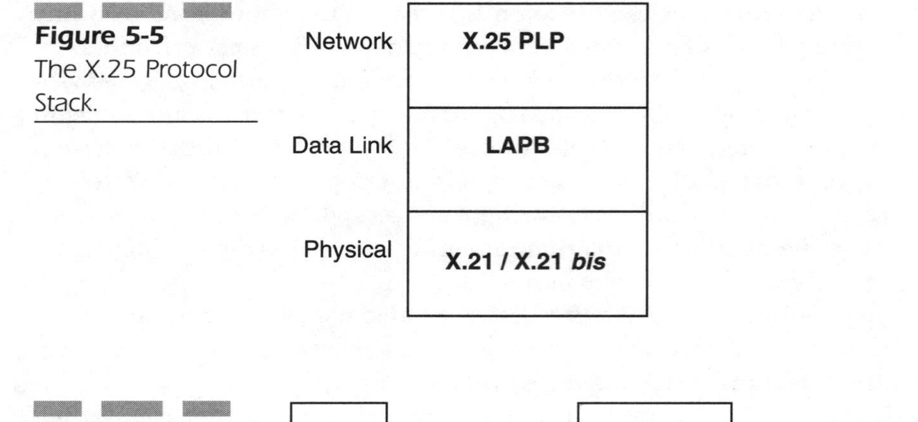

Figure 5-5 The X.25 Protocol Stack.

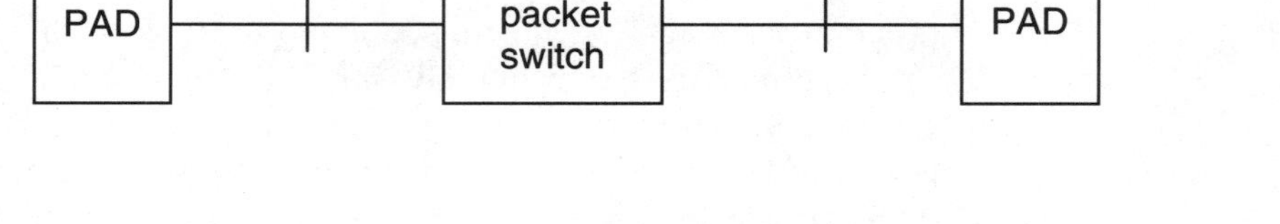

Figure 5-6 A simple application of X.25.

tions for the cable and transceivers that carry the upper layers from node to node through the network. The real work of the X.25 suite is performed by the data link and network layers.

LAPB (Link Access Procedure Balanced—a superset of the venerable OSI-RM HDLC protocol) is the data link layer protocol that connects a packet assembler/disassembler (PAD—essentially an X.25 mux) to a switch or a switch to another switch. This connection is always point to point, and generally exists over a single physical line. In most cases, an end-to-end X.25 connection is formed between two PADs over a series of these point-to-point LAPB links. Customer equipment, such as computers, terminals, and printers, connects to the PADs via asynchronous serial links or other local-area networks (LANs), often unaware of the presence of X.25. Figure 5-6 shows the simplest representation of such a network, with two PADs connected by one switch to form a single host-to-terminal connection.

Of course, this is not a typical public network, because there appears to be only one link (not much networking going on here). In fact, usually the PADs connect to separate switches, which in turn are connected to each other via trunk lines or still other switches. The connections between the switches and the switches themselves form the network "cloud" that service providers of all types refer to regularly. In fact, X.25 essentially invented the *network cloud*. Note that the cloud is a cloud precisely for the reason that its contents are unknown to users. Users (customers) have no way of knowing what physical or data link technologies are in use between the two switches. However, since the network is a PSPDN, users do know that it can support X.25 circuits, regardless. And what's more, no one really cares about the contents of the cloud. This is true because as long as customers see X.25 in and X.25 out, user data will get from one end to the other. In fact, in most PSPDN services, the cloud network is a proprietary one, custom made for the purpose of connecting X.25 switches to each other.

Unlike the physical circuits in the PSTN, the circuits that X.25 offers are *virtual* circuits, meaning that they appear to be true discrete circuits to those who use them but they are not truly separate in any but a logical sense. Virtual circuits may be set up manually by the people who run the cloud, or they may be set up and broken down as needed, on a dynamic basis, by the users of the network. These are permanent virtual circuits (PVCs) and switched virtual circuits (SVCs), of course. However, PVCs are just as "switched" in the network node sense as SVCs are, and SVCs may stay provisioned for years (and can therefore be quite permanent), but this is networking terminology in action.

In X.25, PVC provisioning is a simple, but lengthy process involving the manual configuration of each switch along the path from end to end, as well as of the PADs that delineate the ends of the circuit. When complete, however, the user has a custom-built virtual circuit (VC) that, given proper network engineering and design, has been configured with his or her application in mind, making use of the network links and switches that best fit his or her assumed traffic characteristics. Thus, if a user wishes a VC with the lowest possible end-to-end delay, he or she could specify this to the X.25 service provider and thereby receive a path that spanned only the fastest switches and links in the network.

SVC setup must occur dynamically, so there is no time for a manual provisioning process. Instead, when the user asks the network for a virtual circuit, an end-to-end path is discovered by a routing protocol, and then some type of signaling protocol is used to instruct the switches along that path to connect the appropriate links together and thus form the circuit. Once this process is complete, the user is free to pass data, in packets, over the VC. When the user is finished with the circuit, he or she will instruct the network to release the connection, whereupon the entire setup process occurs in reverse, and the switches' and links' resources are released for use by another user.

It is worth taking a moment to discuss some of the services that such an arrangement can provide. In the cases of both a PVC and an SVC, the circuit establishment process results in a "reliable" connection. *Reliability* has a specific meaning in data communications—the data will in fact be delivered exactly as it was submitted to the network. The user can *rely* on some quality of service (QoS) from the network. If for some reason this delivery is impossible, the transmitting user will at least be notified of that fact, allowing him or her to take appropriate action.

True reliability therefore must guarantee data integrity, with all the error correction and sequencing such a concept implies. Therefore, each LAPB link over which the X.25 circuit runs ends at a switch, which now has two jobs. The link must both buffer the X.25 packet and check it for errors. If the link level protocol finds no errors, then the switch sends the traffic onto the appropriate LAPB link toward its destination. If one LAPB link does find that the packet has been corrupted en route, it can ask for retransmission from the switch that sent the packet, which keeps a copy for this reason until informed of the packet's successful receipt.

This takes a lot of effort, but providing true reliability is no simple task. It requires a lot of resources—like time and bandwidth—that most X.25 networks can ill afford. However, when X.25 was designed, most network engineers saw little choice but to provide such a service. After all, what chance did the data have of making it end to end without any errors if they were not verified to be error-free at each step of the way? The error rates of even the best physical networks, which were all-copper and ill-shielded from electromagnetic interference, were very high. And if errors are commonplace, it is smart to be on the lookout for them at every opportunity, because there is no sense in forwarding a packet with garbage in it.

This seems like a sensible way to go about the business of data transfer, and it does result in a high-quality stream of data at the receiver—one that looks exactly like the data that were placed into the circuit by the transmitter. Thus, what are the disadvantages of doing things the way that X.25 does?

First of all, the process of setting up a circuit takes time, which delays the start of communications. For example, in X.25, the sending device must begin the setup process by signaling its intentions with a specific type of packet, known as a *call request*. On receiving this, the intended receiver must, in turn, signal its willingness to open the connection with another special *call accept* packet. Finally, the sender must acknowledge the receipt of the *call accept* with a *call complete* message. Only then is the circuit between the sender and receiver considered open and ready for data transfer.

However, open it is, and the two end stations thus connected by an X.25

circuit may transfer data as fast as the physical connection will permit and for as long as the connection stays open. The efficiency of such a process depends not only on the call setup procedure itself but also on the amount of data transferred over the call. After all, if the call is being established to transfer the contents of the Library of Congress, one can certainly see how the six packets required to open and close the connection shrink into insignificance beside the sheer volume of data. However, what if all that the sender wishes to send is a single packet to signal that a key has been depressed or that a door has been opened? Here, we find the other variable in the decision as to whether establishing a circuit is wise or not—the importance of the data itself. Much data are crucial and therefore worth all the protection and reliability they can get—even if the data are just a single press of a key. However, some data are not time critical, are easily resent if lost, and therefore may not justify the trouble—and bandwidth expenditure—of having a circuit established for them. Other traffic cannot accept the delay that circuit establishment imposes on it.

Thus, given the fact that some traffic may not be well served by circuits, what can be done to satisfy the requirements of such traffic? The simple answer is: Just send it. In other words, if node A has a short message for node B that would not end the world if it were lost, why not simply send it to node B without the complication of circuit establishment? It turns out that the only difficulty in doing this is that, as usually deployed, X.25 has no provision for such "circuitless" data transfer, so a way had to be developed to make it possible using capabilities native to all X.25 equipment.

Thus the X.25 *fast select* service was born. In fast select, nodes wishing to send a single small packet of data to another node simply may append that data to the end of a call request message addressed to that node. Of course, the X.25 network sees this packet as an attempt at connection setup by the sender and duly forwards it to the intended recipient, who simply strips off the header information and takes the data from the end of the packet. No setup, no fuss.

Many applications benefit from such an approach. Consider a request generated from a credit card swipe in a retail store. For such an application, the card number and amount of purchase can be appended to a call setup message and the message routed by the PSPDN to the central database that houses the cardholder's account information. When the database has authorized the purchase, it sends another small amount of data back to the store with an authorization number. The consequences of either or both packets being lost are the same: no reply to the point-of-sales terminal. This leaves the sales clerk a number of options, the easiest of which is probably to repeat the process, which, barring a serious network fault, will more than likely succeed the second time.

Fast select filled a need, the need of the users of X.25 networks to have a way to send small amounts of data without a lot of overhead. However, by all accounts, fast select never sat well with the users or engineers of X.25 networks, being as it was an attempt to use the network in a manner for which it was not designed. It is not hard to visualize fast select as a reckless and unreliable way to use the solid, staid, well-mannered PSPDN. After all, the very nature of X.25 and its underlying LAPB layer is to guarantee the delivery of data in an error-free and complete manner. Fast select is like taking city buses that seat 50 and using them as family cars for commuters with only the driver.

X.25 was intended by its designers to eventually become an integrated network of connected PSPDNs, in much the same way that the global PSTN is formed by connections between the world's telephone networks. In fact, X.25, while achieving great success in some markets such as Europe and parts of the Pacific Rim, did not succeed as the integrated global network for one outstanding reason—X.25 was filled with "option sets" that made it hard to interconnect national X.25 networks. There was "French X.25," "Belgian X.25," and so on. This was a disaster for an intended global network. After all, if a connection fails to a city within the same country, a simple telephone call may be all that is needed. However, when connections fail to Turkey from Italy that pass through Greece, who should handle the trouble call? Perhaps a network should first and foremost offer universal connectivity to one and all, without the need to establish end-to-end connections for traffic.

Enter the Internet

While all this X.25 development was going on, engineers from the U.S. government and a number of contractors and universities had been working on their own network, which would one day become the Internet. The genesis of this network was not in the day-to-day concerns of business, but rather in an experiment in network survivability in the event of a nuclear war. The U.S. military was concerned that, should war come, its increasing reliance on computers would be a vulnerability, because the dedicated circuits that linked these computers to each other and to their remote users would be an Achilles heel. After all, if a bomb were to disable or destroy a central connection point for a number of important circuits, it could effectively cripple communications between computers with a single stroke.

The protocols that grew out of this design effort behave in a manner directly opposite to that dictated by the world of circuit switching. Even the idea of "packets following virtual circuits" was considered and rejected because of the difficulty of redirecting the circuit path if one of the switches through which it was routed should fail (or be the target of a nuclear blast).

Instead, the designers of the ARPANet (Advanced Research Project Agency Network), later DARPANet (Defense Advanced Research Projects Agency Network), the great-grandparent of today's Internet, imagined a network where information could follow any one of a number of possible paths to a destination. The sender would take each message and break it into a number of small datagrams, each 1500 bytes or less (sometimes quite a bit less), and then send them to a router, which is a device responsible for locating another router closer to the destination and forwarding datagrams to it. Each router would know the addresses of all the destination networks and the neighboring routers that offered routes to those destinations and therefore could "further the cause" of each packet by sending it to a router that was closer to the destination. In this "hop by hop" manner, all the packets of a message eventually would find their way to the intended destination, be reassembled by the recipient, and their contents read and acted on. While it is certainly possible, and even probable, that all the packets containing a particular message will follow the same path and therefore arrive in the same order in which they were sent, there is no assurance that this will happen, as there is in a circuit-switched network or even a virtual circuit network like X.25.

The routing information necessary for the routers in the ARPANet to forward datagrams originally was propagated manually by regular updates to the routing tables in each of the routers connected to the network. Every time a new network was added to the system, the new routing table that included it had to be created and sent to the administrators of each router, who would replace their old routing table with the new one, thus allowing their devices to reach the new network. After the ARPANet grew to a certain size, however, this rather onerous process was replaced by the creation of routing protocols, which share network reachability information dynamically as new networks are added and old ones retired.

Today e-mail messages, Web pages, and all the other traffic of the Internet are carried in datagrams formatted according to the standards of the Internet Protocol (IP). While IP describes a datagram size of between 20 and 65,000 bytes, most datagrams are 1500 bytes or smaller because most user messages are relatively short. Appended to the beginning of the data in the message is the IP header itself, which contains the addressing and other information necessary for the routers to do their jobs.

Despite protocol-based routing table updates and the addressing imposed by the IP header, it is certainly possible for a datagram to be lost as it travels between the sender and the recipient. This may occur because of router or link failure, corrupted routing tables, malfunctioning equipment, misconfiguration, or any number of other causes. It is also possible, of course, for the data in an IP packet to become corrupted as the packet crosses the network. In either of these cases, we have seen what a reliable protocol such as X.25 does: It causes the offending (lost or corrupted) packet to be retransmitted from the last point at which it was known to be good. With a little luck, the second or a subsequent transmission of the packet will not encounter the same trouble that the first one did, and the data eventually will be delivered.

Faced with a loss or corruption of data, IP acts in a way that horrifies those used to working with reliable protocols such as X.25. IP does *nothing* to retransmit or otherwise deliver the lost data. In the data communications lexicon, IP is an *unreliable* protocol, meaning that it does not guarantee the delivery of, nor the quality of, the data entrusted to it. IP cannot be *relied on* for any QoS at all, except connectivity between source and destination. IP ignores—in fact, it is not even aware of—virtually all transmission errors that affect user data. The unreliability of IP extends even to an inability to recognize that corrupted data are in fact bad. How can a protocol function this way and still allow the systems connected to the Internet to depend on the integrity of the data they receive?

Things are not as bad as they seem, though. In the first place, the networks that make up the Internet and other IP-based networks, depending to the extent they do on fiberoptics and carrier-class routers, seldom introduce errors into datagrams. One could argue that the likelihood of receiving corrupted datagrams is so low as to make the inclusion of error-correction mechanisms into IP not worth the effort. For some types of data, this is not the case.

Second, in certain applications it is sometimes acceptable for datagrams to occasionally become misplaced, because the users (or programs) at the two communicating computers may recognize the mistake and request retransmission themselves. For example, think of a user waiting to download a Web page—if he or she waits for more than a few seconds for some acknowledgment that the request was received, he or she can certainly reenter the address of the Web server and thereby attempt to connect again.

Finally, other "helper" protocols can be used in partnership with IP to provide a fully reliable transmission service when this is required. Chief among these is the Transmission Control Protocol (TCP), which does in fact work hand in hand with IP for many common applications such as electronic mail and the World Wide Web. The use of IP (for addressing and routing) paired with TCP (for reliability) is so common that many people refer to the entire suite of IP-supported protocols as TCP/IP.

The Internet Protocol Suite (TCP/IP) Architecture

IP forms the centerpiece of a huge and complex suite of protocols, each of which has its own particular role in the transmission of data over IP networks such as the Internet. This architecture, and the definition of each of the protocols within it, is described in a set of standards maintained by the Internet Engineering Task Force (IETF), a body that has grown considerably from its simple beginnings as a forum for the general sharing of information about IP suite development initiatives. IETF standards are developed first as *Requests for Comments* (RFCs), which on the one hand simply provides a nice name for a "draft" standard but on the other hand provides a glimpse into the relatively fair-minded and equitable academic methods that have defined the IETF from its inception in the late 1960s. The IETF now comprises well over 100 working groups, each tasked with solving its own particular set of Internet engineering problems. Since the economic stakes have risen over the years to considerable levels, the RFC process has become more typical of that generally ascribed to standards organizations, i.e., somewhat brutish, politically charged, and drawn out. The IP suite remains, however, one of the more "open" protocol architectures available, in that IETF standards are available for free on the Internet, are widely known and adhered to, and are deployed on more networks around the world than any other comparable standards.

Figure 5-7 shows the relative positions of the main building blocks of the suite, seen from the point of view of an OSI-like model. The bottom of the protocol stack, i.e., those portions addressed by the physical and data link layers in OSI terminology, is not explicitly addressed by the standards covering the IP suite. While this may seem like a shocking omission, in fact, it frees IP-based networks to be constructed from

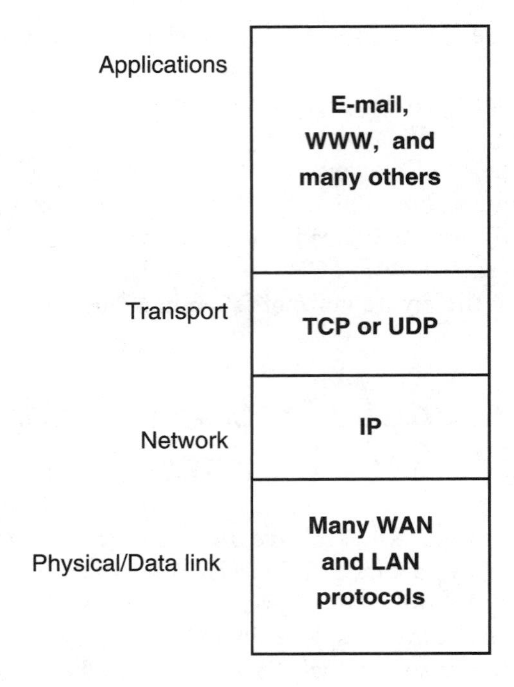

Figure 5-7 Fundamental components of the Internet Protocol suite (TCP/IP).

nearly any network infrastructure, almost regardless of the qualities of that network. Therefore, one can find effective IP networks running over all forms of copper and fiberoptic cable services, wide-area packet switching, wireless services, and even some of the more esoteric technologies such as point-to-point laser. As long as it can support a useful payload size, almost any network can carry IP datagrams. Note that, unlike in X.25, there is no requirement here for an error-free transport mechanism. Since the service that IP provides is known to be unreliable, IP itself does not care if the network over which it is running is reliable or not.

Moving up the protocol stack, there is the OSI-RM network layer and therefore IP itself. Most of the responsibilities of the network layer are handled by IP. These include

1. Unique addressing for networks and the host computers connected to them
2. Defining the format of the IP datagram and its header
3. Routing IP datagrams through routers and across networks to their destinations

4. Fragmenting and reassembling chunks of data too large to be carried across a particular network
5. Passing data contained within datagrams to the layer above, for eventual use by an application

There are two important network layer functions not covered by IP itself. These are the propagation of routing information and the interface to the lower layers. The authors of IP have made the wise decision not to integrate these capabilities into IP because doing so would restrict the ability of the protocol suite to develop over time without the need to rewrite the fundamental IP specification itself.

Routing information is propagated either manually or via one of a number of common protocols that have been designed expressly for this purpose, such as the Routing Information Protocol (RIP), the Open Shortest Path First (OSPF) Protocol, or the Border Gateway Protocol (BGP). The specific operation of these protocols is beyond the scope of this book, but they all achieve more or less the same end: They allow routers to communicate, or advertise, the network reachability information that each knows. As routing protocols operate, each router shares with others the addresses of networks to which it has access, and in this way, every router eventually is told of the existence and reachability of each network.

Since there are so many types of networks over which IP may run, its interface to the data link layer is made on a case-by-case basis and is defined by a number of standards—each of which addresses a particular data link technology. This is necessary because of the different requirements that each type of network makes on IP and the different way that each network satisfies IP's protocol demands. There are standards for communication via IP over LANs such as Ethernet and Token Ring, packet services such as X.25 and frame relay, and cell relay services such as asynchronous transfer mode (ATM), as well as a large number of less prominent networks technologies.

While IP itself is completely peer to peer, and therefore simply concerned with the addressing and transmission of individual datagrams, there is an implicit relationship in the transfer of data from host to host across the Internet. This is the *client* and *server* roles that define the transmissions between any two hosts communicating via the same application or upper-layer protocol. A *server* is a computer whose role is to provide access to a certain kind of data or service for use by other computers or users. A *client* is simply a computer that initiates communication with a server for the purposes of obtaining some data or service. A single machine may be acting simultaneously as a client for, say, both a WWW session and an SMTP e-mail exchange while at the same time acting as an FTP server transferring data to another machine. It is therefore between clients and servers that the transport layer establishes communications, using IP to carry its data.

Transport layer functions within the IP suite are usually handled by one of two protocols: the Transmission Control Protocol (TCP) or the User Datagram Protocol (UDP). TCP is the IP suite's way to provide a fully error-corrected, connection-oriented delivery of information, and it is TCP that is used by the majority of "traditional" user applications, such as the World Wide Web (WWW, or simply the Web) and Simple Message Transfer Protocol (SMTP) electronic mail. TCP sets up and breaks down connections in much the same way as the X.25 protocol does, to provide an end-to-end, acknowledged, reliable virtual circuit for data. UDP is used in cases where the nature of the data makes reliability unnecessary or when the delay associated with TCP's circuit establishment, circuit breakdown, and error correction is not acceptable to the application.

Regardless of whether or not reliability is desired at the transport layer, there is one other important transport layer function that is always necessary: identification of different streams of data to the receiving host for delivery to the proper application. Since any given IP datagram may contain data for any one of a number of possible user programs, both TCP and UDP assign a *port number* to each application's particular stream of data. Hosts maintain a list of the port numbers and applications that are in use at any particular time. When a host receives data from a server's port 80, it knows from the port number that the data are coming from a WWW server and that it should pass the data to its Web browser.

The combination of IP address and TCP/UDP port number is called a *socket*. Sockets are convenient ways for network applications to interface with the rest of the Internet protocol suite. This is so because when writing an application program, a socket can be treated just like a file. A program can create and open a socket (get ready to use the network), write to a socket (send), read from a socket (receive), and close a socket (disconnect from the network). *Winsock* is the term used to describe sockets in a Windows environment.

Sitting at the application layer of the suite are all the user applications that are part of the "traditional" Internet, as well as many newer uses for the services that the rest of the suite provides. Since the Internet, and the IP suite in general, traditionally has been considered a non-real-time network—that is, one in which the end-to-end delay encountered by applications is assumed to be on the order of seconds, if not minutes—such applications generally have been those which could tolerate delay of some magnitude. Electronic mail, for example, carried via the SMTP protocol, functions just as well across slow networks as fast ones. File Transfer Protocol (FTP) sessions are similarly tolerant of delay, since their mission is usually to transfer data files from host to host for use by the receiver at a later time. Faster networks will speed up the transfer process, but the success of the application does not depend on the speed of the network.

The situation with the World Wide Web is somewhat different. WWW sessions, since they are initiated by a user who then waits impatiently for the requested page to arrive, are less tolerant of delay than either of the preceding protocols. Most of the time this wait is on the order of seconds, but large and complex pages, network or server congestion, significant numbers of errors (which must be corrected by TCP), or simply slow access links between the server and browser can cause minutes to elapse between the time a page is requested and the time it is eventually received. While this delay was never an issue in the pre-WWW Internet, its reduction or elimination has come to be the defining cause for improvement of the Internet. It also has helped redefine the Internet's role from that of a *non*-real-time network to a *near*-real-time network, a fuzzy concept to be sure, but one for which the end-to-end delay, under normal conditions, can be considered slight.

Users typically do not mind waiting a few seconds for a Web page to appear, but what happens as the applications that everyone wants to use begin to tolerate very small amounts of delay, perhaps only on the order of milliseconds? Similarly, what is the effect of the increasing size of the messages that must be transferred for communications to function, as the programs and protocols that enable communications continue to increase in complexity and size? The evolution of the WWW provides perhaps the best example of the complex relationship between speed, delay, and the requirements of network services as they have continued to evolve.

The World Wide Web

For the scientists, engineers, and academics who have used it from its earliest days, the Internet has always been a compelling resource. This relatively small population, owing to their early adoption of computers and the nature of their work, was willing to make the investment in time and development necessary to learn to communicate via the rudimentary application protocols available in the early days of the Internet. Their justification was the enormous amount of time that could be saved by using electronic mail instead of postal mail, FTP instead of shipments of magnetic tape, and virtual terminal sessions rather than needing to be physically present at a computer in order to use it.

However, the early Internet was an "expert system" in that it required much commitment to derive these benefits. Simply put, it was too difficult to use the Internet, or any other computer network for that matter, to expect the average consumer, or even the average office worker, to be comfortable with it and regard it as a benefit to exploit rather than an expensive toy. It took the happy coincidental arrival of inexpensive personal computers, high-speed communications protocols (including 56-kb modems), and the development of the World Wide Web to make the extraordinary benefits of computer networks usable by the masses.

As originally conceived in the early 1990s, the WWW and its attendant application layer protocol, the Hypertext Transfer Protocol (HTTP), were used to share Web pages comprised entirely of plain text and relatively simple graphics, formatted according to the Hypertext Markup Language (HTML). Given the rudimentary graphics capabilities of most computers of that period, the graphics files embedded in most contemporary Web pages were necessarily of low resolution and color content, resulting in small file sizes. The average Web page of 1992 was probably no more than 5 or 6 kB in size. Even given the slow modem speeds of the time, such a small file typically took no more than a few seconds to download.

During the 1990s, developments in the protocols that make up the WWW, as well as the Web's increasing importance to the corporate world, have vastly increased its possibilities. By 1999, a single corporate Web page may be comprised of dozens of graphics files (each containing thousands of pixels), ten or so separate text files, some executable code for special effects, a music clip of some type, an animation or two, and even some video stored in a file. This growth in the number of services that a Web page may offer has resulted in an enormous growth in page size, from a few kilobytes to hundreds of kilobytes.

The main objective of this development of the WWW protocols, and the causes of the growth of pages and the Web itself, is an increase in the amount of interactivity and variety of media types that a Web site may offer to a person viewing it. It is worth a moment to review some of the developments that have fostered this shift.

Web Graphics

Graphics have always played an important part in the Web, but their role has changed from that of a simple means by which to include diagrams, charts, or simple line drawings to the major element of information in most pages. This change in role has resulted in many pages consisting of almost all graphics, with the text of the page being part of the graphic images themselves. Such an approach allows Web page designers greater control over the look of their pages, since the font, size, and positioning of text elements are "locked in" when the graphics are created rather than being subject to the vagaries of how text formatting information is interpreted by various types of browsers. It also allows hypertext links to be keyed to individual graphic elements rather than to text, which simplifies HTML coding.

Still, graphics are usually stored in either Graphics Interchange Format (GIF) or Joint Photographic Expert's Group (JPEG) file formats; which is used depends on the nature of the image at hand. GIF files, because of the compression technique employed, are usually most suitable for line drawings and images with solid blocks of a small number of colors. JPEG file format is usually used for photographic-quality images. There are other file formats that most browsers can read, but these are the two that most pages employ.

Web Animation

The earliest animation techniques employed on the Web were animated GIF files, a method of sending a number of GIFs to a single position on a page with an instruction to the browser software to "flip through" them, displaying them sequentially. This relatively crude means of sequenced image is quite suitable for small, simple animations, but it allows the designer little control over the speed or duration of the display. Despite this limitation, animated GIFs are still used in many Web pages as an easy way to provide rudimentary animation capability.

To meet the needs of designers who want more control over their work, as well as to allow the creation of larger and more complex animations, a number of animation packages have been developed. As computing power and access link speeds have allowed use of the larger files that such elements require, development environments such as MacroMedia's Director allow designers to create full-motion, object-oriented animations that are downloaded to the user as part of a Web page. Since standard Web browsers do not necessarily understand the code used to control and display these files, typically the user must download and install an additional piece of software, called a *plug-in*, that is used by the browser to interpret and display the animation.

User Interaction

Interactivity has always been a hallmark of the Web, and the immediate control that hyperlinks afford the user is arguably the single most important reason behind the success of the WWW as a service and the growth of the Internet as a whole. Clicking on a hypertext link causes the browser to take an action on behalf of the user; most typically this involves displaying a different page related to the text or graphic that was clicked.

While the ability to allow users to follow their stream of interest along a random path of links is compelling, early Web page designers struggled to create opportunities for users to truly interact, in multiple ways, with services pages were to provide. Through the Common Gateway Interface (CGI, the oldest and simplest type of user interaction), a browser may display an electronic "form" to be filled out by the user, which may then be sent to the Web server and used by a program meant to collect user data, answer user questions, or provide some kind of conditional response to the user. However, while this ability fostered a degree of communication and interactivity between the Web server and the user, the ability to provide customized programs for users remained elusive until mid-1995.

It was then that Sun Microsystems, a pioneer in the development of the UNIX operating system and the manufacturer of the Sun workstations on which most early WWW development took place, developed the Java programming language. Java, based on the C language, which is the *lingua franca* of modern programming, is a fully browser- and computer-independent language that functions by allowing Web pages to instruct browsers to download small programs, called *applets*, from a Web server and run them as discreet applications on the browsing computer. Such programs may do anything, from computing mortgage amortization tables, to displaying stock price tickers, to plotting graphs for barometric pressure. In fact, Java applets are a Web server's way of serving custom applications for use by the browser.

At this point, however, the appeal of user interactivity (and the financial possibilities for the providers of the service) has become so great that software developers have created means other than Java by which to perform customized user functions. For example, Microsoft has developed ActiveX, similar in capabilities to Java, which is essentially their own object-oriented programming language. The battle between Java and ActiveX for domination of the WWW interactivity market is allied to the war between Internet Explorer and Netscape Navigator for control of the browser market, with perhaps greater implications for the future of how the Web looks and operates.

Audio and Video

At least some of the potential of the Internet could never be realized without the capability to offer traditional media over the network. Most marketing professionals probably would agree that no amount of fine page design or Web animation can take the place of audio and video when it comes to getting a message across. The search for analogues to traditional radio and television delivered over the Internet has been hampered by congestion and bandwidth troubles, but in recent years, many solutions have come to light for the provision of high-fidelity audio and full-motion video.

Perhaps the most notable of these have been the RealAudio and RealVideo products offered by Real Networks. Initially offered as two separate products, Real now supplies a single integrated player (RealPlayer) that accepts buffered streaming audio and video for display. Given a decent access rate (at least 33.6 kb/s) and clear network conditions, RealPlayer is capable of delivering high-quality television-like presentations, although in a relatively small window. A number of other manufacturers, most notably Microsoft, offer similar products.

At the provider end of the audio/video stream is the media server, which is generally a software package written specifically to communicate with only the manufacturer's players and uses a proprietary communications protocol to do so. The relationship between the server and the player is a sensitive one, because the main difficulty of delivering audio and video data over the Internet is maintaining the flow of information at a constant rate between them. This is precisely the type of application that is best served by a dedicated physical circuit, because such a circuit is not subject to the vagaries of multiplexing, having a certain amount of bandwidth available to the audio/video stream at all times.

There are a number of ways that this difficulty may be overcome. For simplex (one-way) transmission of audio or video, the simplest is that of buffering, and this is the method employed by the Real Networks and Microsoft products, among others. In such a scheme, the player requests a video or audio clip, and the server begins transmitting information to the player at a rate slightly greater than that necessary for full-motion (or full-fidelity, in the case of audio data) playback. As the stream of data begins to reach the player, it is not immediately displayed but rather is stored in memory until enough has accumulated to ensure acceptable quality on a temporary interruption in reception of the stream of data. In other words, there is always some amount of data outstanding between what the player is receiving and what it is playing at any given time. As the receiver's buffer fills, it requests the server to slow the rate of information transfer until it matches the rate at which the player consumes it. At this point, should a network interruption occur, the player has, say, 10 s worth of video signal for use before the viewer notices the outage, by which time it hopefully will have been resolved. The amount of outstanding data depends on the network conditions; likely and frequent interruptions demand a large amount of buffered data, whereas good connections and high network throughput allow smaller amounts to be buffered. This concept is similar to that employed in "sport" model portable CD players, which hold a few seconds' worth of digital audio in a RAM chip in anticipation of a skip in the CD playback when the unit is jostled.

This is quite similar to the jitter buffer approach for packetized voice. The greatest consequence of employing receiver-end buffering is the delay that it imparts to the initial reception of the audio/video stream. In the case of broadcast television, for instance, this delay causes no trouble, since the viewer is seldom aware that the video he or she is watching actually was sent by the server 10 s ago. At worst, the need to buffer simply introduces a slight delay between the time that the viewer makes a request for some video and when he or she begins seeing it. Most would agree that is a small price to pay for a high degree of stability in the image.

However, this minor consequence for broadcast data is of the greatest import when it comes to duplex (two-way) conversations, like a normal telephone call. It takes only a quarter-second of delay to make communications difficult, and much more than a half-second is intolerable for most people. One need only imagine a telephone conversation in which one party's words were not heard by the other until 10 s after they were uttered to see the problem our buffering model would cause if used for two-way voice conversations. While some buffering is used in duplex audio and video carried over the Internet, the amounts are much smaller than those used for broadcast streams (typically under half a second), and the buffering procedures are controlled by specific rules covered in Chapter 7.

These problems notwithstanding, it is primarily the ascendance of the World Wide Web that has helped pave the way for the pursuit of interactive, telephone-style voice, and even videoconferencing, over the

Internet. As the number of Internet users has grown, and as they have come to depend on (and expect more from) the Internet almost as they do their telephone service, the providers have responded by making the Internet more worthy of their needs and expectations. While the Internet infrastructure has a long way to go to match the reliability and performance of the PSTN, it is clear that this transformation is under way.

An Internet for Voice?

In an ideal world, of course, the Internet connection between source and receiver would be of such high quality and subject to such small amounts of delay that datagrams carrying audio or video data would arrive almost instantaneously and without errors. While this is often the case, the fact that this cannot be relied on has caused many to maintain that the Internet, due to its very nature as a connectionless network, can never equal the performance of the circuit-switched PSTN for voice communications. However, on examination, there are a few converging trends that, when taken together, challenge this assertion.

The first of these is the continual and ongoing improvement in the physical network of the Internet. While one might say that improvements in the reliability, stability, and performance of the Internet infrastructure are the natural outcome of the progress of technology, there is no doubt that they reflect the business concerns of organizations and people who have come to rely on it. And these concerns have multiplied at the fantastic rate at which use of the Internet has grown in the last few years. The result has been a transformation in the capacity, overall speed, and quality of the services that the Internet provides.

These improvements would have proceeded at a much slower pace if it were not for the continual, and now almost complete, privatization of the Internet. From its beginnings as a government-sponsored endeavor, the Internet was heir to the bureaucratic delays and stilted administration of any government agency. At the time, though, demand was light, and the groups of users were much alike. The recent increasing growth and diversification of the Internet community made that administration a much bigger job, one probably not best handled by a government agency. Luckily, the U.S. government recognized this, and by the mid-1990s, the National Science Foundation had committed itself to getting out of the Internet business and turning the provision and administration of the network over to those companies and organizations who were interested in pursuing commercial ends by providing Internet access.

Estimates of the current growth rate of the Internet, whether measured by number of hosts, number of users, or amount of daily traffic, are almost not to be believed. However, major Internet backbone providers profess to be seeing traffic doubling every 6 months, and some claim growth rates of twice this! While it would seem that astronomic rates like this cannot be sustained forever, the growth of the Internet, as both a network and as a service, has amazed analysts from the mid-1990s onward, and there is no reason to believe that anyone will be able to predict accurate numbers more than a year or so in advance anytime soon.

Thus, while the numbers of hosts, users, and networks continue to grow, so does the capacity and quality of the backbone networks. And at least so far, the growth of capacity has managed to stay ahead of the number of users. And it must continue to, because a network that is so slow as to be unusable is one to which it is difficult to sell access. This is a problem that network providers and ISPs are very aware of and on which they spend vast amounts of money to avoid. This process of continual upgrade, growth, and improvement results in a much faster and higher-quality network than a static population would. The growth of the Internet is in some way its salvation.

The second reason that the Internet is becoming worthy of voice traffic is that the development of protocols and procedures for solving all sorts of user issues has come into its own. The pace of application and protocol development has advanced at a rate comparable with the size of the Internet itself. (One indication of this is the number of IETF RFCs published; the period 1984-1988 saw 192 new RFCs, whereas the period 1994-1998 saw 907.) Since the Internet (and internetworking in general) is such big business these days, there is enormous market incentive for developers to attempt to capture a portion of the huge pool of users by coming up with the next "killer application."

The last 5 years therefore have seen significant development in the tools and protocols that users and network architects have to work with. While this portends a growing complexity (and, of course, a sea of three-letter acronyms), there is no denying that the improvements in routing standards, transport layer protocols, encoding specifications, and the like have made it possible to provide services that depend on certain attributes (e.g., throughput, speed, robustness) over the Internet, something that was impossible to do in the past. This development makes the Internet an environment much more compatible with voice and other real-time services, which many believe are that next "killer application" everyone is looking for. The specific protocols that enable this will be explored in Chapter 7.

CHAPTER 6

Voice-Enabled Web Sites

The Internet and Web have become so common that it is often surprising when someone admits not having access to the Internet and Web, at least in the United States. Even public libraries routinely offer free Internet access to the general public, and even homeless people have e-mail addresses (no real mailbox, but a virtual mailbox in "cyberspace"). Some people scorn television for the lack of content and wealth of violence but have Internet access because they see no way to do much of anything, professionally and/or personally, without the Internet and Web.

However, although the Web is interactive, it is usually only *visually* interactive. There is often audio, but frequently of the prerecorded, canned variety. There are real-time radio broadcasts on the Web, but no way to talk back. What this chapter investigates is adding interactive voice to the Web.

Ironically, this chapter easily could have been entitled "Web-Enabled Voice Sites" without changing much at all. A *voice site* is a call center, inbound for 800 number catalog sales and tech support and outbound for telemarketing (often seen as the most annoying activity on the planet). This chapter mostly assumes that the server end of the voice connection, the Web site, is essentially a call center. Web-enabling a call center is adding multimedia Web pages to the voice experience. In contrast, this chapter is focused on adding the voice experience to the multimedia Web pages. The end result, however, is the same: People can both see the Web content and speak to each other at the same time. The question is how to do it.

But maybe the "how" is just as important as the "why." Just because something is possible does not mean that it is a good idea or that people will pay good money for it to happen. The networking attic is stuffed with ideas that never caught on (anyone remember switched digital video?). Thus the reasons why voice-enabled Web sites are a good idea are important ones.

Why Voice-Enable the Web?

People talk on the telephone. People look at the Web. This is the big difference. There is more pure *communication* in the sense of information exchanged in voice communication than in an exchange of visuals. This is why using the telephone seemed more natural to people than writing a letter.

This does not mean that voice is better at everything than the Web. Both methods of communicating have advantages and disadvantages.

TABLE 6-1

Voice and the Web

	Voice Communications	Web Sites
Advantages	Human interaction	Visual orientation
	Flexibility	Unlimited information
	Sophisticated features	Available 24/7
Disadvantages	Hard to describe abstracts	No human interaction
	No visual element	Hard-to-find information
	Costly to use	Static content

The whole idea is to combine voice and the Web so that the strengths are reinforced and the weaknesses are canceled out. Table 6-1 shows the advantages and disadvantages of voice and the Web. The table considers the traditional Web, perhaps with audio, but definitely without newer XML (eXtensible Markup Language) capabilities, which will take a while to appear in force in any case.

As might be expected, the entries form natural couples. For instance, the table notes that voice is, above all, human interaction, whereas the Web experience is not (in fact, the social isolation that Web browsing fostered has been noted many times). On the other hand, while there is no visual element to voice telephony, the Web is almost all visual. However, some of the elements in the table are not so obvious.

For example, voice communication is about as flexible as life gets. Not only does the listener not know what is coming next, but sometimes the speaker doesn't either. However, go to a Web site 20 times a day, and not much will change over that period. People offer more sophisticated features than the Web too. People typically know what they do know and do not know. The Web often doesn't. Try a search one way and it comes up empty. Try it another way (perhaps by just changing a letter), and the user in suddenly inundated with information from the same set of Web sites. The telephony concept of *collocation* is spelled (apparently equally properly) both as *colocation* and *collocation*. However, people are usually consistent in their usage. Thus a search on one spelling of the term turns up perhaps 500 Web pages, and a search on the other spelling turns up maybe 500 *different* Web pages! This could never happen in conversation.

Ever find something useful on the Web and never, ever be able to locate it again? Anything from how to upgrade a BIOS chip to how to fold a paper airplane. It's all there, but there's no one there to simply *ask directions*. (Women may smugly note at this point that the Web was obviously put together by men, since no one can or will ask for directions when they are hopelessly lost.) Information on the Web is almost unlimited, but often it is hard to find. As an example, a 5-minute conversation by one of the authors with a human tech support representative furnished needed information that was available on the company's Web site but not located after a half an hour of Web-site searching by keyword and following dead-end links. The linking that makes the Web what it is often creates a maze. People are much more sophisticated than a Web page.

Web sites offer mainly static content, and the Java and ActiveX applets that can change Web pages on the fly often only serve up flaming logos and scrolling advertisements. On the other hand, the Web is there 24 hours a day, 7 days a week. Staffing a human call center around the clock is possible but very expensive. Some estimate that a call to a human being is 20 times more costly than browsing a Web site (about $0.50 versus about $10 for the human alternative).

Thus the reasons for voice-enabling the Web are numerous and attractive. And instead of a traditional all-human call center, adding the Web can make the call center into a "contact center." The marketing database also can be tied in to create new "customer relationship management centers." The whole idea is to make sure the customer comes back again and again. Customer loyalty is always the number one goal of any for-profit organization.

These are the benefits of voice-enabled Web sites. Consider for a moment the consequences of a totally separate call center and Web site for a business. Perhaps this is a company selling computer peripherals. Initial contact is most likely through the Web site (this is often true even for non-high-tech businesses such as office supply companies). A potential customer makes his or her way through the on-line catalog and finds a disk drive on sale (today only!) for a very low price. If the potential buyer has a question, how can this be asked? The Web page may contain an e-mail address, but who knows when this will be checked and answered (the sale is today only). If the Web site lists an 800/888 number, this may be the only feasible approach.

Assuming the potential customer is not bothered by interactive voice response (IVR) systems with distracting multiple menus and not annoyed by a long wait for representatives who are "busy assisting other customers," the human eventually will get to speak to another human. However, if the representative has no access to the Web site, the potential customer is forced to repeat every step he or she has already taken at the Web site. One of the authors had the odd experience of doing a price check for the sales representative while on the telephone because the agent did not have access to the current prices posted on the Web! Needless to say, this is not the way to foster return business (although it was tempting to invent new, lower prices on the fly).

Thus the question is not whether voice-enabling Web sites (or Web-enabling call centers) is a good idea or not (it undoubtedly is), but just how to do it. There are three main ways to add voice to the Web. The first two methods use the PSTN to complete the voice portion, while the third employs voice over IP (VoIP). The first two differ in the way that the call between customer and call center is set up.

Before examining each in a little more detail, one way of adding human interaction to the Web should be explored. This is adding Web "chat" capability to the Web site. On the Internet, an application called Internet Relay Chat (IRC), a form of interactive e-mail for short messages, has been around for years. IRC is an important part of on-line games and other social "chat room" activities on the Internet. Web-based chat just adapts this application for call center use.

Each Web page has a button labeled "Chat with an agent" that the customer can click at any time. When clicked, a small box with insert cursor usually less than a full screen line long appears where the customer can type his or her message. When the customer types in the message and presses Enter (or clicks on another button), an agent gets the message. The agent can respond at once.

Web-based chat, although interactive, and despite its name, does not involve voice or VoIP at all. It is mentioned only because Web-based chat is so popular and because of the possible confusion arising over its name. This section will only consider voice-based solutions.

You Call Us

A customer using Web-based chat is in the "you call us" mode of operation. That is, the agents are there waiting to help, but unless the customer initiates the conversation, they will be waiting forever. In some early applications of voice-enabled Web sites (which were little more than lab demos, to be sure), the chat application was replaced by a real, live telephone call.

The Web chat button was replaced by a "Call us" button on the Web page. When clicked, the Web browser used a helper application based on computer-telephony integration (CTI) to actually place a PSTN voice call to the company's call center. Naturally, the PC needed access to a telephone line, but if the customer was on-line, obviously one was available.

Of course, this posed an interesting problem. If there was only one telephone line attached to the customer's PC, how could a voice call be placed at the same time? There were three proposed solutions to this problem, none of which was ideal and for the most part eventually lead to the abandonment of the whole scheme. First, the customer could be required to have a second line attached to the PC (which had to have speakers and a microphone, of course). This was not a popular plan, because of both the added expense and the utter lack of privacy, although headsets could be used.

Second, the CTI software in the browser application actually could disconnect the call to the Internet service provider (ISP), and redial the call center number. This was not popular because the customer now lost the ability to reference and link to another page while speaking to the call center agent. And there was always the issue of reconnecting to the ISP when the call was over.

Third, and most promising, was when ISDN was used to access the ISP. Residential ISDN service allowed two B-channels at 64 kb/s to be active at the same time. Even if both were used to access the ISP at 128 kb/s, one "call" could be suspended and used to connect to the call center. Internet access continued at 64 kb/s, although the same privacy issue was raised. This method, which depended on the widespread availability and use of ISDN, did not go far. This whole "you call us" approach was more or less abandoned just as digital subscriber line (DSL) solutions were appearing that allowed analog voice calls to be made at the same time that the line was in use for Internet access (in fact, ISDN was the first DSL). Unfortunately, DSL spread as slowly as ISDN.

All these methods required a multimedia PC with speakers and a microphone, in addition to the special Web software and the telephone line. Not everyone at the time had these. However, almost everyone who had a PC had a "regular" telephone. Why not use this approach?

We Call You

In contrast to the "you call us" methods, the "we call you" approach actually has made it to market through a number of companies that offer the software required. All of them are based on the idea that many on-line shoppers still want to speak to a human to "close the deal." A high percentage of e-commerce "shopping cart" activities fail to complete because frustrated shoppers abandon the process or confused shoppers do not complete the transaction properly. There is often no choice given the

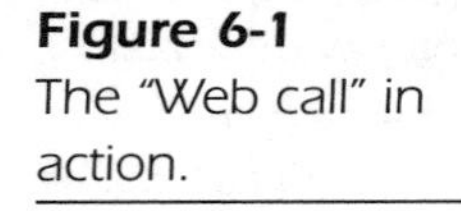

Figure 6-1
The "Web call" in action.

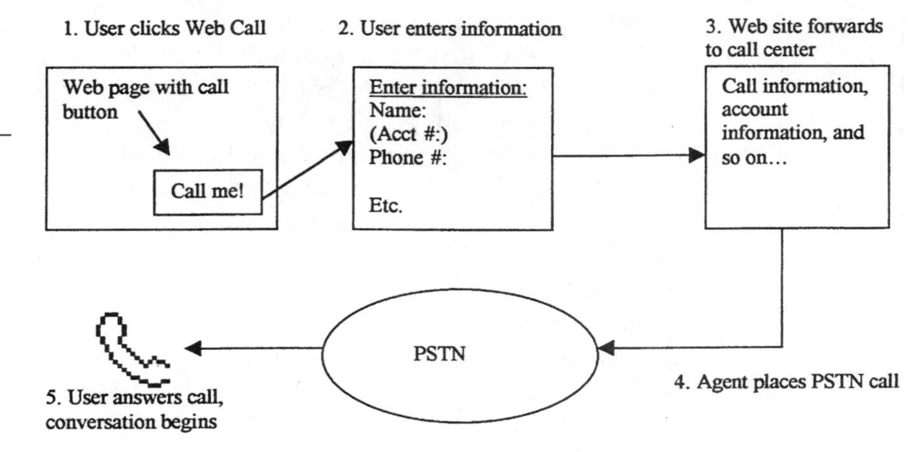

shopper except "Do it" or "Exit." If people are not comfortable for any reason at all, they leave. Even a commit decision cannot be honored if needed information is missing.

What the "Web call" adds to the equation is shown in Figure 6-1. Now, the user is back in control. Here is how it works. Presumably, the user has access to a telephone as well as a PC. While browsing a Web site looking for some product or another, the potential customer spots an unbelievable bargain, too good to be true. So good that their first thought is "maybe this is a typo." However, his or her second thought is "got to check this out." This is where the Web call button comes in. If the Web call button is there, it says basically, "Call me!"

It is important to realize that the Web call approach requires no special software or hardware on the client PC. Clicking does not allow talking. Instead, the second step requires the user to fill in a Web form, no major shock to a Web user. Information is the basic name, account number (if known to a returning customer), telephone number to call, and so on. This information is passed by the Web server (where special software *is* required) to the company's call center.

The company call center is where things get interesting. All the information that the customer entered appears on the call center agent's screen, perhaps even with the whole Web page that prompted the inquiry. The agent places the call to the user over the PSTN, using the number that the user provided. The telephone rings, and the conversation proceeds—maybe.

Calling on the PSTN while the user is attached to the Internet can be a problem whether the user initiates the call or the company initiates the call. If there is only one telephone line capable of only one conversation, the same problem of how to both talk and enjoy Web access arises. There are two workarounds employed by Web call vendors, assuming that ISDN or DSL is not in use (a good assumption). One approach is to prompt the user for a time to call *later*, when he or she is off-line. This approach sacrifices the interactive intent of the whole process, but it works. The alternative is to prompt the user on the Web form for his or her *cell phone number*. After all, hardly anyone accesses the Internet over their cell phone, so it should always be available for a voice call. When questioned about this strategy of presuming that everyone has a cell phone, a vendor of Web call software replied, "We'll *sell* them a cell phone!" The vision of the Web salesforce obviously includes unbounded confidence.

As stated, various Web call packages including the forms and the pass-along to the call center have been available for the past few years and used successfully in some cases. However, by now it should be obvious that the drawback of these schemes is the need to place a second call, in addition to the ISP call, for the voice communication. Isn't this what VoIP is all about?

We Have VoIP

Perhaps the best way to integrate voice and Web on one network is to use VoIP for this purpose. As long as voice can be packetized successfully (by no means a given), then VoIP will be the best way to integrate data packets for the Web and voice packets for people on the same physical link at the same time.

The biggest advantage to voice-enabled Web sites featuring VoIP is that there is no longer any need to use the PSTN for the voice portion of the conversation. The PSTN just provides the bit pipe to the ISP, the same as it did before. The link just transfers the packet. Some of the packets are data and others are voice, but the network does not care, only the users.

Tying voice to the packet stream eliminates the problem of establishing a second connection for the voice portion of the interaction. And since IP is the same no matter where it is used around the world, there are even fewer interoperability problems than there are in ISDN. (However, there are plenty of mutually incompatible VoIP methods that will make VoIP interesting.)

Once VoIP comes into play, the "Web VoIP call" works as shown in Figure 6-2. Note how much simpler and cleaner the VoIP arrangement is.

Figure 6-2
The Web VoIP call.

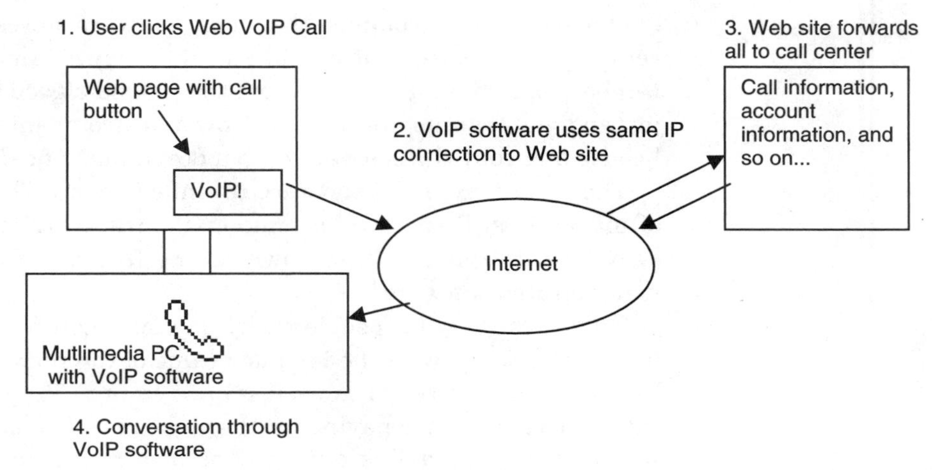
1. User clicks Web VoIP Call
Web page with call button
VoIP!
2. VoIP software uses same IP connection to Web site
3. Web site forwards all to call center
Call information, account information, and so on...
Internet
Mutlimedia PC with VoIP software
4. Conversation through VoIP software

The agent side can get all the information it needs directly from the browser (as long as the user has entered the information properly). No telephone number is needed, just the IP address of the customer. The Internet is used for everything of consequence. A multimedia PC is needed, but these are more or less a given today. The software for the client is usually freeware, so cost is no longer a real issue.

One of the first applications of integrated VoIP and the Web is expected to be in the area of computer equipment and software technical support.

Calling Tech Support...

It has been said that all software today is really free. That is, the cost of reproducing the software is minuscule. But what about the cost of developing the software in the first place? Application generation tools have made the process much easier than ever before, and almost no one writes Windows applications from scratch anymore anyway. It's all point and click, drag and drop, object and modules. The user interface is as important, if not more so, than what the program actually accomplishes in some cases. How many word-processing packages does the world need? Anything that performs basic computer functions such as spreadsheets and e-mail is probably downloadable over the Internet from somewhere absolutely for free.

And how common is the practice of copying software? CD-ROM burners have become so common that some software stores will not even take returns on software that is still shrink-wrapped, since the shrink-wrap can be replaced with the proper materials and a good hair dryer. It is not just individuals who are at fault. How many companies buy a single-user license and copy the software onto a dozen machines?

The point here is not software revenue threats. The point is that a lot of money is still spent on commercial, shrink-wrapped software today. Why isn't everybody either downloading freeware from the Internet or using pirated software?

Only because what people really pay for when buying software today has nothing to do with the intrinsic value of the software. (The boxes can be attractive, but they are useless once empty.) They are buying technical support. They are paying for the right to call someone when the software balks or crashes outright or otherwise does not perform as expected. This is the reason for the registration cards in the boxes and the serial numbers on the CD-ROMs and even the key numbers needed to install the product in some cases. It all has to do with who can call for help and who cannot.

All this has been pointed out before. But few take the next logical step. If it is true that software buyers are really paying for technical support and little else, then among similar products, *the one that offers the better tech support is the better product*. This is now true more or less by definition. The discriminating factor could be the availability of an 800 number, or more knowledgeable tech support representatives, or something else, but it all revolves around tech support.

It is not only in the software field that tech support counts for much. Users adding a scanner, second hard drive, external backup device, uninterruptible power supply (UPS), or some other peripheral often have occasions of head scratching and wondering just what to do next. The next step is often a call to the tech support center.

What has all this to do with voice-enabled Web sites? Frequently, the problem is an outdated installation guide or lack of the latest revision of the software. A visit to the vendor's Web site often can solve many of these problems. All the documentation is there, as well as the latest downloadable software versions. As long as the installation or upgrade has not rendered the computer useless, the Web site can be a good source of information and often the only one needed.

In fact, the traditional tech support call center has come to rely on the easy availability of Web-based information almost too much. Commonly, the first question asked of a caller to the tech support center is, "Have you read the FAQ on the Web site?" The frequently asked questions

(FAQ) page on the Web can be a help, but most sophisticated users have been there, done that. If the answer is out there on the Web site, it is hidden behind such a maze of links that the user feels like Theseus in the Labyrinth with the Minotaur waiting around the corner. And at least Theseus had a trusty trail of string to follow out. All that is available today are the "Back" and "Home" buttons.

However, once the caller has avoided being deflected back to the Web, there is usually a human to speak to at the tech support center. These are often cleverly concealed behind the bewildering menus of an interactive voice response system, which often leads not to the human agents but to a fax-back service that often faxes you the same FAQ Web page that did not help in the first place.

Even then, the user is typically placed on hold until a customer representative is available. A recording informs callers that "all representatives are busy assisting other customers" as if they were chained to their work locations. In reality, some are out to lunch, some are taking a break after an exhausting hour-and-a-half call, and some are even actively in the process of avoiding work. On hold, the caller is usually treated to some form of music, anything from Bach to Marilyn Manson (no kidding). Periodically, the music may be interrupted to inform the caller how important this call is to the company (but not important enough to add resources to the call center to handle it right away).

It is easy to criticize the process, which is universally derided and mocked. But the very best sites inform the caller of the approximate wait time (usually a little high; no one complains that the wait was less than expected) and let them decide for themselves whether to call back at some other time. One famous site of a software vendor treated callers on hold to the early history of the Internet. They had the odd experience of callers asking to be placed *back* on hold so they could listen to more.

If the Web site is to be the focus of the tech support group in terms of aid and information, then why not voice-enable the Web site with VoIP? This makes a lot of sense and avoids some of the problems that plague voice-only or Web-only tech support call centers.

Some of the benefits of voice-enabled Web sites for tech support centers are

- *The conversation is interactive.* Submitting a support request by e-mail is fine, but the request often needs clarification, leading to a second round of messages with no progress at all.
- *Problems can be handled more flexibly.* Talking to someone is an exploratory process. The agent can tell immediately the technical level

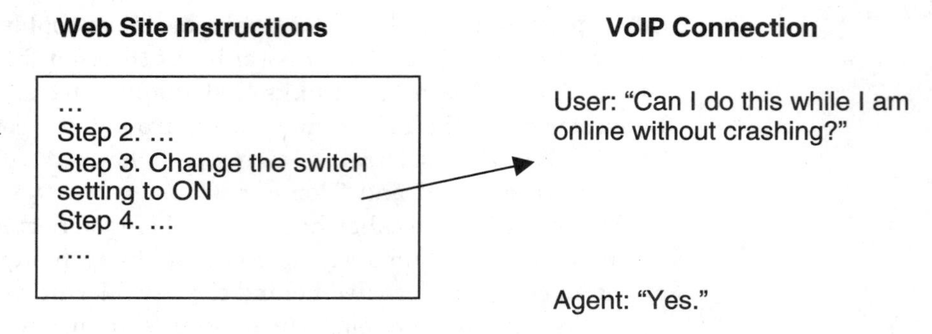

Figure 6-3 Voice-enabled tech support.

of the caller (the caller also can quickly assess the ability level of the agent as well). It is hard to tell an FAQ, "Look, I tried all that already."

- *New problems are known immediately*. The human can gather information on a newly discovered problem much more quickly than by relying on e-mail postings.
- *Cost savings*. There is no need to maintain a separate voice network for all support calls. Not all calls may be handled through the Web, but just a call on hold can cost the support center anywhere from 6 to 8 cents per minute. The VoIP equivalent of "on hold" costs much less.

How a voice-enabled Web site can aid tech support is illustrated in Figure 6-3.

As tech support becomes more and more the way that software and hardware vendors distinguish themselves, voice-enabled Web sites will become more valuable than ever. However, tech support is not the only Web activity that can benefit from VoIP integration. Even commercial sites marketing goods and services outside the high tech field can benefit.

Can I Get That in Red?

More people use the Internet and Web to buy everything from cars to flowers than use the Internet and Web to connect tech support. Traditional catalog sales companies have been scrambling to retool their businesses to take advantage of this growth of Internet and Web use. Not only does this integration benefit from the capabilities of the PC, which are above and beyond that of the simple telephone, but it saves money as well. There is better customer segmenting (needed when a company offers many types of goods and services) and customer retention and loyalty. It is even possible to offer entirely new customer service programs ("Based on your previous purchases, we recommend this book..."). Add-on sales are seen more as concern for customer well-being than as just another pushy sales effort.

Retooling the business for the Web means retooling the call center for the Web. This means adding voice to the Web site. Traditional voice-only call centers' uses for sales have unpredictable and uneven service levels, and there is a real fear of overwhelming the customer with choices (who wants 79 choices on the voice menu?). It has been hard to effectively add technology to the voice call center, yet customers demand better service that is hard to provide without a technology basis. Above all, competitive pressure is increasing in this market.

Customers have grown to expect certain things from these call centers. They want faster response times and the ability to talk to an agent whenever they want. They expect the company to know they are a returning customer and are no longer put off by voice-prompt systems (although some people still will not leave a recorded message on an answering machine). As much as customers like and use the Web, they still want to be able to choose the form that the ordering process will take. Some people will gladly give a credit card number to a human agent but hesitate to enter the same information at the Web site (of course, the agent is just typing the number into his or her own Internet-attached computer).

There is some degree of computer-telephony integration (CTI) in the modern call center. More details will be presented later in this chapter. But what exists now, before VoIP-enabled Web sites, is a piecemeal approach to the situation. In many cases, customers can order goods and services through the PSTN with a PBX and automatic call distribution (ACD) device, at the Web site, through e-mail, or even with a fax. All are separate applications, which use separate databases and verification procedures. The service levels depend on the means used to access the company. There are no integrated ("blended") agents or applications at all. This situation is shown in Figure 6-4.

Voice-enabling a call center for sales is a little trickier than adding VoIP to the tech support process. Even Web chat can get a message to the call center agent. But what if the message, text or audio, is, "Can I get that in red?" What, exactly, is "that"? The customer sees the product, but the agent does not. Thus the next step is to pass not only the message but also the Web page to the call center agent. This is rarely necessary in tech support applications but crucial in on-line catalog sales situations.

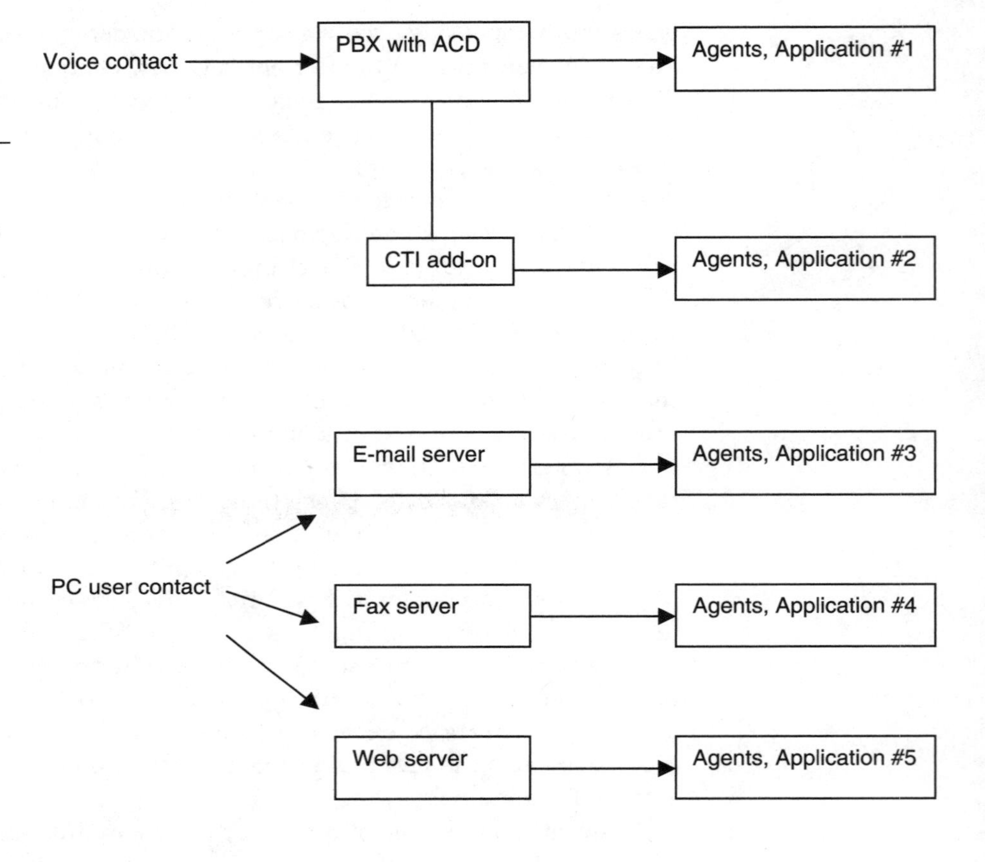

Figure 6-4 The traditional call center.

There is one step beyond even "shared browsing" that allows the agent to follow the user through the Web site while talking. This is *page pushing*, which allows either browser, at agent or user end, to guide both through the process and share the experience. Page pushing involves some delicate positioning to do properly, and this means human considerations, not technical ones. If the agent is too "pushy," in both senses of the word in this context, the user feels that he or she has lost control of the situation. How would a person feel in a "real" store, while trying to look at sports shirts, if the salesperson kept steering him or her over to dress suits? If the agent is too "hands off," then the potential customer feels abandoned.

Adding VoIP to the call center requires a voice-enabled multimedia PC and Web browser on the customer end, of course, just as in the tech support application. This is a given.

The bottom line is that the customer has to be able to choose the level of interaction needed. Contact could be made anywhere, anytime, and either over the telephone or through the Internet. Customers will decide how to contact the company. If one type of method is favored over another, it is up to the company to provide an incentive for customers to use the approach that the company wants them to ("10 percent off everything sold using VoIP").

Call Centers and the Web

By now it should be obvious that there are many reasons to voice-enable a Web site. The high tech industry and sales examples used are by no means the only industries interested in voice-enabled Web sites. Any activity that traditionally includes both written documentation and verbal interaction is a prime candidate for this integration. The text made its way onto the Web site a few years ago. Now the voice will follow. Businesses that will use VoIP to add voice to their Web sites include

- Airline reservations ("Can I connect through Philadelphia instead?")
- Hotel reservations ("Is that room on the ocean side?")
- Banking ("What would the payment be if the interest accrues?")
- Insurance ("How can this policy be converted?")
- Ticket sales ("Can I get four seats together in another section?")
- Stock trading ("Will I make the split requirements?")
- And many others

While all these queries can be handled over the Web without voice interaction (usually with a form or Web chat or e-mail), the voice alternative is quicker, easier, more flexible, and more acceptable to customers.

Levels of Call Center Integration

It has already been pointed out that when it comes to integrating voice and the Web, it makes little difference where the process starts. The end result is the same. As long as both user and agent have access to both Web browser and VoIP, it does not matter if this process began by voice-enabling a Web site or Web-enabling a call center. This section looks at the process from the starting point of the call center.

In order to integrate VoIP with the call center, two main things are needed. First, an IP telephony gateway must be attached to the Web server and the ACD device. Any database information available to the agents must be available through the integrated system as well. Second, the call center agent must have a Web browser available (he or she has voice already), and the customer's PC must have VoIP software (and perhaps hardware) installed (he or she has the Web already).

Even so, the transition to Web-based call centers is far from simple or painless. The fact is that most call centers have some degree of voice and data integration already. However, the data portion of the call center is just a local-area network (LAN) and does not extend to the customer at all. Agents use the database instead of written documentation to answer questions, access customer account information, and even to see what to say to irate customers. This arrangement is shown in Figure 6-5.

In the figure, there are dual networks for the voice and data operations of the traditional call center. These have been around since at least the mid-1980s and include a PC at every agent location. The LAN connects each location to a database, which has all the information about the voice call that the agent needs to know. The information differs greatly between inbound and outbound (telemarketing) call centers, which will be discussed in more detail in a later section of this chapter. But now voice flows on the LAN, and the PBX/ACD are purely voice devices. There is a link between the PBX/ACD and the database, but it is of little use other than to shuttle telephone numbers and some other call information back and forth between the voice and data worlds.

There are three drawbacks to this arrangement. First, as has been pointed out already, the whole thing required two distinct networks, one

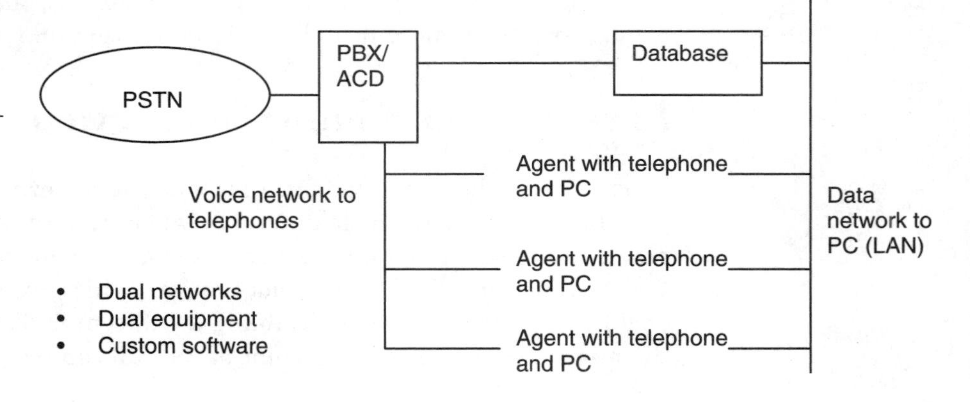

Figure 6-5 The voice/data call center.

for voice and one for data. Not only did each of these networks use separate hardware, software, and protocols, but the physical networks used for each (the wiring and other equipment) had to be totally isolated. Maintaining both required two separate staffing functions, management, and areas of expertise. Second, dual-agent equipment was needed. Every agent had a telephone (usually not a handset, but a headset to keep the hands free) and a PC or terminal to access the database. If an agent's telephone was out of service, the PC did no good. The same was true when it was the PC that was unavailable. Backup work locations needed *both* so that an agent could shift work locations and keep on being productive. Third, this arrangement required custom software to make the whole thing work. For instance, there was no standard application or protocol that allowed the PBX/ACD to send the caller's telephone number to the database and look up account information. This had to be written by hand from scratch. And while the database query language on the agent side could be a standard, this did not define how the information appeared on the agent's screen. In many cases, the PBX/ACD could supply the telephone number to the database application, but there was no "push" application or protocol that could make the caller's account information fetched from the telephone number appear on the screen of the agent that eventually took the voice call. Thus the agent had to type in the caller's name and number over again, even though the database had that information already.

Just this simple step caused a revolution in call center operation and perception. For example, when a major computer vendor first used a database to supply information to callers instead of printed manuals in the late 1980s, customer satisfaction went through the roof. It was hard to see why, since the information given out was exactly the same. It took a while to figure out exactly what was going on.

It turned out that callers did not like to hear the agents flipping through the documentation when they called. They all thought, "I didn't call up just to have them read the manual to me." Of course, in many cases this was exactly what was happening. (Not only do users not read the manual, apparently they do not want anyone else to read the manual either.) However, when the documentation was on the database, all the caller heard was the clicking of keyboard keys as the agents typed in the query to fetch up the proper document. Callers now thought, "Wow! They have a computer right there. They are really trying to help me." Customers were so pleased that they did not realize that the manual was still being read to them.

The next step is to Web-enable the call center, which is the same as voice-enabling the Web site. What is achievable today is shown in Figure 6-6.

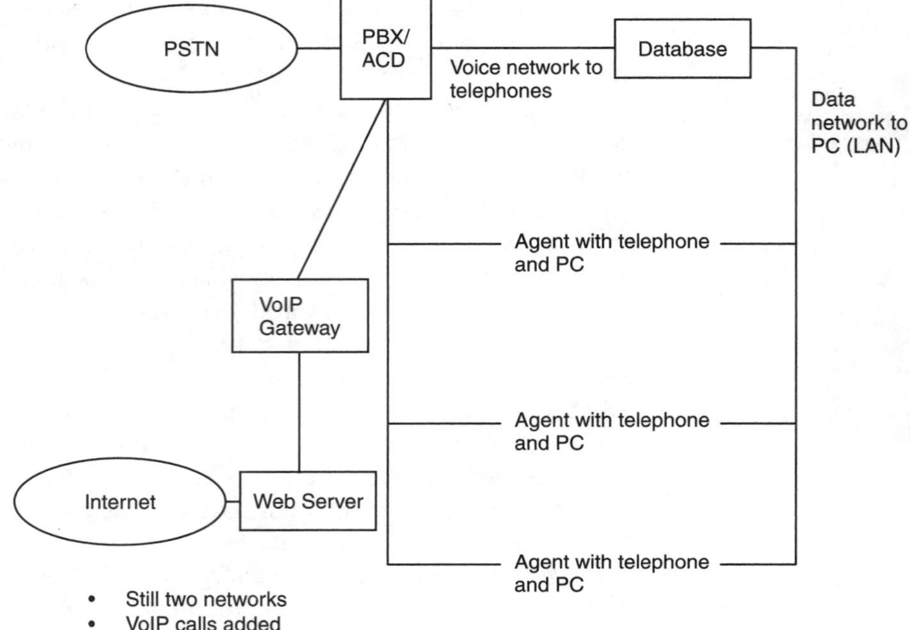

Figure 6-6 The Web added to the call center.

The major difference between Figure 6-6 and the preceding figure is the addition of the Web server representing the Web site (in reality, this could be a whole group of servers) and the VoIP gateway. The Web server is the main target for the activities of users on the Internet, naturally, including users with VoIP capabilities. The "Web VoIP call" button now links to the VoIP gateway that enables voice communications with these VoIP Web users, but only through the same PBX/ACD arrangement as before. What essentially has happened is that VoIP has been added to the call center. However, all VoIP calls go through the same PBX/ACD as "normal" voice calls. On the other hand, the agents do not necessarily need VoIP, although access to VoIP and the Web site would be a good idea for inbound call centers so that agents can see what the caller sees. Note that the Web server could reside on the same LAN as the agents and the database, but this is perceived as somewhat of a security risk today. More likely, the public Web portion of the network is isolated from the intranet portion of the network with the agent's database(s). It is also common today to replace the database with a Web-based version of the same thing and give the agents a simple Web browser interface, but this is not always done.

Note that there are still two networks for voice and data in this configuration. The main thing is that VoIP calls have been added, but really only for an inbound (e.g., tech support) scenario. There is no real way to generate outbound calls over the Internet to potential customers with VoIP. Clearly, for high-tech telemarketing organizations, which always seek to reach such leading (bleeding?) edge customers, this is a serious limitation.

And unless the agents also have transitioned to Web-based applications (by no means a given), custom software still abounds in this configuration. What has been added is a way to connect calls from the Internet to the call center. This may be enough, but VoIP can do more, much more, for the call center. Figure 6-7 shows a much better integration between the call center and VoIP.

Figure 6-7 The integrated Web site and call center.

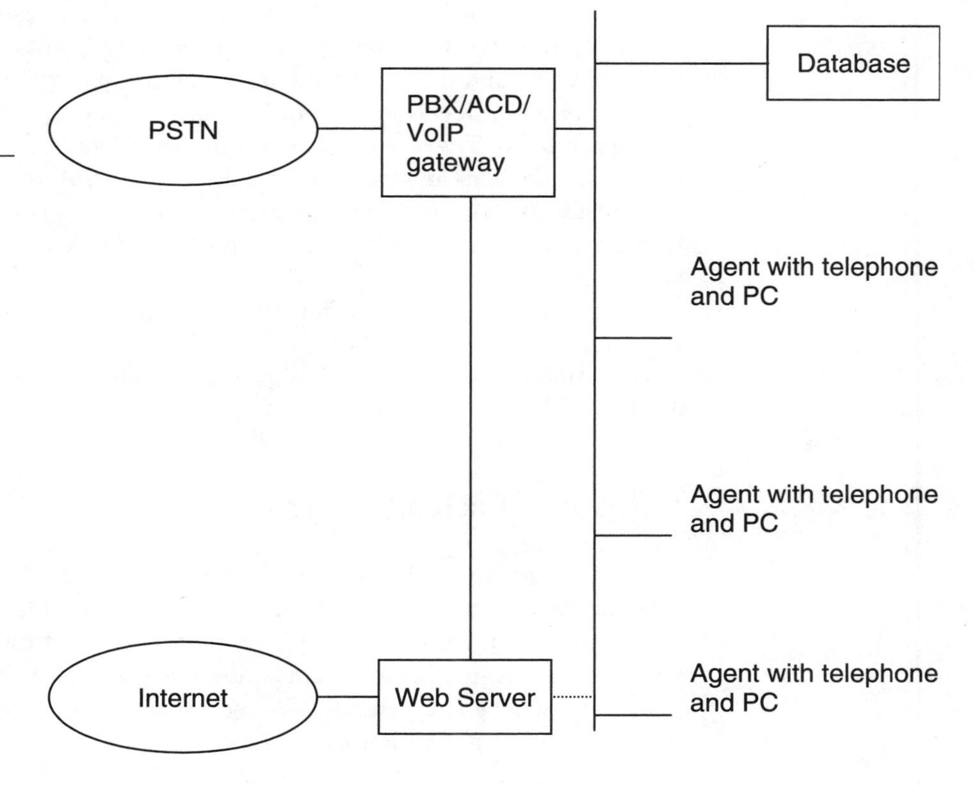

The figure shows complete integration between Internet/Web data operations and PSTN/Internet voice operations. The agents still have a telephone headset and PC before them, but now the PC with VoIP handles *all* calls, not just Internet calls. The PBX/ACD is not housed in the same device as the VoIP gateway and in actuality is probably just another server on the LAN. Several LAN-based VoIP PBXs and ACDs are currently available. The PBX/ACD/VoIP gateway can still take and make calls over the PSTN, but these calls are all handled as VoIP calls to the agent locations. Internet/Web VoIP calls require no real conversion. The call center database is now accessed with a simple agent Web-browser interface. The Web server can be on the same LAN as the call center database, as before. However, as the dotted line indicates, security now becomes a paramount concern. The call center now has exposed itself to all the risks of the Web site, for better or worse.

The main benefit, however, is that now there is only one network used for both voice and data. This means one staff, one wiring infrastructure, and easier agent location backup (just plug a spare PC into the LAN connection and go). There are not two sets of hardware, software, and protocols, just one. This means that standard, plentiful Web and IP software and protocols are used throughout the operation, not just for the data portion. Any problem is just an IP problem, even on the voice side of the operation.

More than 30 vendors offer hardware and software that make the voice-enabled Web site shown in the figure possible today. The major issue is the degree of conversion or outright change required. But all can do the job.

Inbound Call Centers

There is a major difference in function between the operation of an inbound and outbound call center. This difference in function will extend to call centers that feature VoIP, so a clear understanding of the differences in important. Inbound call centers are wrapped up with 800 calling packages, where the company called pays the bill, not the caller. The most common application of inbound call centers is some form of customer services. However, many traditional catalog sales companies that still distribute printed catalogs employ 800 numbers for sales as well. But this is just another form of customer service.

As people buy more and more on the Web, the human touch recedes into the background. VoIP can help to bring personal service back into the limelight. The number one reason that people always cite for avoiding Web-based shopping is lack of confidence in a system that they really do not understand. Consumers are typically asked to fill out a form to complete the on-line transaction, a form in which personal information is sent over the Internet without the consumer really knowing who is on the other end or if he or she will really get the item he or she just paid for.

The best way to calm these common fears is to just let the customer hear a human voice. With VoIP, inbound call centers become easier to use and more efficient, and consumers get the best of both worlds: automated real-time assistance over the Web, and all over one telephone line. This is all tied in with the concept of customer relationship management (CRM), a key idea behind call centers. A call center agent can now not only talk to the customer but also view the same Web pages.

VoIP-enabling the Web site should be coupled with Web-enabling the call center database. This is the key to making the whole integrated system perform properly. Just what is the function of the inbound call center database? Some generalities regarding the function of this database have been mentioned, but this section will supply a few more details. The call center database takes the telephone number of the caller (this will not work if the caller has blocked the caller ID function, of course) from the PBX/ACD and uses it to look up key information that the agent can use to best deal with the customer concern. The customer concern could be anything from "Where's my order?" to "How do I hook up your printer?" to "Can I get six of those at the sale price?"

Usually, the database will supply some basic information about the customer, if the database possesses the information, right on the screen of the agent taking the call. If the caller's number is supplied by the PBX/ACD, then the agent can even say, "Good morning, Mr. Smith. What can I do for you today?" before the caller has spoken a word. If the caller's number is not available to the database, the agent must prompt for the caller's name and/or number or some other database record key such as account number (which callers usually do not know). Then, based on the nature of the call's request, another screen can furnish further information. Typical information supplied by the call center database is shown in Table 6-2.

Notice that some of these types of information can be made available to the caller safely and easily directly from a Web site (the printer hookup). Other types of information are needed for the agent to assist the caller, but not necessarily the kind of information a company might want to freely share even with the individual concerned (the sale). Why should the caller have to know that the decision may rest on the caller's promptness in paying bills or the size of previous orders.

TABLE 6-2

The Call Center Database

Where's My Order?	How Do I Hook Up Your Printer?	Can I Get Six of Those at the Sales Price?
■ Order date	■ Hardware requirements	■ Item and price
■ Order content	■ Software requirements	■ Sales price
■ Shipping date	■ Installation steps	■ Sales dates
■ Shipping method	■ Troubleshooting	■ Availability
■ Tracking number	■ Power?	■ Order history
■ Payment status	■ Cable?	■ Payment history
■ Order history	■ …	

This is why the solution cannot be simply the Web or the call center all by itself. Adding VoIP solves all the issues at once.

However, adding VoIP to a call center brings up a number of issues on its own. A company has to look at its client base. Will human assistance make a difference or not? Sometimes VoIP makes a difference in surprising circumstances. For example, a Web-based Internet game site became even more popular when it began to offer VoIP for participants to converse with each other over the Internet while playing the game. Sounds odd, since there was really no "call center" involved. However, users liked VoIP because they did not have to stop using their mouse and joystick to type in "chat" messages or take their eyes off the screen to read messages at the risk of getting splattered.

Outbound Call Centers: Telemarketing

Inbound call centers spend their days taking calls. Outbound call centers do the opposite. Outbound call centers are used for telemarketing. Although telemarketing has gotten somewhat of a bad reputation for annoying calls, disrupting dinners, and so on, this is big business. After all, if no one bought anything from telemarketers, they would all quickly go out of business.

Outbound call centers differ from their inbound siblings not only in the direction of the call but also in the form and function of the database used by the agents. The most common purpose of the database is to feed the telephone numbers to be dialed to the PBX (there is no ACD in an outbound call center, of course, since ACDs deal with incoming calls). Most also supply the agent with the "script" to be used during the call. The script usually fills in the called party's name and sometimes town and even street address. Sophisticated packages will tap into Internet information on weather and so forth as well, the better to maintain the illusion that the telemarketing concern is a local business.

The telemarketer can ask the identity of the person who answers the telephone by name and then add appropriate comments. If the customer is one known to the company from previous orders, then some additional personal information may be available as well.

The most important part of the call is the first 10 to 30 s. Most people will give the caller at least that much time to intrigue them, if only out of a need not to appear rude. The energy level of the agent means a lot here. Who wants to hear a bored person obviously reading from a script? Some telemarketing calls last less than 10 s, usually because the caller does not know how to pronounce the person's last name correctly. (Telemarketers seem convinced that the authors' last names are pronounced "gor-a-LAS-ki" and "ko-LON.") There is a fortune to be made from software that can show a telemarketer how to pronounce last names. The best telemarketers sound like they are long-lost friends who just called to say hello. All this points out a firm rule of sales: People buy things from people they like.

Sometime during the first 30 s or so of the "pitch," the telemarketer has to make a decision about where to go next in the script. The conversation can take a nasty turn, so there are menu picks for "hostile" and "silent" as well as "wants to hear more." Based on the agent's choice, the call can now branch off in a number of different directions, sometimes even adjusting the product offered. For example, a remark about a small kitchen can lead to home improvements instead of the pitch about roofing that was intended. Naturally, the sponsoring company must be able to offer this variety of services.

The outbound call center database therefore consists of a chained series of script pages that can be accessed in a number of ways. If this does not sound like a perfect application for the hyperlinked Web, then what is? The general idea behind the outbound call center database is shown in Figure 6-8.

The whole point is that it only makes sense for the outbound call center script database to be a Web application housed on a Web server. In the figure, the underlined words are the ones supplied by the database and differ for each and every call. The links at the bottom are Web

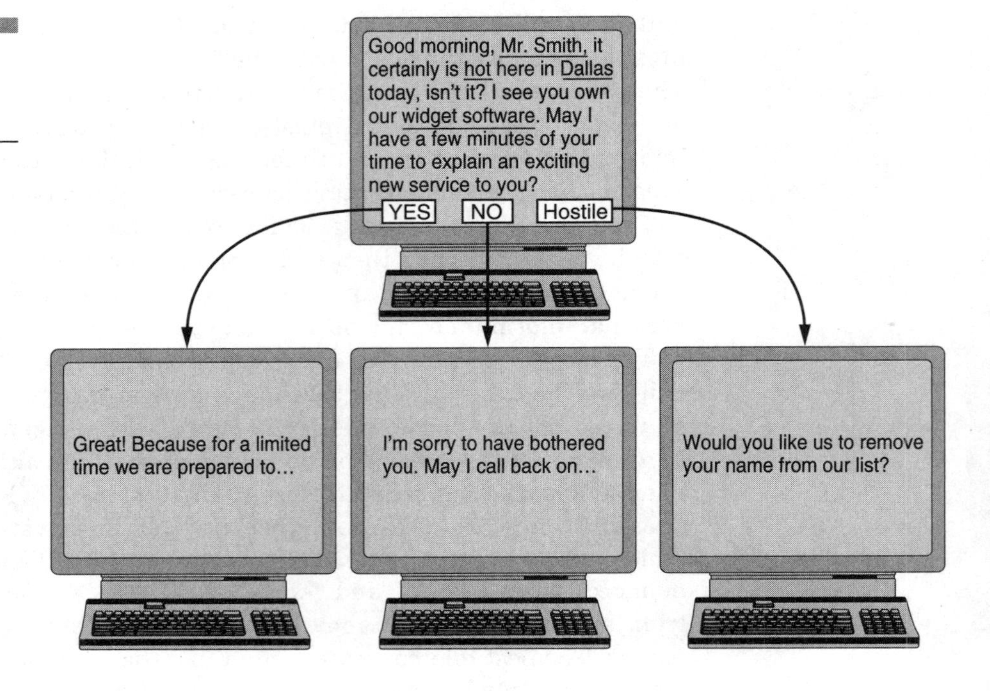

Figure 6-8 The call center script.

links, naturally. The chosen link depends on either a response to a question or just the agent's feeling about the call. Some express surprise that hostile reactions are accounted for and usually result in a "Don't call back" flag being set on the customer record (the record is usually not deleted outright). But why waste valuable time and effort annoying people who obviously will not buy anything? The art of focusing the lists of people called is all part of customer relationship management as well.

So VoIP and the Web are naturals for telemarketing. VoIP is used to trigger the calls onto the PSTN, although conceivably there could be telemarketing calls to VoIP-enabled PCs (selling new VoIP software and hardware?). The scripts become standard Web forms and are easily modified by a number of tools. Telemarketing scripts are frequently changed and tuned both in response to consumer feedback and to keep the agents from becoming terminally bored and sounding too rehearsed. Agents can even work at home with a PC as long as they have VoIP and secure access to the telemarketing Web server where the scripts reside.

A Distributed VoIP Call Center

What would a *distributed* VoIP call center consist of? What would it look like? How does one go about building one? There are many guidelines that can be followed, but implementing such call centers is not so routine that the rules are hard and fast. Sometimes these new types of call centers are called *virtual call centers*, but they all feature VoIP prominently.

Once the leap to VoIP at the call center is made, some amazing things can be done with the basic call center setup. For example, the fact that everything is now LAN-based, including the PBX and ACD, means that the call center can be spread out among several LANs in several locations. Network routing can shift calls from East Coast to West Coast or the other way, depending on load, time of day, or other factors. The multiple sites can still be managed centrally and seen as a single entity. VoIP offers universal call transferability if it is done correctly. There can be many small centers or fewer large ones, whatever the situation warrants.

Agents can now work anywhere there is a VoIP-enabled PC and headset, even from home. Home workers make the same salary, of course, but do not need parking spaces, lunchrooms, or buildings at all. There is less overtime as well for larger groupings, since most calls can be handled during normal business hours. The agents are now location-independent and yet completely supervised using well-established TCP/IP techniques. The scheme also offers redundancy and backup. If a flood or storm knocks out one "center," the others can pick up the slack, as long as the proper database information is available. Callers are not even aware of any changes. Also, specialists need not be gathered in one place but can be spread around and take calls as needed. Not only that, great telemarketers or problem solvers who can only work part time, or who do not want to fight their way into the office, or who want a flexible schedule can be accommodated easily.

Distributed workers do not need to be compensated in the same way as others who spend 8 hours day after day in an office. Salaries can be based on piecework, such as making or handling so many calls per day. For high-tech call centers, even a retired expert can be called into action once or twice a month, if need be. The cost of having such an expert in a call center all the time would be prohibitive. Just give him or her VoIP at home (which might be a houseboat in Florida).

A key piece of this system is a secure network between sites. There needs to be dedicated security equipment added to the equation to form a virtual private network (VPN), but this is beyond the scope of this discussion. The

VPN capability could be added to the site router, but only at the risk of a performance bottleneck, especially when VoIP needs are factored in.

The Internet Watch Company

Consider the activity of an agent assisting a customer buying something (a watch, perhaps) from a Web site using VoIP as well as the Web in the call center. Sometimes discussions of technology such as VoIP tend to focus on the technology itself and not on the human aspects. This is a good time to put this all in perspective.

1. The agent arrives for work (or sits down in his or her home office) and prepares for work by firing up the PC and donning the headset. If this is a typical call center, there are calls waiting for immediate attention. The agent sees this as a pop-up window that identifies the first call as coming from the PSTN. This information comes from the call center application and is needed because the caller may not have visited the Web site at all. Then again, the caller may have gotten the telephone number from the Web site and just likes the human touch. The caller's name is there on the screen, as well as the general nature of the query gathered from the interactive voice response prompts ("Press 1 to buy a watch..."). There is also any other information about the caller gathered from the call center database, which is now just another Web server, through customer relationship management software ("They don't like rubber watches").

2. The agent speaks with the customer, possibly addressing him or her by name (watch out for pronunciation woes!) and helping with his or her concern. All this happens over the LAN with VoIP, even though the other end is just a person with a "regular" telephone. The VoIP gateway takes care of any conversion needed. The agent ends up transferring the call to a watch specialist in another city and is confident that the customer records will be forwarded there as well. There will be no need to repeat information to the new agent handling the call, and the original agent can add any annotations needed.

3. Once the first call is over, another pop-up window signals the presence of a VoIP call coming in over the Internet. Again, this is useful information because this customer may be looking at the Web page at that very moment. A few clicks of the mouse, and the agent can now see the page (if any) that the customer currently sees and can even put a frame on the customer side of the conversation that includes the agent's name and even photo. If the customer has filled up a shopping cart, the agent sees the contents.

4. The second call proceeds just as the first, with the agent being as helpful as possible. It is easy to "upsell" to a better watchband verbally than through a Web site. Such upgrading to an incrementally better item could be prompted by genuine concern on the part of the agent ("You say you are a construction worker. Could you use a more sturdy watchband?") or just part of the regular sales effort ("Do you need a dual-time-zone model?"). Upselling is much easier with voice. It is easier for panhandlers to gather spare change by asking for it than just by displaying a sign that says "Deposit spare change here."

5. If the customer is skeptical about the utility of the upgrade, the agent can "push" the page with more details out to the caller over the Web. This cannot be done with a PSTN call. If the customer agrees, the agent can even complete the ordering process and push out a final copy of the shopping cart contents, the invoice. This way the transaction is always completed correctly.

It is important to point out that all the circuit switching, the routing, the queuing, the transfer, and so on are all transparent to the caller. This is the way it should be. It does not matter if the call comes in from the PSTN or the Internet, the same agent, hardware, and software handles it all. The amount of information may differ slightly, but the level of customer service remains high. VoIP makes it all possible.

Voice-Enabled Web Site Issues

Like anything else in the telecommunications industry, adding VoIP to the Web is not painless. A fair percentage of users who attempt to download and install VoIP freeware on their PCs do not succeed the first time. The configuration process is awkward, and one study has claimed that it takes about four tries to get the configuration right, even when the other end of the connection is configured correctly already (and compatible). Some people do not even know if their PC has speakers and a microphone, let alone if their machine is "H.323 Internet phone capable."

To offer successful VoIP packages for the call center, vendors will have to do three things. First, they will have to focus their efforts. There must be working relationships between vendors of ACD, VoIP, Web, and other hardware and software components. The environment must be thoroughly integrated and blended. Second, the vendors have to prepare and educate customers for what the future call center with VoIP holds. No one likes surprises. Third, vendors will have to help with the VoIP standardization effort. This point will be raised again at the very end of this chapter.

Some further issues are cultural in nature. For almost 30 years, call centers have revolved around the basic telephone call. It is not only the companies that have to transition to the new environment, but the potential users as well. Adding Web chat, callbacks, and VoIP to a call center can bewilder the agents as well as the customers. As the nature of the work changes, PSTN and VoIP calls may receive differing levels of service. This may be unavoidable in the short run but cannot be allowed to continue for any length of time.

If the voice people in an organization run the call center and the Internet people in the same organization run the Web site, who controls the integrated whole? Weddings between voice and data have been promised before, only to break up after a short period of bliss or heartache. While technology can blur the distinction between voice and data, technology cannot do anything about the distinct differences between voice and data *users*. Voice users generally are more laid-back ("It's busy. I'll call later.") than data users ("It's busy! I need to go on-line *now*!"). Voice and data support groups tend to mimic the same philosophies. If a link between PBXs is down, the voice support group tends to shrug and tell everyone to dial out on the PSTN. If a link between routers is down, the data support group scurries around like maniacs. They know that heads will roll if the link is not restored in 4 hours or so. When mixed together, such "integrated" environments are almost always less productive.

There are also technology issues. How much integration between Web and call center is possible? What kind of network management is available for the "blended" environment? All transactions, whether PSTN- or VoIP-based, must be tracked the same way.

Once again, there is the issue of VoIP standards. Standards are needed for the SS7-VoIP gateway signaling integration as well as for the end-user software. No one will like downloading and installing different types of VoIP software for every Web site they access. But the issue of VoIP standards is important enough to deserve a chapter of its own.

CHAPTER 7

Standards for IP Telephony

One of the most remarkable attributes of the public switched telephone network (PSTN) is the global connectivity it offers. With few exceptions, which are usually due to political considerations, anyone with a telephone can pick it up and dial a number that will ring any telephone on Earth. This ubiquity and reach are due to the voice service providers' realization that the value of a telephone itself is next to nothing. It is the qualities of the network to which it is attached that determines the value of a telephone to people. Thus, from early on, telephone companies have gone to great lengths to extend the boundaries of their networks and to connect their networks to others.

It is in these interconnections between systems that the real difficulty of all types of networking lies, regardless of whether it is voice or data that is being carried. Telephone systems, for example, are all built to do more or less the same thing: carry voice conversations from one telephone to another. This single problem of carrying voice signals has many possible solutions, and telephone systems differ greatly in how they do it. Even when restricted to more or less the same equipment and physical infrastructure, there are many choices that network architects must make when designing a telephone network, revolving around issues of, for instance, capacity versus quality. Systems need not differ that much to be incompatible, and differences can require quite complex adaptation mechanisms if the users of one system ever want to speak to the users of another.

This issue of internetworking has been addressed, and made considerably easier, by the introduction of global standards for telephony. Usually promulgated by the International Telecommunications Union (ITU), this body of standards performs two important functions: It helps limit the number of different types of telephony systems, and it defines the common attributes that phone systems must have to connect. The introduction of Internet Protocol (IP) telephony systems broadens the scope of the ITU's coverage and adds complexity to both these areas. This is so because the still-vast array of possible telephone systems now has IP-based telephony added to it. Any IP-based telephony must, of course, be connected to "standard" or normal telephone networks if all users are ever to be able to speak with one another.

The ITU's Telecommunications Subcommittee (usually called the ITU-T), never one to shrink from a challenge, has stepped up to the plate with a full suite of standards for all the functions of IP telephony—from audio-signal encoding, to call establishment, to routing, and even extending to interconnection with the PSTN. This chapter explores the ITU-T's suite of protocol standards (as well as some "competing" standards), describes their various responsibilities, and examines how products designed around these standards function and interoperate. This chapter also explores the attributes of protocols that fall outside the ITU-T's jurisdiction, such as those routing and transport protocols which are the responsibility of the Internet Engineering Task Force (IETF). Finally, this chapter discusses the proprietary means by which some vendors accomplish the same ends as the standards-based products and examines the consequences of such an approach to the various problems of IP telephony.

Following an IP Telephony Call

Before getting into the protocols themselves, it is perhaps useful to follow an example of IP telephony from beginning to end. This example will be used to illustrate some of the various tasks that must be addressed by different protocols along the way. No single example can realistically encompass all the possible functions of an IP telephony system, especially as development and refinement of existing methods continue. However, the example below outlines a call that not only is possible in today's environment but also is perhaps typical—at least until a time when the public IP telephony network becomes more distributed and ubiquitous. Should it occur that the entire PSTN becomes packet-based over time, many of the "problems" of IP telephony will be simplified by the lack of necessity for circuit-switched network internetworking. If the all-packet future ever comes, however, that day is a long time away.

Figure 7-1 presents an example of what is today the most commonly employed application for IP telephony, called *toll bypass*. More on the toll-bypass applications of voice over IP (VoIP), especially for international calls, will be discussed in Chapter 11. Dave, who lives in Los

Figure 7-1 Toll-bypass Internet telephony.

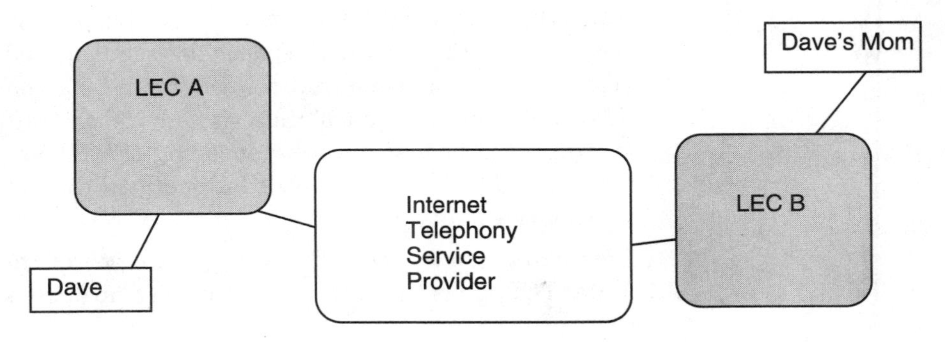

Angeles, wants to call his mother in New York. Dave has subscribed to one of the ever-increasing number of Internet telephony service providers (ITSPs) to handle his long-distance telephone service, wishing to take advantage of their low rates for interstate calls.

When Dave picks up his telephone and dials his mother's number, his local exchange carrier (LEC), realizing the long-distance nature of the call from his first dialed digit (1), hands the call to his chosen interexchange carrier (IXC). The IXC's job is to carry the call to Dave's mother's local phone company (LEC B) so that the other LEC can ring her phone and connect the call to her. Since Dave's IXC is an ITSP, the normal circuit-mode switching of the call ends there. Dave's ITSP must now convert his audio circuit to a stream of packets and send those packets to New York. Once there, the packets must be converted back into an audio stream on a "normal" voice circuit for transit across the traditional circuits of his mother's LEC to her phone.

Some of the functions that must be accomplished in order for this to happen are as follows:

- *Address resolution.* The gateway between Dave's LEC and the ITSP must somehow find the IP address of the gateway between his mother's LEC and the ITSP. This can be done by associating the called number with the gateway that has responsibility for her area code.
- *Routing.* The ITSP network must send packets from the Los Angeles gateway to the New York gateway by the best available means, keeping latency low while balancing traffic across the network so that links are not overloaded.
- *Ensuring quality.* Depending on the services provided by his ITSP [it also may be an Internet service provider (ISP), which carries ordinary data traffic as well as telephony traffic across its network], Dave's call must coexist with other traffic. If there is data traffic sharing the links, the ITSP must have some means of isolating or identifying voice traffic from other traffic that does not require the low latency of voice signals. This is true so that the voice can be given priority and expedited on its way. Even if there is only voice traffic on the network, there should be some way of ensuring that Dave's call will never be overcome by other voice calls that overload the network so much that all callers' quality suffers.
- *Predictable transport.* The transport layer of the Internet Protocol suite (TCP/IP) is tasked with ensuring some degree of reliability to the

stream of data, ranging from sequenced, error-free delivery to simple encapsulation with no error detection or other guarantees. In the case of real-time traffic, perhaps the most important function of the transport layer is to ensure the regular delivery of audio information. When predictable transport is not possible (as it almost never is in the connectionless world of IP), the Los Angeles gateway can as least stamp data with its relative time of origin so that the New York gateway may reassemble it with a similar timing.

- *Call signaling.* Not only must the ITSP network carry Dave's call to the correct gateway, it must somehow signal that gateway as to the eventual destination of the call, in this case Dave's mom's telephone. Thus there must be a way for the New York gateway to learn the called number from the IP stream, in order that it may dial across the New York LEC to Dave's mom and connect the IP call to the LEC call.
- *Voice encoding.* As encoded by his LEC, Dave's call is presented to his ITSP as a stream of digital data running at 64 kb/s. Much of the advantage of IP telephony over the traditional PSTN is its ability to squeeze this stream into a stream running at only 32 kb/s, 16 kb/s, or even smaller in size. Naturally, IP packets are used to convey this stream.
- *Error handling.* If something unforeseen should happen to the call, such as the failure of a gateway, a problem with one of the LECs, or an interruption in the network service of the ITSP, there must be some way of recovering from the error as gracefully as possible. At least there should be a way of ending the call on both ends so that their phones are returned to them.

These are the fundamental problems that must be addressed by any IP telephony system. There are others, such as the type of network the IP runs over, the means of bringing audio to the phone or terminal, and the myriad of human considerations such as billing, management, and maintenance. These other considerations are not addressed by the ITU-T, since they do not directly affect the interoperability of systems, which is the primary reason for the standards. They are therefore left to the designers of systems and equipment to innovate for themselves.

A graphic representation of some of the protocols that fulfill some of the functions just described is presented in Figure 7-2. In keeping with the IP-centric nature of any IP telephony system, it is displayed with reference to the Internet Protocol suite model.

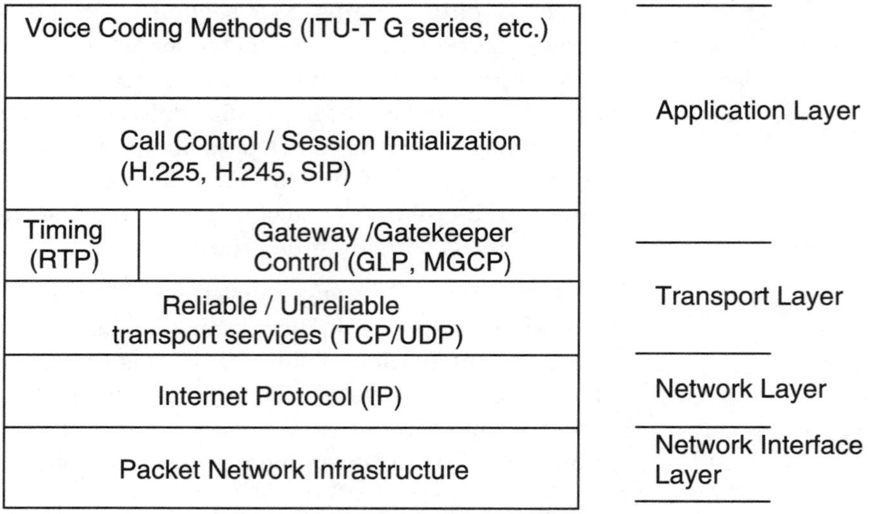

Figure 7-2 Issues/protocols in IP telephony networks.

Voice Coding Methods (ITU-T G series, etc.)		Application Layer
Call Control / Session Initialization (H.225, H.245, SIP)		Application Layer
Timing (RTP)	Gateway /Gatekeeper Control (GLP, MGCP)	Transport Layer
Reliable / Unreliable transport services (TCP/UDP)		Transport Layer
Internet Protocol (IP)		Network Layer
Packet Network Infrastructure		Network Interface Layer

The H.323 Framework

When it comes to IP telephony, the international standard most often mentioned is H.323. This recommendation, first issued in early 1996 and last released in February of 1998, is the keystone specification of most contemporary voice over IP products. A broad and far-reaching recommendation, H.323 is less a specific standard for any one technology than an umbrella under which the actual VoIP protocols and standards fit. In fact, it is quite possible to design an H.323-compliant voice telephone system without employing IP at all. The recommendation only specifies a need for a "packet-based network interface" on the terminal. It is somewhat ironic that H.323 was intended for public X.25-type packet networks, and later ATM networks, but now has found a home on the Internet and TCP/IP, while there is precious little H.323 running directly on X.25 or ATM networks.

As can be seen by the ITU-T title for H.323, *Packet-Based Multimedia Communications Systems*, H.323 actually defines ways of interconnecting systems that do much more than simply transmit and receive a voice audio signal. The multimedia communications systems referred to in the title are expected to support audio communications but also may optionally support video applications such as teleconferencing and "data conferencing" applications such as electronic whiteboarding, file transfer, and the like. However, despite this wealth of different uses for H.323 devices, the primary focus by the market on this recommendation has been for the audio capabilities that make VoIP possible. Thus another H.323 irony is that a full-blooded multimedia specification is being used today for one of the most basic audio applications only—voice. This has been pointed out many times: Perhaps there are much better ways to support VoIP than with H.323. This whole idea will be explored fully later in this chapter.

H.323 Architecture

Since VoIP uses only a portion of the entire H.323 architecture, it is a good idea to examine the full H.323 architecture in overview before moving on to explore the pieces of H.323 used in VoIP networks. The full H.323 architecture is shown in Figure 7-3.

The figure shows that H.323 can be used comfortably on a local-area network (LAN), or on a wide-area packet network. Any unreliable (no guaranteed quality of service), high-delay packet network can be used for H.323. At the top of the figure, a LAN is shown with four key types of H.323 devices. The users all have H.323 terminals, which are typically multimedia PCs that can take advantage of all H.323 features, including multipoint video conferences. All multipoint communications use the H.323 multipoint control unit (MCU). Of course, H.323 capabilities can extend across a WAN, as long as connections are set up among the H.323

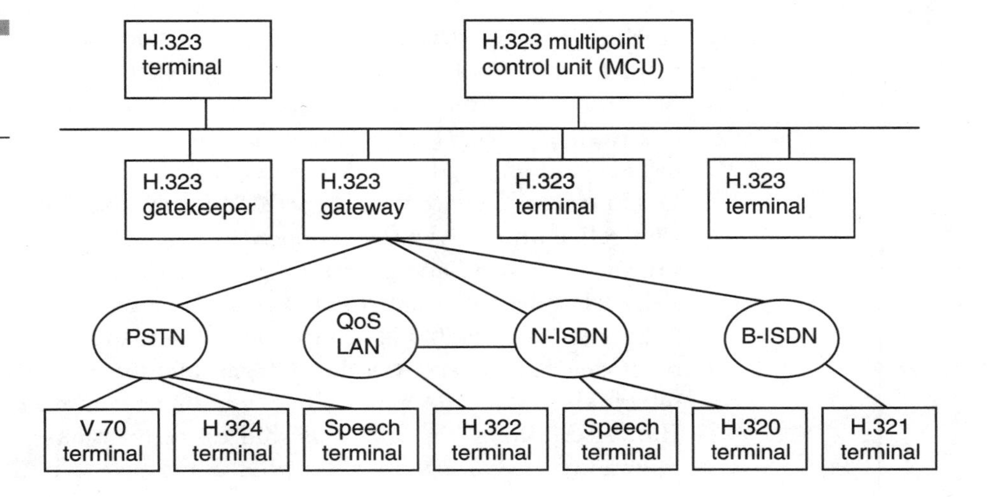

Figure 7-3 The H.323 architecture.

devices. This is the main function of the H.323 gatekeeper device, which is optional in H.323. If the gatekeeper is not present, all devices must be able to generate their own signaling messages directly. All the WAN links are handled by one or more H.323 gateways. Technically, whatever lies beyond the H.323 gateway is not covered by the H.323 recommendation itself, but the H.323 gateway can interoperate with various types of devices on several kinds of networks.

H.323 can be used with the global PSTN, a narrowband ISDN (N-ISDN) network (which runs at or below 1.5 or 2.0 Mb/s), or a broadband ISDN (B-ISDN) network employing ATM (which runs above 1.5 or 2.0 Mb/s). Even a telephone, or speech terminal, can engage in an H.323 conference, but only with audio capabilities. More functions are found in V.70 terminals, which support both digitized voice and data over a "regular" telephone circuit, and H.324 terminals (H.324 terminals can carry real-time voice, data, and video or any combination, including videotelephony, over V.34 modems running at a maximum speed of 33.6 kb/s). Usually, an H.324 terminal is just a PC with some special software.

When H.323 is used with narrowband ISDN, ISDN telephones or H.320 terminals can be used. H.320 describes a generic terminal arrangement for narrowband ISDN visual telephone systems and terminal equipment. These are usually used for videoconferencing and videophone services. If there is a LAN attached to the ISDN that guarantees certain quality of service (QoS) parameters, then an H.322 terminal can be used. An H.322 terminal is a terminal attached to a network in which the transmission path includes one or more LANs, and each of the LANs is configured to provide a guaranteed quality of service (QoS) equal to that of N-ISDN. H.322 was more or less intended to run on an extension to Ethernet called *isochronous Ethernet*, or *isoethernet*, which added a number of 64-kb/s channels to the basic Ethernet architecture. However, for a number of reasons, isoethernet never caught on in the industry.

Finally, B-ISDN networks based on asynchronous transfer mode (ATM) can employ full video/audio H.321 terminals. H.321 applies the H.320 terminal concept to B-ISDN. B-ISDN also can employ a configuration called an *H.310 terminal operating in H.321 mode*. H.310 terminals are sort of "super" audio/visual terminals that can take full advantage of B-ISDN and ATM networks in terms of services and signaling.

In summary, H.320 is used to define four different types of terminals (user devices). There are H.321 terminals for B-ISDN and ATM, H.322 terminals for QoS LANs, H.323 terminals for conferencing and more, and H.324 terminals (for 33.6-kb/s dial-up connections). When used for IP telephony, H.323 covers VoIP calls made between H.323 terminals or between an H.323 terminal and an H.323 gateway. This does not rule out the use of VoIP for other members of the H.320 family; this just focuses VoIP on the H.323 terminal type.

H.323 for IP Telephony

The entire H.323 architecture is overkill for VoIP and even full IP telephony networks. Only a subset of H.323 is needed to run audio terminals (PCs or telephones) over an IP network. Thus, when used for IP telephony, only the H.323 components of the whole H.323 protocol stack shown tinted in Figure 7-4 are actually used.

Only the audio and control portions of the full H.323 protocol stack are used for IP telephony. The audio stack handles all VoIP functions, and the only audio required for H.323 is G.711 (64 kb/s). Of course, in most VoIP configurations, it makes more sense to implement lower-speed digital voice, perhaps G.728 (16 kb/s), but especially G.723 (5.3 or 6.4 kb/s) or G.729 (8 kb/s). Because of the delays introduced by the error checking and retransmissions employed by the Transmission Control Protocol (TCP), the User Datagram Protocol (UDP) is used. H.323 specifies that a Real-Time Transport Protocol (RTP) header be added to the UDP data. To manage the quality of the voice on the networks, some implementations use the Real-Time Control Protocol (RTCP) as well as RTP itself. More details on RTP and RTCP operation will be given later in this chapter.

The control portion of H.323 also can use UDP to quickly set up connections between the H.323 terminal device and the H.323 gatekeeper. The H.323 gatekeeper is basically the remote-access server (RAS) of the H.323 network. H.225 is also used for call control with TCP to set up,

Figure 7-4 H.323 for IP telephony.

<table>
<tr><td colspan="2">Video</td><td colspan="2">Audio</td><td colspan="3">Control</td><td>Data</td></tr>
<tr><td colspan="2">H.261
H.263
(video coding)</td><td colspan="2">G.711
G.722
G.723
G.728
G.729</td><td rowspan="2">H.225
Terminal to gatekeeper signaling</td><td rowspan="2">H.225
Call signaling</td><td rowspan="2">H.245</td><td rowspan="2">T.120
(Multipoint data transfer)</td></tr>
<tr><td>RTP</td><td>RTCP</td><td>RTP</td><td>RTCP</td></tr>
<tr><td colspan="5">Unreliable transport (UDP)</td><td colspan="3">Reliable transport (TCP)</td></tr>
</table>

maintain, and release VoIP connections. H.245, which can be used with all H.320-type terminals, also can be used with TCP. All these somewhat confusing "H" recommendations will be discussed in full later in this chapter.

While some manufacturers of VoIP equipment and software depend on proprietary components and protocols, most use at least a subset of the H.323 recommendation—and many try to stay entirely H.323-compliant for purposes of multivendor interoperability. This fact is what makes H.323 an excellent model for the purposes of discussing—and implementing—the components of a VoIP system. Figure 7-5 shows the functions that are most commonly associated with VoIP systems that follow the H.323 model. For simplicity's sake, only those functions which are essential to the operation of VoIP are shown. As protocols and procedures are discussed, they will be referenced against this figure, using it as a framework against which to plot their features, functions, and capabilities.

Perhaps the first important bit of orientation that one can do with regard to H.323 is to realize what it does and does not cover. As shown in Figure 7-5, H.323 does nothing to define the mechanism by which voice signals get into the H.323-based VoIP system, nor does it provide a facility for hearing them once they arrive at their destination. Similarly, it does not define the characteristics of the network over which the packetized voice will travel or the network interface between the voice terminals and that network. Since its specification is so focused, the role of the H.323 terminal or gateway is very well defined. What H.323 and most VoIP products must do is shown within the dotted rectangle in Figure 7-5.

VoIP Terminals and Gateways Among its other responsibilities, H.323 defines the roles of the components of various types of VoIP systems. Perhaps the most important of these are terminals and gateways—collectively referred to as *end points*. These two pieces of equipment perform the same essential task: They mark the beginning and end of

Figure 7-5 Protocol functions of VoIP systems.

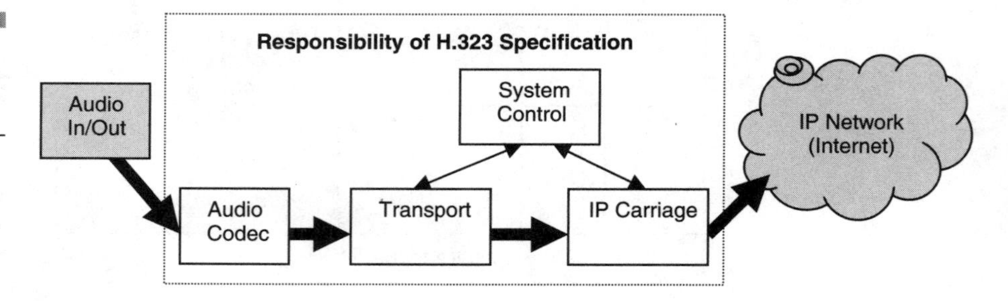

the IP portion of a voice call. When two users of multimedia PCs call each other with a software application resident on their computers, using the microphones and speakers integrated into their PCs, the entire call can be made over IP between the two H.323 terminals. In the case of a long-distance call from telephone to telephone using a VoIP interexchange carrier, the voice goes from PSTN to packet and then back to PSTN, with the transitions from PSTN to IP network taking place at H.323 terminals. The H.323 terminals form the end points of the VoIP portion of the call. In either case, the primary functions of the device remain the same—each describes a point at which voice is put into or taken out of a series of IP packets. H.323 refers to such a series of packets as an *information stream*.

The primary component of a VoIP information stream is the voice, and the H.323 specification allows it to be generated in a number of different ways. In order to comply with the specification, for example, all H.323 terminals and gateways must be capable of encoding audio using the ITU-T G.711 codec (coder-decoder), which is one of the older methods of turning analog audio into digital data. In summary, G.711 uses *pulse-code modulation* (PCM) as the technique for digitizing the voice signal, producing an information stream 64 kb/s per second in size. This is exactly the same technique that has been used by telephone companies on their digital lines for many years.

While not particularly efficient, G.711 at least forms a lowest common denominator that all compliant devices can fall back on to achieve some measure of connectivity, should they prove to be incompatible when attempting more efficient methods of encoding. In addition to G.711, H.323 defines as acceptable a number of other coding techniques, including but not limited to

- *G.728*. An ITU-T specification for a 16-kb/s encoding using a low-delay algebraic-code-excited linear prediction (LD-ACELP) algorithm.
- *G.729 and G.729a*. ITU-T 8-kb/s conjugate-structure algebraic-code-excited linear prediction (CS-ACELP) speech compression algorithms. G.729a is a reduced-complexity version of G.729, which results in faster execution. Of the two, G.729a has received the most attention from gateway manufacturers because of its lower algorithmic delay, which is on the order of 15 ms.
- *G.723.1*. This specifies two bit rates: 5.3 and 6.4 kb/s, both of which must be supported by all compliant coders, because the protocol requires the ability to switch between rates every 30 ms if necessary. In addition to the two noted bit rates, which produce 20- and 24-byte

packet sizes, respectively, G.723.1 also defines a 4-byte packet used for silence suppression and the transmission of "comfort noise" information when the sender is quiet. Compared with the G.729 codec, G.723.1 resolves DTMF ("touch tone") signals better, which is an advantage for in-band telephone signaling.

While the digitization techniques used by each of these codecs differ with regard to their mathematical models, there are three attributes of any coding algorithm that determine its suitability for use in a VoIP system: the output data rate, the amount of delay associated with its use, and the perceived quality of the audio signal resulting from it.

An examination of the specifications of the preceding codecs will reveal that there is an inverse relationship between the quality of an audio experience produced by the operation of a particular codec and the data rate it produces. For example, 64-kb/s PCM is the standard for quality against which all others are judged, producing what most people consider to be "toll quality" digitization. As different algorithms are substituted for it, most listeners perceive that their own subjective rating of the quality of the audio signal decreases, resulting in less intelligible speech.

As mentioned earlier, quality is by its nature difficult to measure, but it is helpful to have some quantitative way by which to compare differing encoding schemes. When it comes to evaluating digitized voice, engineers have settled on two criteria that provide yardsticks against which to measure quality: delay and mean opinion score (MOS). The relationship between data rate, delay, and the MOS is shown in Figure 7-6.

Figure 7-6 Relationship between data rate, delay, and MOS.

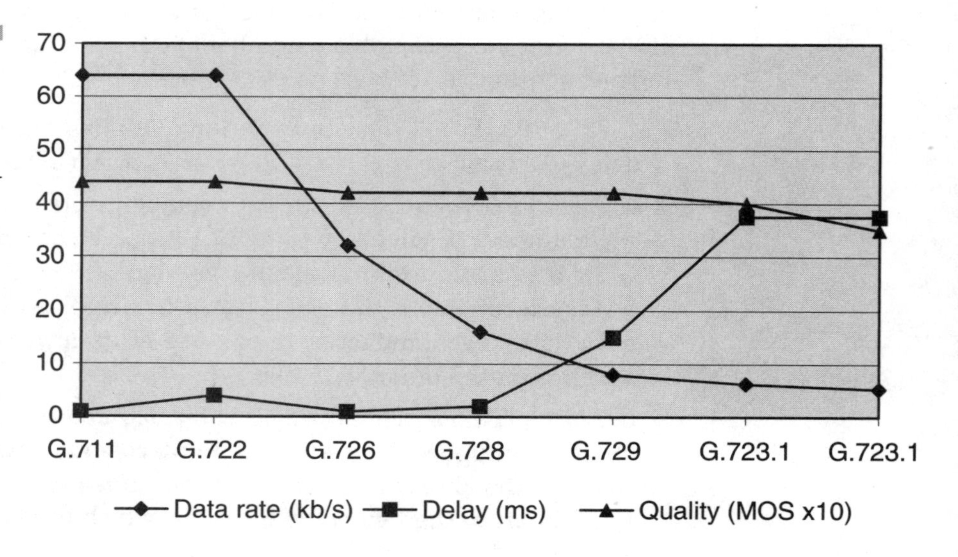

Delay is the amount of latency that the processing required by a particular digitization scheme imparts to the audio signal being processed. As the complexity of the algorithms necessary to compress an audio signal increases, so too does the amount of time it takes even a very fast microprocessor to perform the operation. Gateway and terminal manufacturers attempt to decrease this latency by using application-specific integrated circuits (ASICs) and field-programmable chips in their designs, but the inverse relationship is still there.

Delay is the one component of codec performance that everyone agrees about, and all agree that it is bad. For example, it is the unavoidable delay that comes with spanning the great distances between geosynchronous satellites and the Earth that is primarily responsible for the distressing "echo" associated with overseas telephone conversations before undersea fiberoptic cables became commonplace. While anyone who has ever made such a call will agree that between 1/2 and 1 s of delay (common in many satellite telephone systems) makes conversation difficult, most people are sensitive to a fraction of that and are bothered by much more than 2/10 s of delay.

Things would not be so bad if the delay between sender and receiver were limited to that shown in the diagram, but actually, the delay inherent in the codec is only one component of that experienced by the audio signal. Far greater amounts are introduced by the network processing nodes (routers, switches, and the like), the media access delays inherent in many networks, network congestion, and even propagation delay (the physical limit to transmission speed imposed by the speed of light). Thus the challenge that codec designers truly face is not simply to fit high-quality voice into as small a data stream as possible, but to do so quickly.

While delay itself is important, the second criterion that protocol designers and network engineers use to compare codecs is perhaps the most important one: that of *mean opinion score* (MOS). MOS is a completely subjective measurement of audio fidelity, as determined by a panel of listeners. As one compares the various codecs, perhaps the most striking figures are those for MOS, because the values for the most aggressive of the codecs are astonishingly good, providing 80 percent of the fidelity in 8 percent of the bandwidth.

Of course, equipment manufacturers are under no compulsion to use one of the codecs mentioned earlier, in fact, there are a number of gateway developers who brag that they have a proprietary voice digitization method that preserves the best qualities of voice conversations while managing to produce a lower bit rate than the best G.7xx codec. Nevertheless, virtually all manufacturers do include support for the most popular ITU-T codecs, because the consequences of being non-standard-based, and therefore incompatible with any but their own equipment, are so negative.

Other Application Layer Issues

While coding voice into an efficient digital signal is of paramount importance in transporting voice over IP, it is only one of the responsibilities of the gateway or VoIP terminal under H.323. Many control functions also must be handled by the VoIP system, including the placing and termination of calls, the negotiation of call parameters (e.g., which codec will be used during the call), the measurement of delay times, and even maintenance of the link with "keepalive" messages during silence. Within the H.323 framework, these issues are addressed by the H.225 and H.245 control subsystems.

Gatekeepers

These control functions may be communicated between the H.323 terminals or gateways directly, or they may be "passed through" another device whose sole responsibility is the administration of call control services in a VoIP system, called a *gatekeeper*. An H.323 gatekeeper is not required for H.323 compatibility, but in systems of any significant size, network designers and vendors agree that it makes sense to centralize the administration of these functions in a single unit. Gatekeeper functions include

- *Address translation.* A gatekeeper may allow the local use of proprietary addressing schemes (called *alias* addresses in H.323), such as mnemonics, nicknames, or e-mail addresses, and "translate" these into the IP addresses needed for establishment of IP communications.
- *Admissions control.* Since they play a fundamental role in the establishment of VoIP channels, gatekeepers may control the setup of VoIP calls between the terminals or gateways they are responsible for and the rest of the world. Access may be granted or denied based on authentication, source or destination address, time of day, or any other variable known to the gatekeeper. In this way, the gatekeeper performs a security role in the VoIP network.

- *Bandwidth management.* H.323 gatekeepers have the ability to request H.323 terminals or gateways to change the communications parameters of calls they are taking part in so as to manage the use of bandwidth that may be shared by multiple stations at any one time. In this way, the gatekeeper can act as a clearinghouse for the efficient and coordinated assignment of bandwidth resources.
- *Zone management.* The gatekeeper also can coordinate the interplay of the preceding functions across the network of devices for which it is responsible. For example, zone management may require that no more than 25 calls may be permitted across a particular low-bandwidth link to avoid a degradation of quality. Management of this type also may allow functions such as automatic call distribution (ACD) and other call center-like services.
- *Call signaling.* A gatekeeper may act as a call signaling "proxy" for the terminals (or less likely, gateways) it represents, thus relieving them of the responsibility of call control protocol support. Or it may simply serve as in initial "point of contact" for callers, which, upon admitting the proposed call, prompts the two gateways or terminals to signal each other directly.

A gatekeeper and all the end points under its control form an H.323 *zone*, a logical group of devices under a single administrative authority. This is shown in Figure 7-7.

In addition to the preceding functions, the H.323 gatekeeper may play other roles in future VoIP systems, which are under study and development by manufacturers. In any case, it is important to remember that since the gatekeeper is an optional component of any H.323 system, the signaling procedures defined under H.225 and H.245 have been designed

Figure 7-7
A simple H.323 zone design.

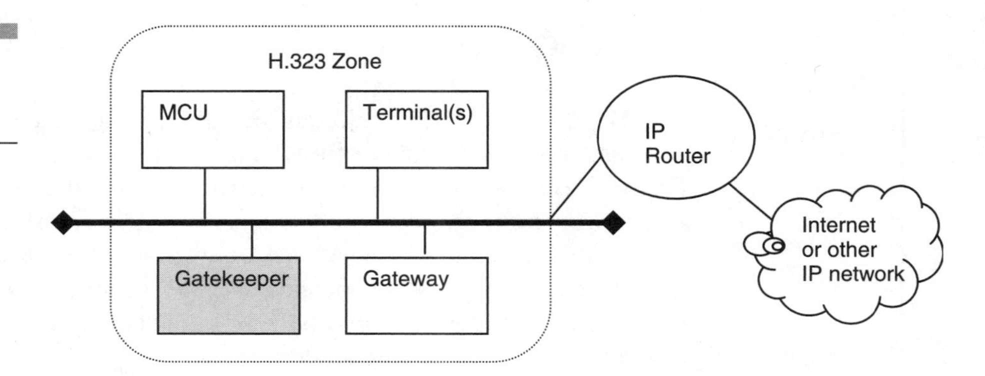

to function both with and without it. Systems without gatekeepers simply allow end points to signal each other directly, putting the responsibility for the management and control functions that the gatekeeper provides onto the end points themselves.

In order for a gatekeeper to "take responsibility" for a particular terminal or gateway, it must register that entity in its own database and note its addresses (both IP and alias) and other information (such as channel bandwidth limits) necessary for the gatekeeper to act on behalf of the end point. This registration can be accomplished manually through the use of gatekeeper configuration files or via the H.323 gatekeeper discovery and registration procedure.

In gatekeeper discovery, end points broadcast to the multicast IP address 224.0.1.41 a *gatekeeper request* message, essentially asking any gatekeeper that can act on its behalf to identify itself as such. This request will be answered by all gatekeepers that receive it with either an accept or reject message, which corresponds to an offer to register the end point or a denial of registration services, respectively. The end point may then choose, from among the gatekeepers that replied affirmatively, which gatekeeper it wishes to register with. The end point then addresses the gatekeeper of its choice with a *registration request*, which forms the formal beginning of the registration process itself. On receiving a confirmation from the gatekeeper, the end point and gatekeeper may then proceed to exchange information about their relationship. For instance, the gatekeeper may dynamically assign an alias address to the end point, or it may offer registration services only for a certain amount of time, after which the end point must reregister.

Gatekeepers simplify the deployment and use of VoIP systems by centralizing and coordinating the administration of call signaling, thereby serving as a "control node" for all the devices in their zone.

H.225 and H.245

The H.323 specification calls on another ITU-T specification, H.225, to perform the signaling for call control. H.225 is entitled, *Call Signaling Protocols and Media Stream Packetization for Packet Based Multimedia Communications Systems*, and like H.323 itself, it defines a much larger set of capabilities than those used in systems concerned only with voice traffic. H.225 itself uses messages defined in H.245 to establish and terminate individual logical channels for audio (or, in the case of a true multimedia terminal, video or other medium).

The initial use of H.225 procedures in the placement of a VoIP call is in the notification by the originating device to the receiving device of its desire to establish a call, which is done on the H.225 registration, admission, and signaling (RAS) channel. To call the RAS facility a "channel" is a bit of a misnomer, since it is defined as an "unreliable" resource and therefore uses the User Datagram Protocol (UDP) as a transport, with no explicit acknowledgment. But the RAS channel does define a common logical "meeting place" that all devices can monitor for the first sign of an incoming call.

An RAS request to establish a call may be answered by the gatekeeper responsible for the called device or by the device itself in the absence of a gatekeeper. In either case, the response will include the IP address and TCP port number of the called device, allowing the caller to attempt to establish a reliable TCP connection between it and the recipient, over which the call signaling itself may occur. During a call, the signaling proceeds through five phases:

Phase A: Call setup. During this phase, the calling end point notifies the called end point of its desire to open an audio (and conceivably video or other media) channel. The setup phase also defines a message for the purpose of informing the caller that the called end point has been alerted of an incoming call. The exact call setup signaling sequence varies depending on the network configuration, in particular on the presence and location of a gatekeeper or gatekeepers in the signaling relationship. In all cases, however, the calling end point initiates the connection with a setup message, to which the called end point responds with a connect message containing the H.245 control channel IP address for the purposes of setting up the media channel with H.245 messages.

Phase B: Initial communication and capability exchange. On successfully completing the call setup phase, both end points proceed to phase B, which is concerned with establishment of the H.245 control channel via the exchange of information regarding the capabilities of each of the end points in the call. In this case, *capabilities* refers to the types of media channels supported. For example, all H.323 gateways must support audio channels, but many will not support the video or whiteboarding media channels defined under H.323.

Phase C: Establishment of audiovisual communication. At this point, the end points are free to establish the actual logical channels that will carry the information streams of the call. For audio information, each end of the conversation opens its own unidirectional channel, because there is no requirement that the same codec or bit rate be used in both directions.

Phase D: Call services. Call services, which are alterations to the parameters of the call as agreed on during the prior three phases, can include adjustments to the bandwidth required by the call, the adding or dropping of participants during a multipoint (conference) call, or the exchange of status or "keepalive" messages between gateways and/or terminals.

Phase E: Call termination. The device wishing to terminate an H.323 call need simply discontinue transmission of information streams (first video, then data, then audio, if elements other than audio are present) and then follow a call clearing message exchange similar to the call setup exchange used to begin the call. As in call setup, call termination procedures vary depending most notably on the role of the gatekeeper in the call. When a gatekeeper is present, it must be notified of the termination of the call so that it may adjust its bandwidth availability tables.

H.323 messages used for signaling are a subset of those used in Integrated Services Digital Network (ISDN) telephony and therefore draw on the Q.931 signaling protocol defined for ISDN. H.225 essentially adapts Q.931 to the "reliable packet mode" environment (e.g., TCP) and does little else.

Multipoint Controllers and Processors

In the same way that traditional telephony systems, both private and public, have developed to support "value added" services such as call forwarding and answering, so too does the H.323 specification allow for the provision of these nontraditional telephony applications, which are in fact much more easily supported under a packet-switched network than a circuit-switched one. Perhaps the most requested of these is the ability to perform three-way calling, or "conference calling," which is a staple of most business environments today. H.323 handles this requirement through the definition and use of multipoint controllers (MCs) and multipoint processors (MPs).

A multipoint controller serves as a central point through which end points may exchange information about the capabilities each has and which may "broker" the possible audio and other modes that are common to all participants in the conference. An MC function may be integrated into a gatekeeper, gateway, or even a terminal. The audio and other information streams making up a conference may be centralized or decentralized depending on the end points' or MCs' capability to support common modes.

In a centralized conference environment, all end points taking part in a conference establish point-to-point communications with the end point containing the MC, called the *multipoint control unit* (MCU). The MC is then able to manage the conference directly, since it forms the center of communications as well as control signaling. Decentralized conferences require the terminals or gateways to directly communicate with each other via multicast, using the MC only for conference control. A decentralized conference puts less of a load on the MCU but requires that the end points (and their networks) support multicast communications and handle the mixing of the various incoming audio streams themselves. In addition, decentralized environments require that all end points support common modes.

While an MC can provide a definition for a particular conference's selected communication mode (SCM), it does nothing to reconcile the differences between end points that may not allow them to communicate. For example, some devices wishing to join the conference may not support the SCM. For this function, H.323 defines a *multipoint processor* (MP) that may form a part of an MCU for the purposes of mixing, switching, or otherwise reconciling the various audio streams that form the conference under the control of the MC. The MP can take audio streams from a number of different sources, each encoded using a different codec, and convert them to analog audio for the purposes of mixing them into a composite audio signal. This signal can then be recoded according to the capabilities of each of the end stations (again, perhaps in a number of different formats) and distributed to each of them. It must do this in such a way as to omit each station's own incoming audio stream from the stream that is transmitted back to it.

H.323 is an effective and popular way to implement VoIP and IP telephony networks. However, H.323 is not the only method that can be used to accomplish this marriage of IP networks and voice networks.

Session Initialization Protocol

The ITU-T is not the only standards organization to weigh in with a proposal for establishing IP telephony connections and packetizing audio for them. The Internet Engineering Task Force (IETF), which more than any other body in the loose world of Internet standards manages (or at least legitimizes) the development of the Internet Protocol suite, has its own proposal for VoIP systems. This is called the *Session Initialization Protocol* (SIP).

Proponents of SIP claim that H.323, arising as it does from the worlds of ATM and ISDN signaling, is ill-suited for controlling VoIP systems in general and Internet telephony in particular. This claim rests on the assertion that H.323 is inherently quite complex, supporting many functions that are not needed for most IP telephony and therefore overhead-intensive and inefficient. For example, H.323 defines three separate methods for interworking H.225 and H.245: on separate connections, tunneling H.245 through the H.225 connection, and the expedited "fast connect" method of integrating the two protocols. Although most implementations probably would do well to support only fast connect, H.323 compatibility concerns require the support of all three methods.

At the same time, those in favor of SIP also claim that H.323 lacks the extensibility required of the signaling protocol for a rapidly maturing technology like VoIP, which will no doubt develop to support unforeseen services and features. Experience with protocols such as those used in Internet mail (SMTP) have provided much of the philosophy that has informed the development of SIP, with much attention being given to adaptability of the signaling set to future uses. Coming as it does from the IETF, and therefore designed with the Internet in mind, SIP ostensibly avoids both the complexity and extensibility pitfalls.

Beneath the "official" arguments between H.323 and SIP proponents is a much more basic issue. Some people cannot understand why ITU-T specifications such as H.323 have been embraced by some in the Internet community. For years, the ITU-T turned a deaf ear to IETF proposals to make all or part of TCP/IP an international standard. Thus who knows IP better than the IETF? Certainly not the ITU-T. Some observers have seen SIP as just an "un-H.323" approach to VoIP and IP telephony, a needless complication in a world where standards are a requirement for widespread deployment and customer acceptance.

SIP borrows much of its design philosophy and architecture from the Hypertext Transfer Protocol (HTTP), the information-exchange protocol of the World Wide Web. It is defined as a client-server protocol, in which requests are issued by the (calling) client and responded to by the (called) server, which may in itself be a client for other aspects of the same call. SIP uses many of the message types and header fields of HTTP, identifying the contents of an information stream with entity headers (content-type descriptions) and allowing for authentication using methods similar to those used on the Web.

SIP defines INVITE and ACK messages that are similar to the setup and connect messages defined in H.225 in that they both define the process of opening a reliable channel over which call control messages may be passed. Unlike H.225, however, this channel does not depend on TCP for reliability but rather handles its own acknowledgment and handshaking. Proponents of SIP claim that this integration of reliability into the application layer allows a tight coupling of timing values (for retransmission and acknowledgment) to the application, which may be optimized for VoIP rather than being subject to the "general purpose" values of TCP.

Finally, SIP relies on the Session Description Protocol (SDP), another IETF standard, to carry out a negotiation similar to that held under the capabilities exchange mechanism of H.245. SDP is used, for example, to exchange codec identification for use during a call using a simple textual description. SDP also is used to carry information elements for the Real-Time Signaling Protocol (RTSP) to negotiate multipoint conference parameters and to define a common format for information of many types when carried in SIP.

The debate about which set of standards, H.323/H.225 or SIP, is better suited to Internet telephony is academic. There are advantages and disadvantages to both approaches, and while H.323 has a considerable advantage in terms of its market presence (having appeared well before SIP), SIP benefits from the momentum of the IETF, which is among the fastest-moving of standards organizations. For the purposes of studying the possible protocols and solutions to the problems of VoIP, it is probably most useful to consider both the H.323 and SIP models together, as one "meta-model," with little regard for the almost religious attitudes of those who attempt solutions with only either in mind. The most useful of the solutions, whether appearing in real equipment or just proposed, do just that.

Addressing and Naming

Of course, session initialization or call setup is only useful if the caller is able to accurately specify the party he or she wishes to call. Then that specific party's identification can be resolved by the system into a valid IP address with which a VoIP session may be established. Currently, VoIP systems depend largely on manual configuration of the end points with other end point addresses. For instance, a gateway that forms part of a system of 10 gateways (or 10 gatekeeper zones, for that matter) may have addressing configured manually by associating IP addresses with area codes. On receiving a direction to place a VoIP call to a given number, the gateway may compare the specified area code with its list of area codes and initiate a VoIP session with the associated gateway. Of course, such a scheme works best for systems of only a few nodes and does not scale at all to support fully distributed VoIP systems in which a large number of end points call each other directly.

Furthermore, the addressing and naming systems necessary for successful VoIP networking differ depending on the architecture of the particular system, since VoIP networks may be built according to so many different models. For example, in a fully distributed environment in which all users occupy terminal end points and therefore can contact each other directly via IP, a single IP address equates to a single user. In such a system, a user could be identified merely by the DNS hostname of the VoIP terminal at which he or she is sitting. DNS would then perform the address resolution required to obtain the IP address of that end point. Unfortunately, the fully distributed model is not the one that the majority of current VoIP implementations follow, and many individual workstations capable of acting as VoIP terminals do not even possess DNS hostnames. Current users of PC-to-PC VoIP packages must either rely on proprietary directory services or share ahead of time the addresses they need to establish communications. In any case, such a system would require that users keep each other notified of their whereabouts, in much the same way that telephone callers usually tell each other where they may be contacted at a particular time, in accordance with their schedule. However, the need for such a manual process is hardly in keeping with the supposed benefits of VoIP. It would be much better if the called user's physical location did not have to be known to the caller.

In the gateway-to-gateway model, a single IP address resolves to a number of users, each of whom is identified by their phone number or extension. In that case, finding the gateway at which the called party resides is the first problem. The IETF iptel (IP telephony) working group has proposed Gateway Location Protocol (GLP) toward such an end. While not fully defined, GLP will work by exchanging gateway identity information in much the same way as some routing protocols exchange network layer reachability information, with *location servers* (LSs) being defined as the "gateway representative" for a particular group of gateways in an administrative domain. It is worth noting that this capability is mentioned under "gatekeeper characteristics" in the H.323 model, although its inclusion is still under study.

The Media Gateway Control Protocol (MGCP), an initiative of the IETF Media Gateway Control Working Group, defines a method by which a calling gateway, on receiving an E.164 address (PSTN phone number) from the calling PSTN node, may signal the called party's gatekeeper. Together, GLP and MGCP will allow the gatekeeper to (1) identify to the caller the gateway or terminal that has responsibility for that number, (2) provide to the gatekeeper the number to be called, and (3) establish a session (or allow a session to be established) with that end point for the purposes of completing the call. The exact operation of MGCP and GLP has yet to be specified, but all proposals contain (in addition to the gateway location and called number information) bandwidth restriction information and codec negotiation.

In Internet VoIP networking there must be a provision for naming the called party or user that is compatible with the DNS. A number of possibilities have been suggested by the SIP and H.323 communities, with most tending toward an e-mail-like representation, such as *user@domain.com*. Using this already established format would allow existing database code to be reused, as well as permit the inclusion of URLs such as *h323://user@domain.com* or *sip://user@domain.com* in HTML Web pages, thereby making the initiation of a call as simple as clicking on a hypertext link.

Directory Services

A naming service such as DNS is a relatively static entity that contains no provisions for being dynamically updated with changes between IP addresses and domain names. As a result, DNS—in its current implementation—is useful for finding gateways that have been associated with hostnames but much less so for user VoIP terminals. Terminals introduce the difficulty of IP addresses that may change on a regular basis, either because they change domains altogether (when travelers leave their corporate LAN, for example, and connect to the Internet via a different ISP when in a hotel) or because of the trend toward dynamic address assignment such as that used in the Point-to-Point Protocol (PPP) or Dynamic Host Configuration Protocol (DHCP).

Therefore, designers of IP telephony systems where the end points' locations change unpredictably must draw on a true directory service rather than a simple naming service such as that provided by DNS. A directory service differs from a naming service in that it can access information in much the same way that a relational database does, retrieving records based on a number of search parameters, each of which may be a description of some attributes of the entry. For example, a directory service user could ask of the directory, "What is the terminal IP address of *user@domain.com*, right now?" Of course, since the value of *right now* is continually changing, the directory will answer according to the values in the record, with one address during the work day and another in the evening, in whatever way the directory entry has been written. In the same way, a user could ask the directory for the "work" or "home" gateway or terminal addresses of a particular user. Currently, the leading contender for use as an Internet VoIP directory service is one designed around the Lightweight Directory Access Protocol (LDAP), which is itself a simple (thus the *lightweight*) client-server means for accessing information contained in an X.500 directory database.

Network Requirements

While H.323 and SIP both provide common (and as yet incompatible) languages by which the end points in a VoIP conversation may signal each other, interoperate with gatekeepers, and code and decode audio signals, they are comparatively vague when it comes to the network facilities over which telephony services should run. For example, H.323 states that Novell's IPX protocol is as valid a network layer as IP for voice, and even the IETF mentions this as a possibility. These network layer protocols are intended to run over frame- and cell-based networks such as Ethernet, Token Ring, ATM (particularly the AAL 5, unreliable service), FDDI, and the like.

However, while the network itself (meaning the data link and network layers) is expected to be unreliable and "best effort," the application that is responsible for transmitting and receiving the voice data must be able to glean at least some types of information from the network service that is delivering it. That is, the application would like to "keep its eye on" the network in order to make adjustments to its own operation, or its interaction with the network, to achieve the best possible service. To this end, both H.323 and SIP specify the use of the Real-Time Transport Protocol (RTP) and the Real-Time Control Protocol (RTCP) defined in the IETF RFCs 1889 and 1890.

RTP and RTCP

RTP was designed to provide a number of services important to the delivery of information with real-time requirements, particularly voice and video traffic. More than any other single protocol, RTP attempts to bring some regularity and predictability to the applications that use the Internet for the transport of time-sensitive traffic.

In particular, RTP can provide

- *Timestamping.* The variable amount of delay between a sender and receiver in a packet-based network such as the Internet invariably results in packets arriving at the destination in "clumps." That is, a number of packets arrive closely together in time, followed by an absence of packets for a period, followed by another number of packets, and so on. The effect created is called *jitter*, and it is a far from optimal environment for the provision of a service to replace a dedicated circuit, in which data arrive with some fixed amount of delay. RTP addresses this problem by including a 32-bit field that is used to identify the instant in time that the beginning of the data field was sampled, using a random identifier for the first packet. For subsequent packets, the timestamp value is increased in a linear manner according to the number of "clock ticks" that have passed since the last packet.

  While this does nothing to guarantee the timely arrival of data, when such a scheme is used in combination with a receiver's *jitter buffer*, the received data can at least be fed to the receiving application with the same interpacket delay characteristics as it had when leaving the transmitter. The presence of the buffer will allow the receiver, with the timestamp information delivered by RTP, to impart additional delay to packets that arrive quickly, thereby "smoothing out" the delivery of data.

- *Sequence numbering.* Most packet networks allow packets belonging to the same data stream to cross the network following different paths, spreading the load of the transmission across many different links. Since these links may have somewhat different delay characteristics, it is quite possible that packets will arrive at the receiver in an order different from that in which they were transmitted. The RTP sequence number assigns a random number to the first packet of a stream and then increments the number by one for each subsequent packet sent. This allows the receiver to (1) order the packets in the sequence intended and (2) detect loss of one or a number of packets. While RTP has no facility for requesting retransmission of lost packets (indeed, real-time or near-real-time delivery of data does not usually allow this), the receiving application can at least be notified of a gap in the audio stream.

- *Delivery monitoring.* RTP can provide valuable feedback to the sender of an audio stream via its companion protocol RTCP. RTCP defines both sender report (SR) and receiver report (RR) packets that contain information such as interarrival jitter, number of packets lost, total numbers of packets and octets transmitted, and other data useful in the diagnosis, monitoring, and correction of some types of network error conditions. For example, adaptive encoders might switch to smaller, lower-quality packets when the amount of end-to-end delay or jitter increases to a point where it threatens the fidelity of the audio more than the lower bit rate would.

RTP also leaves open the possibility that RTP header information could be acted on by network service providers, which, while not involved as an end point, may adjust priorities, queuing, or even routing based on what it considered acceptable limits—for example, to stay within the tolerances imposed by a service level agreement with a customer.

- *Payload identification.* It is essential not only that audio data delivered via RTP be interpreted with the correct timing characteristics but also that it be decoded according to the same rules by which it was encoded. RTP therefore identifies the information it carries by assigning a payload type identifier to each packet. Payload types (essentially, specific codecs that may be used for audio and video) are identified in RFC 1890, but additional types may be defined by petition to the IETF or dynamically by software.

The ITU-T approach for RTP implementations is for RTP to be very tightly integrated with the application layer it supports and not stand on its own as a separate protocol layer. RTP therefore becomes more of a "protocol framework" for the support of whatever functions the application requires. Those not supported in the protocol header may be added by application developers, although how this will affect multivendor interoperability is not something the ITU-T addresses.

Neither RTP nor RTCP include multiplexing information and therefore have no way of distinguishing different streams of data from one another. Since a participant in a multiuser conference who may simultaneously be receiving multiple streams of audio data must be able to tell them apart, RTP packets are encapsulated in the User Datagram Protocol (UDP), which is itself encapsulated in IP. UDP forms the unreliable transport layer of the IP protocol suite and provides identification of port number but no error correction or sequencing.

The whole struggle between H.323 and SIP groups is just heating up. H.323 has an early lead, but SIP has many powerful IP groups behind it. However, when all is said and done, the limiting factor in widespread

VoIP deployment may not be H.323 or SIP but just the quality of service available over IP networks.

Quality of Service

A full discussion of QoS, networks, and IP is contained in Chapter 12. For now, the issue at hand is the specifics regarding QoS present in the current versions of IP. The issue of QoS in IP networks was addressed in the current version of IP (version 4) by the type of service (ToS) octet. This second octet of the packet header is divided into two sections. The first is precedence, which serves as a 3-bit indication of the priority that routers should give the packet, from routine (low) to network control (high). The second is service parameters, 4 bits that indicate a request from the sender that the packet be routed with concern for any combination of delay, throughput, reliability, or monetary cost.

While most current router implementations honor the precedence subfield when possible, applications' use of the field in the manner in which it was intended is spotty. Furthermore, the service parameter bits have never been implemented on the Internet because of a lack of options for treating traffic differently. After all, the ability to honor the "low delay" directive presupposes a knowledge of a "low delay" path in addition to the "normal" path. Routing protocols that support this ability have been developed, and the IETF's Open Shortest Path First (OSPF) and Cisco System's Interior Gateway Routing Protocol (IGRP) both support the differentiation of network paths based on link- and path-quality information. Neither protocol, however, has been implemented widely in ways that take advantage of this capability, a fact that has made the IETF, for example, drop the requirement for service parameter routing from the specification.

On the Internet, even flawless implementations of OSPF and IGRP would lose their capacity to choose effective path qualities. This is so because larger ISPs are organized into routing "clouds" known as *autonomous systems* (ASs). Once the packets the protocols are handling leave the AS within which the routing protocol under the control of a single ISP is operating, unless the neighboring AS should happen to implement them in exactly the same way. Thus, while it is possible to differentiate between separate qualities of service in a single AS, preserving this capability across AS boundaries requires a degree of cooperation that would be difficult to achieve in a much less tumultuous place than the Internet.

At least part of the reason for the spotty support of IP precedence and the lack of a solid protocol for IP service parameter routing is that there remains a fundamental disagreement among network service provider engineers about whether any such thing is necessary. After all, IP and its ToS field were developed at a time when common WAN speeds were 9.6 kb/s or less, while in today's networks, a 1.5-Mb/s DS-1 line is considered slow. With the increasing network speeds that result from advances in optical-fiber technology, digital subscriber lines, wavelength division multiplexing, and the like, some engineers think that distinguishing between different classes of traffic adds needless complexity to the transport of data, which should be simple in order to be as fast as possible. Some pundits refer to this dispute as the "stupid network versus the smart network" debate. To some, the solution to all networking problems is just to add bandwidth until the problem goes away.

Those who defend the network's need to keep some of its intelligence to supply IP network QoS can point to the unbelievably rapid growth of the Internet in the last 5 years, a growth rate that continues to increase. In such an environment, these people contend that demand will always outstrip capacity, and therefore, there will always need to be a way to send some traffic "first class." For IP networks, two approaches to this problem currently receive the most attention: the *integrated services* model and the *differentiated services* model. Mercifully, for a world overburdened with three-letter acronyms (TLAs), they are known by their abbreviations *intserv* and *diffserv*.

Intserv and RSVP

The integrated services model seeks to provide a way for applications to choose among several levels of "delivery service" for their data streams. Intserv is best described as an architecture consisting of two primary components:

- *Routers and hosts that understand, and can differentiate between, different levels of service.* These devices must be able to take direction as to how packets fitting a certain description should be treated (particular routes, priority queuing, etc.). This function is sometimes referred to as the *QoS control function* of intserv. Examples of QoS control functions include controlled load and guaranteed, which are outlined in RFCs 2211 and 2212, respectively.

  Controlled load service uses admission control to identify and prioritize packets fitting a certain description, making that traffic experience conditions similar to those on an unloaded network. Under controlled load, however, the network service still seems "best effort" with regard to delay and reliability; i.e., it remains essentially unpredictable. Under guaranteed QoS control, the network elements agree to provide a service that introduces mathematically predictable amounts of delay into the data stream.

- *A method by which users or applications may inform the nodes of their requirements.* This is the function assigned to the Resource Reservation Protocol (RSVP) in RFC 2205 and can be called generically the *setup control function* of intserv.

The intserv model depends on the interplay of these two functions and their application to *flows*, which are streams of packets that are related to each other in some way and therefore classifiable as a unit. For example, a host on the Internet may establish a session with another host for the purposes of conducting a VoIP call, and all the packets with the characteristics of that particular call could be seen as a single flow. This is so because the packets share the same destination and source IP addresses, the same transport layer ports, and perhaps other attributes. If, at the same time, the two hosts also conducted a file transfer of non-real-time traffic, the packets belonging to the file transfer would not be part of the same flow (having different port numbers, at least). Therefore, the file transfer could be distinguished from the VoIP traffic by network nodes.

Flows are essential to the operation of RSVP and intserv in general because the intserv framework uses flow descriptions as the means of identifying traffic to routers (the setup control function) for treatment according to specific QoS control parameters (the QoS control function). Taken together, the intserv process follows these basic steps, illustrated in Figure 7-8.

1. The sending application tells its RSVP process about the traffic it expects to generate, providing it with a description of the flow characteristics that the traffic will exhibit. The application also provides to RSVP a description of the requirements that the flow will need.
2. An RSVP setup message containing these two elements is forwarded along the path to the intended receiver and is modified with information about the capabilities of each of the nodes as it is relayed hop by hop, creating a path description. For example, part of the flow's requirements may be a maximum transmissible

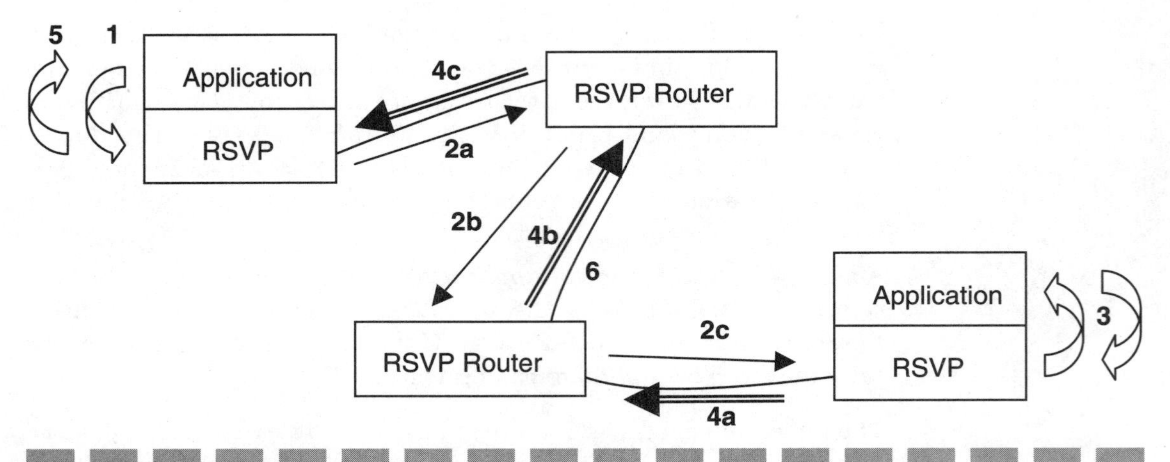

Figure 7-8 RSVP reservation process.

unit (MTU) size of 1 KB, and each router along the path would state its ability (and willingness) to grant that MTU to the flow.

3. When the setup message arrives at the receiver, RSVP gives the path description to the application, which can use it in selecting reasonable path attributes to request. Once the receiving application has indicated its needs, it hands that request back to RSVP.
4. RSVP transmits the request upstream along the path to the sender, requesting those attributes from the routers.
5. When the sender's RSVP process receives the message, it informs the sending application of the existence of a path with the specified characteristics, open for use.
6. Routers honor the "reservation" by matching the flow characteristics mentioned in the sender's setup message with the attributes indicated by the receiver's request.

While the intserv process is somewhat complex, it certainly is worth the effort of reserving a certain set of network characteristics for a particular flow—at least from the point of view of the individual application. It is not difficult to imagine all the devices on the Internet being capable of asking the network for exactly the level of service required for whatever they were doing at a particular moment or not making any request at all when only best-effort service was required.

The intserv model has a limitation, however, that is more related to the rapidly expanding scale of the Internet than to the design of the protocol. While it may not be difficult for a server to transmit the hundreds or thousands of RSVP requests it must make on behalf of the clients simultaneously connecting to it, the routers at the core of the network potentially would have to maintain information about the state of all the flows that were passing simultaneously through them, involving not one server and its clients but rather many servers and all their attendant clients. This could involve, for certain routers in high-throughput locations, keeping track of the status of millions of conversations at once, in addition to routing the packets that make up the conversations.

RSVP therefore has the disadvantage of all so-called stateful protocols, in that the effects of traffic persist in the nodes even when the traffic itself does not, requiring the node to devote processing power and memory to the anticipation of traffic and to the maintenance of that state information. Backbone routers have enough work to do simply routing packets and maintaining network reachability information (running routing protocols) without incurring the additional overhead of processing state information, which would only increase intrarouter delay. As Internet networks increase in size and speed, the number of traffic states that would require maintenance increases proportionally.

Another limitation to the wide-scale deployment of RSVP is the interprovider coordination that would be required to allow RSVP flows to be registered, tracked, and acted on along a multiprovider path. Asking all applicable providers to support the same QoS control function, compatible service levels for given flows, and sufficient capacity for a given number of flows would strain the cooperative abilities of Internet service providers.

As a result of these two limitations, intserv and RSVP are seen by many as useful at the edges of paths, for the individual portions of VoIP calls that cross a given provider's network and for single-provider networks dedicated to VoIP and other real-time services—as long as they are under a single administrative authority. However, intserv is clearly not the answer for large-scale support of VoIP or other connection-oriented, real-time services on the Internet as a whole.

Diffserv and the ToS Octet

Having examined solutions to Internet QoS like intserv and RSVP, attention in the ISP and router communities has returned to the little-used second octet of the IP header, the ToS field. In December of 1997, the

Integrated Services Working Group of the IETF "spun off" a number of members, who formed the Diffserv Working Group, with the express intent of creating a scalable method of "service differentiation." In contrast to the relatively complex intserv/RSVP solution, which had resulted in the sophisticated identification and classification scheme detailed earlier, the diffserv approach uses a "stateless" packet marking approach—allowing routers to decide per-packet forwarding behavior based on the contents of a single field.

Recall that as originally specified, the IP ToS field contained two subfields: the 3 precedence bits to provide eight levels of priority and 3 or 4 bits meant to solicit certain types of path attributes such as low delay, high throughput, or high reliability. The exact number of bits in the subfield depends on the time of implementation. The diffserv initiative recast this octet to contain a single 6-bit field, leaving the other 2 bits unused. The 6 bits are used to select a *differentiated services codepoint* (DSCP, or simply CP) that is used to indicate to routers and other nodes the per-hop behavior expected at each node. Such use of the field is completely incompatible with the "traditional" use of the field as defined in RFC 791—a demand that is not seen by most as problematic, given the spotty current use of the field.

The diffserv model is shown in Figure 7-9. The diffserv model defines a *DS domain*, which consists of a number of interconnected routers that understand the same definitions for a set of per-hop behaviors (PHBs). It is assumed by the model that this domain will be a single service provider's network or some other network under the control of a single administrative authority over which common PHBs can be defined. Allowing access into and out of the DS domain are routers known as *DS boundary nodes*, which "classify" traffic coming into the domain based on

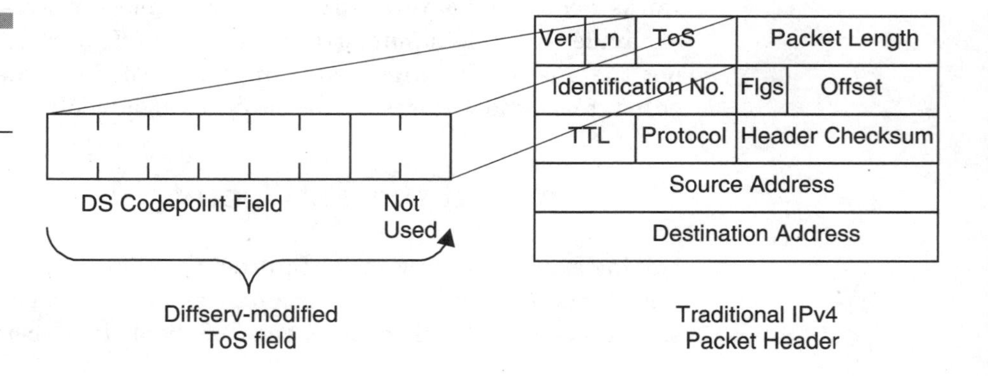

Figure 7-9 Diffserv and the ToS field.

a traffic profile defined by the network administrators. Traffic fitting a particular profile may have a particular CP value that indicates that a certain PHB is to be followed by the routers in that domain. This value may or may not have the same (or any) meaning to routers outside that particular DS domain.

For example, traffic entering the DS domain network from a customer paying for low-delay service could be classified by the ingress boundary nodes with a codepoint meaning "low delay" to all that domain's routers. The boundary nodes can classify traffic based on a number of IP header attributes, including IP address, protocol type, and the like—even on the prior value of the DS field, which may be set by the host originating the packet or the host's network's routers. In this way, any number of observable characteristics of traffic can be reduced to a single codepoint indicating the way that traffic should be treated by the network.

The differentiated services architecture depends on a common understanding of PHB rules and how they relate to a specific QoS that is expected from a network. To this end, diffserv expects that traffic crossing DS domain boundaries will have to be covered under contracts such as service level agreements (SLAs) and traffic conditioning agreements (TCAs). While some PHBs will be standardized by the Internet assigned numbers authority (IANA) for consistency's sake (such as the "default" PHB, similar to current "routine" Internet traffic today), providers will be free to define their own codepoints and PHBs to offer levels of service that interest them and their customers.

QoS under IPv6

The next version of the Internet Protocol, version 6 (IPv6), addresses the problem of QoS both explicitly and implicitly. First, the IPv6 header defines a 4-bit priority field that can indicate 16 different priority levels in much the same way that the IP ToS octet did. Half these values are intended for use with non-real-time traffic such as file transfers and "uncharacterized traffic," and the other half define 8 levels of priority reserved for real-time traffic (presumably this could be used to differentiate between simultaneous audio and video feeds). While implementation guidelines for the use of this field by IPv6 routers are not firm, the field is so similar in nature to the ToS field of IP version 4 that its success as a network prioritization mechanism depends entirely on the willingness of router and application vendors to support it. Given the lack of implementation of ToS features in IPv4 platforms, one must wonder if the priority field of IPv6 will not suffer the same fate. One reason for optimism is the new emphasis on QoS that the data communications industry is currently placing on IP, which will surely concentrate efforts to provide priority.

Priority labeling is not the only way that IPv6 addresses the issues of QoS in the IP networks of the future. The basic IPv6 header also includes a 24-bit flow label that can be used by originating applications to mark a stream of packets as belonging to a particular IP flow so that routers may classify them without having to look up addresses, application ports, and other information for that purpose. Having flow ID information available within the IP header itself allows routers to reduce the amount of effort, and therefore time, it takes to assign a particular QoS to a packet. Note that the flow label does nothing to address the provision of the QoS itself—RSVP and other reservation and provisioning protocols must still be employed.

Of course, the impact of the powerful new features of IPv6 can only be felt when significant portions of provider networks implement it, and this date seems to be proceeding farther and farther into the future. Current methods of "retrofitting" IPv4 to provide next-generation services such as flow labeling, address assignment, timing, and even the expansion of the address pool seem to be extending the life of IPv4 considerably. The likely path for the eventual conversion of the Internet and other networks to IPv6 will most likely involve backbone networks first, with a slow migration of the outlying networks next, followed by corporate and private networks much later.

The key element in all these standards, from H.323 to SIP and QoS schemes such as diffserv and beyond, is the VoIP and IP telephony gateway. In most cases, the VoIP and IP telephony gateway stands at the boundary of the IP world and the PSTN world. It is time to take a closer look at the form and functioning of these gateway devices.

CHAPTER

IP Telephony Gateways

Many have said that whether speaking of technology, culture, components, purpose, or even intelligence, the public switched telephone network (PSTN) and the Internet are fundamentally irreconcilable except on very basic levels. Even the most avid voice over IP (VoIP) proponents will readily acknowledge that the differences between the two networks are immense and that while very simple "integration" (such as modem calls over the PSTN or simple PC-to-PC IP telephony) is well understood and commonplace, full-scale interoperability—with seamless billing, services, and feature sets—is still far over the horizon.

There are many components in each of these two enormous networks that must be modified to achieve wide-scale integration and allow for seamless transfer of calls from PSTN to Internet and back again. For example, the Internet—and even private Internet Protocol (IP) networks—must adopt a way or ways to provide the voice network's guaranteed, reliable service, while the PSTN must build in the means to allow IP access to all its many services. Exactly where one network's responsibility ends and the other's begins is an issue that will remain open far into the future, but there is no doubt that the major portion of the work toward reconciling these two worlds is accomplished by what most people think of as a single device: the *IP telephony gateway*.

On the face of it, the basic function of a VoIP gateway is easy to define: It is the device that converts "traditional" analog voice data from a telephone into digital data and sends that data over an IP network like the Internet to another gateway, where it is converted back into analog data and sent to another telephone. While correct in the simplest of terms, this definition belies the huge number of issues that must be resolved in all but the very simplest of systems. This chapter explores these issues from the point of view of VoIP gateways and other "helper" devices and examines some of the solutions that have been presented to what are now the telephone system's greatest integration problems.

Since the boundaries between the roles of various devices are so dependent on which vendor or specification is being consulted for definitions, the term *gateway* will be used in this chapter to refer to all the functions necessary for VoIP-to-PSTN integration, with distinctions between different devices being made where necessary.

Gateways and Their Attendants

As the VoIP industry develops, and as developers and hardware manufacturers address integration issues one by one, the full complexity of the gateway function is becoming apparent. Specifications such as H.323 and proprietary vendor schemes identify various roles such as gatekeepers, multipoint controllers, telephony switches, and IP phones, but these roles are rarely consistent across manufacturers or standards. While marketing and development efforts from competing vendors obscure the clear definition of various component roles, it is clear that casual use of the term *VoIP gateway* usually refers to the device or devices that perform the following functions.

Destination Lookup

Gateways typically receive the telephone number of the called telephone in the form of dual-tone multifrequency (DTMF) tones, as generated by the caller's telephone when he or she dials (actually "keys in") the number. In a normal PSTN environment, that 10-digit number (using U.S. long-distance calling as an example) has a very specific meaning.

In the example outlined in Figure 8-1, the caller's local exchange carrier (LEC) PSTN switch knows that a long-distance (interexchange carrier, or IXC) number is about to be dialed by the presence of the 1 at the beginning of the dialed number. The switch therefore contacts that caller's chosen IXC. The region that is being called over that IXC is indicated by the area code 212—very important to the IXC network, which must route the call setup to that area based on its knowledge of area code mappings. The IXC network must now contact the switch within the 212 area code that is responsible for the 555 exchange and request that the 555 switch attempt to connect the call through to its line number 1212. Added to this process is the additional digit that must be dialed (typically 9) at the beginning of the number when the call is begun from behind a private telephone switch such as a PBX, resulting, perhaps, in a 12-digit dialing string.

In a VoIP environment, the gateway that receives the DTMF tones indicating the desired called number must perform a similar role in the process of connecting the call but must attempt to do so over a network

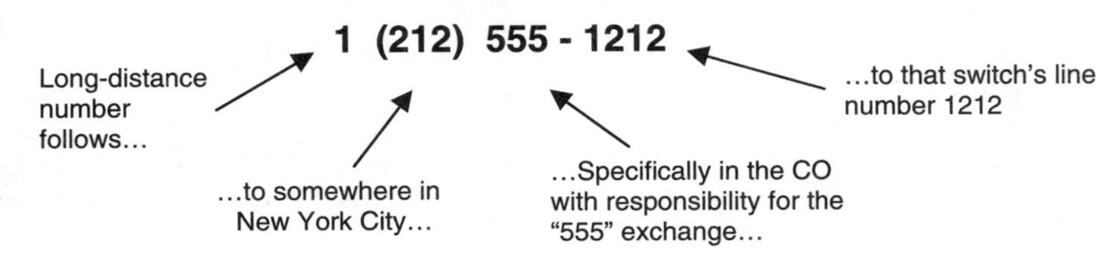
1 (212) 555 - 1212
Long-distance number follows…
…to somewhere in New York City…
…Specifically in the CO with responsibility for the "555" exchange…
…to that switch's line number 1212

Figure 8-1 Segments of a North American telephone number.

in which the DTMF tones and PSTN signaling methods have no inherent meaning. The gateway therefore must be able to associate a called number with the IP address of another gateway that is willing and able to terminate the IP portion of the call and connect to the called number through the PSTN. In a well-designed, geographically distributed VoIP network, there will be a gateway "close enough" (in PSTN terms) to the called number to complete the call within the called party's LEC. That gateway, however, is not able to be identified by the same means as the LEC switch—the number of the exchange that it covers is not likely to have any relationship to its IP address.

In small networks, this function can be performed by a simple table lookup in the originating gateway by statically mapping the called area code or exchange to the IP address of the gateway providing coverage for the called area. For an example of a partial listing, see Table 8-1.

A table such as this states that

1. Calls intended for numbers in the 899 exchange in the 802 area code should be routed via the 192.168.16.23 gateway.
2. Calls intended for numbers in the 542 and 341 exchanges within the 212 area code should be routed via the 10.34.56.78 gateway.
3. All calls for the 203 area code, regardless of exchange, go to the 10.64.253.10 gateway.
4. All other calls go through the 192.168.16.47 gateway.

Depending on the size of the network and the coverage area of each of the gateways, a table like this can range from short and simple to very large and complex. The limitations of statically configured lookup tables are that they are difficult to maintain in any but the smallest and simplest networks and that the maintenance difficulty is likely to grow exponentially as the number of gateways increases—which increases both the

TABLE 8-1

A Simple PSTN/IP Routing Table

Area Code(s)/ Exchange(s)	Gateway IP Address
(802) 899	192.168.16.23
(212) 542, (212) 341	10.34.56.78
(203)	10.64.253.10
...	192.168.16.47

number of table entries and the number of devices that must be kept up to date. Their ability to dynamically adjust to changes in network configuration or failure of a particular gateway is also essentially nil, making this approach less than robust.

To simplify the process of gateway location, a number of protocols can be employed to help convert PSTN numbers to IP addresses. The simplest protocol-based solution uses the Domain Name System (DNS), long employed in the Internet for client lookup of servers, to "name" gateways with identities that align closely with the area codes or exchanges they cover. For example, *voipprovider.com* may identify its Manhattan gateway as *212.voipprovider.com*, allowing other gateways to associate the DNS hostname (*212*) with the area code of dialed numbers destined for that area. Using DNS in this manner provides some flexibility, allowing IP addresses of gateways to change without affecting gateway operation, but leaves important details of implementation (such as hostname conventions and security) proprietary.

A standards-based protocol that promises some assistance in this area is the IETF's draft Gateway Location Protocol (GLP), having been designed from the ground up for the association of PSTN area codes, exchanges, and numbers with particular VoIP gateways. GLP will be capable, given a particular phone number, of resolving either a gateway or a signaling server that is able to direct the calling gateway to the correct address. GLP borrows heavily from the Internet standard for interdomain routing, the Border Gateway Protocol (BGP), in that it uses a combination of link-state and distance vector procedures to propagate area code and exchange (rather than IP prefix) reachability information from VoIP domain to VoIP domain. Call routing information thus shared is stored in each domain's location server (LS), which then provides interdomain call routing information to gateways as needed. Vendor support for GLP, however, is awaiting the protocol's advancement through the standard track. Various vendors, not content to wait for an open standard, have developed a number of proprietary methods for accomplishing the same function.

Despite the attraction of a protocol-based gateway location protocol, however, there is one enormous advantage to the static configuration approach: table lookups are very fast. Since all protocol-based approaches depend on the network to some extent for gateway identification, they are at the mercy of the servers that provide the translation information and the links that carry it back and forth, and network congestion or busy servers can slow the protocol operation considerably. The result is that while a table lookup in even a very large table is unlikely to take more than a fraction of a second, the DNS or GLP operation may take 5 or even 10 s under adverse conditions. This may not seem like a large amount of time, but it forms only a single part of the overall connection delay, and users waiting for a call to go through are likely to revolt if connection times are consistently even a few seconds longer than they experience in the PSTN.

IP Connection Management

After learning the identity of the gateway at the "other end" of the IP portion of the call, a gateway must make a VoIP connection to the destination gateway over which the packetized voice may flow. Gateways may, in cooperation with their gatekeepers (when there is a gatekeeper present in the system), use a number of different protocols to establish, maintain, and—when the call is through—break down this call. Currently, there are two distinct but slightly overlapping standards-based protocols toward this end: the ITU-T H.323 *Packet-Based Multimedia Communications Systems* specification and the IETF Session Initiation Protocol (SIP). In addition to the standards-based protocols, there are also the proprietary methods that gateway vendors have developed for use with their own products.

Connection management is a complex task, and both H.323 and SIP reference a number of other standards to complete the suite of functions they must provide to make calls. Among these functions are capabilities exchange and the negotiation of compression and digitization schemes. Where VoIP networks use gateways employing both H.323 and SIP signaling, a "gateway gateway" may be necessary to allow SIP and H.323 services to interact.

As stated earlier, while the term *gateway* is defined in standards documents, many of the functions that gateways often perform may just as often be offloaded to other devices in the VoIP network. Signaling is one such function; gateways can use the services of gatekeepers to perform call establishment and maintenance for them, thus allowing the gateway itself to concentrate solely on the actual call data.

Compression/Digitization

The conversion of the voice data from analog or standard pulse-code modulation (PCM) digital format to a low-bit-rate stream is perhaps the most notable job of the gateway and is its definitive responsibility—one that is never handled by an assisting device. The conversion from analog to digital data is not in itself remarkable; after all, voice networks have used digital trunk lines for decades. However, even today's PSTN primarily uses digitization methods that result in data rates of 64 (or less often 32) kb/s, which is far below the state of the art for today's microelectronics yet persists for reasons of backward compatibility.

Much of the business case justification for IP telephony is provided by the compression function inherent in the gateway, yielding data rates for a single call of between 24 and 5.3 kb/s. The gateway therefore can fit up to 10 times as much voice data over a given digital channel as can be achieved in standard PSTN encoding, allowing for much higher utilization rates and therefore a large degree of savings over traditional telephony. (Actual rates transmitted by the gateway vary according to the compression mechanism employed, as well as the overhead contributed to the data stream by the various headers of the transport protocols.)

It takes a significant amount of computing power to perform this digitization and compression, however, so the gateway designer must strike a careful balance between the competing requirements of bit rate, gateway processing power (and therefore hardware expense), and audio quality, which maintains an inverse relationship with bit rate. There exists a wide variety of audio codec DSPs (digital signal processors) yielding myriad combinations of these variables, each with its own suitability for a particular type of network. Since gateway vendors usually wish their products to be suitable for a number of different network applications, most gateway platforms are configurable (either in software or through the installation of optional DSP modules) to employ a number of different techniques to meet the particular needs of their deployers.

IP Packetization and Transport

After the gateway converts the voice data into a low-bit-rate stream of digital data, that data must be transmitted across the network in a manner that, as closely as possible, mimics the transport qualities of the circuit-switched PSTN. While the service offered by the PSTN and that provided by an inherently connectionless IP network could not be more different, there are a number of procedures and protocols that allow the packet-switched environment to retain enough of the qualities essential to good voice carriage to maintain good voice quality. Chief among these is some method of timing control, and both SIP and H.323 gateways employ the IETF Real-Time Protocol (RTP) to allow a receiving gateway to reconstruct an incoming audio data stream with its timing intact, compensating for the variability of delay inherent in packet-switched networks. RTP depends in turn on a system of jitter buffers in the receiver to allow for the short-term storage of audio data while it awaits its proper time for presentation to the audio codec, when it will be decoded back into an analog signal.

Once the audio stream is placed inside RTP packets and stamped with a time value, it must be delivered to the destination gateway with as little delay as the network allows. In networks that are designed to carry solely VoIP traffic, the number of calls allowed at any one time can be limited so as not to overwhelm the bandwidth available, thus providing a de facto method of quality control. However, since much of the interest in VoIP is directed toward the eventual convergence of voice and data traffic on the same network, the construction of purpose-built IP networks for voice only is a very expensive solution and one that does not allow users to leverage their investments in network technologies very well.

Instead, devices transmitting voice traffic must have a way of obtaining "expedited service" for their packets. This ability is often associated with the transmitting gateway, since it is the only device that is explicitly aware of the contents of the packet and thus the only device capable of classifying voice data as needing special treatment. The ability of IP hosts (like VoIP gateways) to request a certain quality of service (QoS) from the network, while considered in the design of IP from the beginning, had not received a great deal of attention until the mid-1990s, with the result that a number of different methods for defining, providing, and obtaining QoS are competing for attention and implementation in various types of networks. Single-vendor solutions are already relatively mature and functional, although of most use in networks under a unified administrative control—which is, ironically, where they are least needed, because traffic in such networks is centrally managed along with network design.

VoIP gateways will be able to use heterogeneous multiprovider networks like the Internet most effectively when significant portions of networks implement new, standards-based protocols for QoS such as diffserv coding and the intserv Reservation Protocol (RSVP). The gateway itself can play a pivotal role in the operation of each of these protocols: coding each VoIP packet with a bit pattern indicating the QoS desired from the network (in the case of diffserv) or contacting routers along the path to the destination to reserve the necessary amount of bandwidth and other resources necessary for its call (in the case of RSVP). To this end, most vendors either have announced support for either or both of these protocols or have already implemented some prerelease subset of QoS functionality in their gateways.

Advanced IP/PSTN Signaling

The PSTN has evolved enormously over the last 100 years, from a simple manually connected, circuit-based network for carrying audio from one static location to another to an enormously complex set of facilities providing services such as 800 number service, number portability, caller ID, call forwarding, call waiting, and others. The ability to provide these services has in large part depended on the construction of a separate call control network integrated with but separate from the voice-carrying network itself. This "intelligent" network is based on a highly reliable, packet-based protocol called Signaling System 7 (SS7). SS7 is the means for turning simple voice carriage into true telephony service, with all the intelligent features that customers have come to expect from the "plain old telephone" network.

In the PSTN, SS7-capable switches identify calls that require intelligent services and query one of a number of signaling control points (SCPs) for directions on how to handle the call—for instance, transferring it to a number different than the one dialed (call forwarding) or connecting instead to a voicemail system (call answering). These queries are routed through one of a number of SS7 signaling transfer points (STPs), which act like routers to direct the query to the SCP that contains the desired information. SS7 can then instruct the necessary switches to set up the call to the proper end point. Since SS7 signaling occurs over a network separate from the voice network itself, database lookups and setup messages consume no voice bandwidth, no ports on the telephony switches, and far less processing power than they would if they occurred in-band.

In order for consumers—especially businesses that depend on advanced telephony features—to begin wide-scale adoption of VoIP, gateways must be able to take advantage of the capability of SS7 to use the PSTN as the intelligent network it is and not simply as a collection of circuits over which to feed audio into the VoIP network. Conversely, VoIP telephony providers want to find a way to provide these services not simply to keep customers happy, but because intelligent network services account for a significant portion of telephony revenue—by some estimates, up to 50 percent.

As yet, true integration of SS7 into VoIP is elusive, since no accepted standard for the transport of SS7 messages over IP yet exists. Current

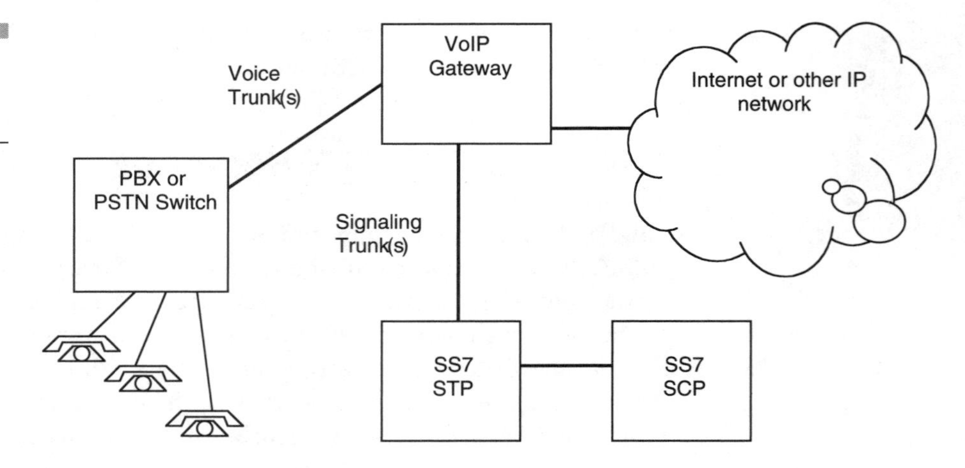

Figure 8-2 Gateway/SS7 connectivity.

solutions are proprietary to particular gateway vendors and depend on simply tunneling SS7 messages across the IP network to the remote switch, which requires that switch to use traditional PSTN methods of providing the service required. In a fully integrated network, the originating gateway will be able to use the SS7 network as though it were a PSTN switch itself, querying the SCP for directions and providing the requested service over the IP network (Figure 8-2). In large part, this distinction determines whether the VoIP gateway is seen by the PSTN as a switching node itself or simply as another end point outside the services of SS7.

Debate is ongoing regarding how best to provide true integration between SS7 and VoIP gateway and gatekeeper equipment, with most of the focus on two protocols: the IETF Media Gateway Control Protocol (MGCP) and a yet-to-be-numbered protocol from the ITU-T, referred to in process as H.gcp (for Gateway Control Protocol). Both these protocols seek to make VoIP gateways appear like traditional voice switches to the SS7 network yet use IP to carry out the directives provided by SCPs. Coming as it does from the Internet domain, MGCP is seen by many as having the momentum necessary to achieve interoperability standard status, and support for it is planned by most gateway vendors.

Authorization, Access, and Accounting

In addition to providing voice connectivity to callers, PSTN switches are responsible for security (e.g., disallowing control input from certain telephones), access control (allowing only the use of services contracted for), and of course, accounting (keeping track of services used so that they may be billed for). In the PSTN, this functionality has been highly tuned over the decades—in contrast to the IP world, which has only begun to seriously address the needs of electronic commerce or billing for anything but access in the last 5 years.

True integration of VoIP equipment into the provider environment demands that VoIP services be billable in a manner similar to current telephone billing practice: that the customer receives a single bill with the details of each call (time and date, called number, length of call, etc.) enumerated for scrutiny by the customer. Currently, each of the carriers of a telephone call (the LEC, sometimes an IXC in the case of long-distance calls, and perhaps an international carrier when required) is responsible for generating a call detail record (CDR) for each call and delivering it to a central billing center, usually provided by the caller's LEC. The result of a customer's accumulated CDRs over a given period of time, charged according to rates provided by the various carriers, is that customer's telephone bill (Figure 8-3).

To make this possible in a VoIP network, gateways (or more likely the routers, access servers, or gatekeepers that assist them in call connection) must be capable of providing the same services to the provider network that PSTN voice switches currently do. The exact methodology to be used will depend on the billing model used by the provider, something that is only being defined now as VoIP providers decide on how to offer VoIP services. For example, voice calling over IP may be offered in a

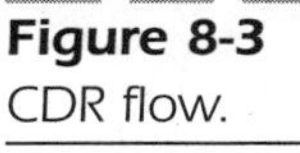

Figure 8-3
CDR flow.

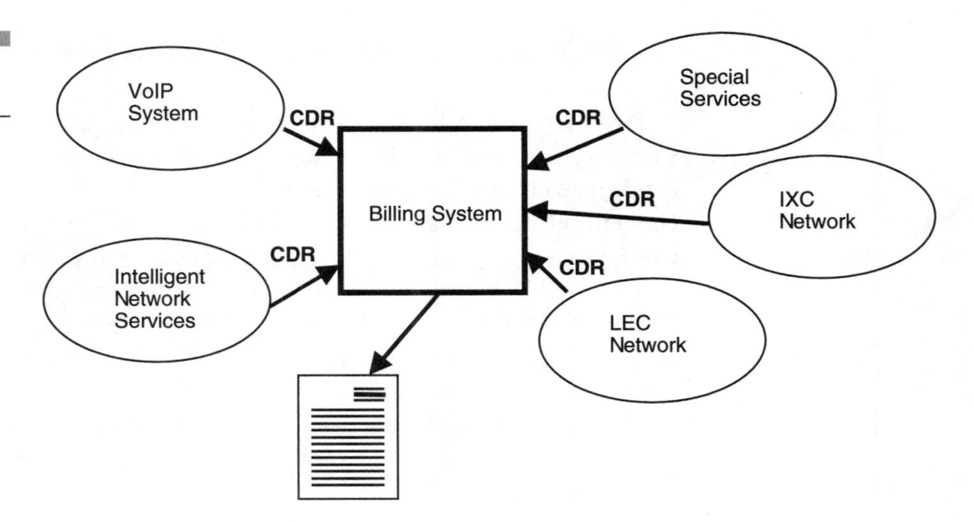

manner similar to how it is currently offered over the PSTN: according to a subscription model as part of a monthly agreement between the provider and the customer. Or it may be that the capabilities of VoIP networks lend themselves to the definition of new billing models, such as a "pay as you play" system in which callers may choose the service (and even the provider network) they wish to use on a per-call basis. In either case, authentication, authorization, and accounting (AAA) protocols such as those defined in the IETF draft specifications for Common Open Policy Specification (COPS) and the Diameter Protocol must be supported by the gateways in the network to provide the ability to generate a standard CDR for use by the billing center.

Necessary Functions

With all these functions and others needing tending to, more comprehensive terms such as *IP telephony switch* and *media gateway / controller* are coming to replace the more common concepts of gateway and gatekeeper as defined in the standards. There are many applications for VoIP, and each has its own requirements in each of the areas outlined earlier. This results in a bewildering array of gateway products, even within a single vendor's product line, as hardware manufacturers attempt to produce platforms that fulfill the needs of a particular user application.

Are Gateways Always Necessary?

Before attempting to categorize the different gateway hardware and protocol architectures, it is useful to examine the most minimal gateway configuration of them all: not having one. While much of the industry concentrates on the integration of the PSTN and IP network worlds, there is an undercurrent of support for those who would drop most access to the PSTN altogether and provide calling over pure IP networks (Figure 8-4). While the idea of allowing connectivity options as comprehensive as those achieved by the PSTN is still years away, it is an attractive idea—although daunting in scope.

What is needed for such a dream to be viable? The primary requirement is for a global IP network—which exists to some extent in the form of the Internet—to replace the global PSTN. Only recently has the

Figure 8-4
"Gateway-less" VoIP.

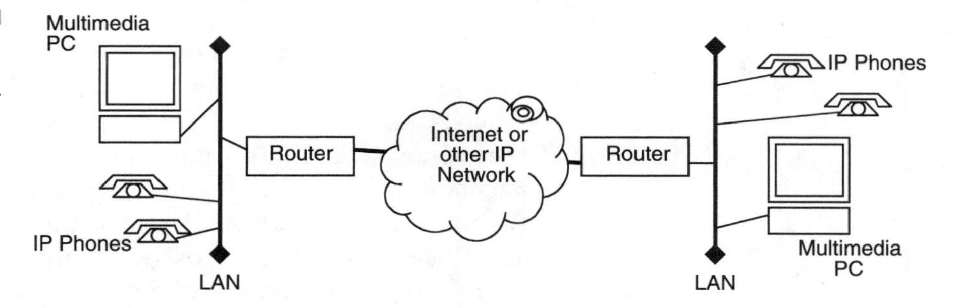

Internet begun to address the issues of reliability, ubiquity, and accountability that are the hallmarks of the PSTN, and as the continual progression of failures and brownouts makes clear, it will be some time before the Internet is ready to face up to the traditional telephony network on these issues. Even assuming that the reliability problems of the Internet are being solved by providers eager to use it for the provision of voice services, the unique characteristics of a network that carries both real-time traffic like voice and less urgent data traffic require additional capabilities such as QoS provision and security procedures that are only in their infancy.

At some point, however, it seems inevitable that the Internet will come to provide, among its other services, a reliable, secure, quality-assured networking service that offers connectivity similar to that provided by today's PSTN. At that point, there is nothing to prevent Internet users from using multimedia- or simply audio-equipped PCs to place VoIP calls directly across the Internet, with the PCs serving as H.323 or similar terminals and providing the compression, packetization, and call control mechanisms themselves. This capability is available today—at least nominally—in software packages such as Microsoft NetMeeting and VocalTec Internet Phone.

When used over private or semiprivate networks, such as a local-area network (LAN)- or wide-area network (WAN)-based IP intranet, these simple packages are capable of reproducing astonishingly good voice quality. In networks such as these, the challenges of the Internet are not a factor, because private networks are designed—usually by a single administrator—to exhibit exactly the characteristics the Internet lacks. Attempts to use them over the Internet at large, however, are subject to widely variable and unpredictable quality. Only when both parties in the call are connected to the Internet via an uncongested link and the network path between them is also uncongested and relatively short can calls be conducted with anything resembling PSTN quality.

And quality, important as it is, is still only half the battle. Users must be able to identify and address the other party they wish to contact in some Internet-friendly way, something for which traditional phone numbers are completely unsuitable. The caller therefore must obtain the IP address of the callee—something that is often changeable because of address management protocols like the Point-to-Point Protocol (PPP) or the Dynamic Host Configuration Protocol (DHCP), which assign addresses on an as-needed basis. Finally, even between the best-connected users, the today's Internet is still the Internet, subject to outages, slowdowns, and congestion.

Therefore, if users of multimedia PCs and "Internet phones" wish to call only those with which they share a high-quality network, terminal-to-terminal VoIP is a viable reality right now. Other calls, to those on other networks, across the Internet, or to persons whose only network connectivity is to the PSTN, must use a gateway to "get out" to the wider world.

Gateway Architectures

As mentioned in the beginning of this chapter, the market for VoIP gateway and other equipment is a rapidly growing one, with upwards of 25 vendors producing everything from software-only solutions for implementation on a Windows NT Intel–based server platform to PSTN/VoIP "mediation devices," which are said by their manufacturers to be capable of fully replacing a class 5 PSTN switch. In order to examine and compare gateway solutions fairly, it is useful to classify these devices according to the markets they are meant to serve.

In some cases, especially among the larger devices, call control functionality is passed off to a gatekeeper process, which may or may not be an integral part of the gateway itself. Since gateway-gatekeeper interaction is usually handled according to the H.323-specified H.225 and H.245 protocols, there are many cases in which gateways from one vendor may interoperate with another vendor's gatekeeper, but not with that vendor's gateways.

Finally, the categories listed below for different "classes" of gateway are not clearly defined by either vendors or the industry at large. There have been many attempts to define such categories, usually using metrics such as the number of calls that may be handled simultaneously or the typical markets targeted by gateway manufacturers. The categories below are a general indication of the hardware platform on which gateways are built, and it must be remembered that there are many hybrid models and others that may span more than one category.

PC Platform Gateways

These typically consist of one or more add-on boards meant for the backplane of a server PC or other general-purpose computer, as well as software to control the board(s) and to manage the calling efforts of the gateway. Each board usually contains one or more analog or digital interfaces to the PSTN side—and sometimes to the IP side—of the network, as well as hardware specific to the compression and digitization of voice; this is typically a digital signal processor (DSP) chipset. The PC platform provides the backplane environment for the cards, the power supply, the management processor (the main PC processor), the physical box in which the boards are housed, and sometimes the IP network interface. The end result is a configuration that looks something like Figure 8-5.

An architecture built around a PC platform is limited by the PC specification itself, in that there is relatively little industry-standard support for many of the features that would be desirable in a gateway. For example, PC backplanes typically have between three and six open card slots available for the installation of DSP and interface cards, making it difficult to scale this architecture beyond the smallest configurations. Zero-downtime features such as hot-swappable backplane buses are rare and proprietary in the PC server world, although hard drives and power supplies fare better in this regard. Finally, coming as it does from a general-

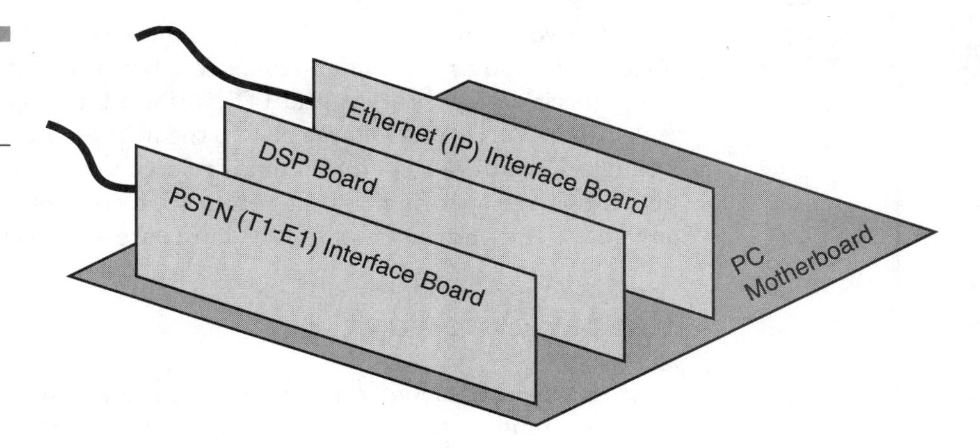

Figure 8-5 The PC platform configuration.

purpose specification to provide a platform for thousands of different applications, the PC platform is not optimized for the software or hardware necessary for VoIP. In small-scale applications, this may not be a limiting factor, but it does affect how large such systems can grow and how well they perform under high loads.

For the very smallest low-density applications, a number of vendors offer software-only solutions meant to run on Pentium- or SPARC-based workstations, performing the voice processing and all other functions in the workstation processor itself without the aid of DSPs. All hardware necessary for this solution is provided by the workstation, excepting the interface to the phone system.

Whether software- or hardware-based, PC or workstation gateways usually offer fewer of the enterprise features that the more custom solutions do. The primary advantage of the PC-based gateway approach is low cost. By using the PC platform for the nontelephony operations of the gateway, the need to design, test, and build custom hardware beyond the interfaces and chipset is obviated, allowing the cost of a simple board-and-software solution to be much less than that of a comparable standalone purpose-built hardware device. With even high-quality PC servers available as a commodity item, the platform cost itself is quite small.

The PC platform gateway model is followed primarily by vendors who

- See the VoIP gateway as a software device with a small amount of custom hardware.
- Do not wish to create full-blown custom hardware platforms themselves.
- Maintain a low cost of production by borrowing the standard PC design and adapting it with custom hardware to their own needs.

As the VoIP market matures—lowering the price of gateway hardware in general—and as even small-scale users begin to consider VoIP another mission-critical application, the PC platform gateway market probably will shrink. This is due in large part to the poor reputation of the platform for high reliability—something that is related to both the use of a general-purpose server operating system rather than one custom built for telephony and to the limitations of the PC hardware architecture itself.

Router Gateways

Vendors who traditionally have produced routers and other connectivity hardware tend to see the functionality of the gateway as being largely hardware-driven, a view that fits their product line and business model well, unlike the PCs on which routers are designed to provide connectivity to the outside world by connecting local networks (usually data LANs) to WANs as well as the Internet. Inasmuch as there is any "generic" router design, it is inherently much closer to providing the function of a VoIP gateway than a PC.

Routers come in a huge variety of sizes, from small monolithic access routers with two ports to enormous expandable chassis with hundreds of high-speed ports. The router models that have been borrowed or adapted for use as VoIP gateway platforms typically exhibit the following characteristics:

- They come from the middle range of the router product line, being inexpensive enough to provide a relatively low-cost platform on which to build a gateway yet being capable of providing the fast traffic forwarding and advanced features (custom queuing, traffic management, etc.) that are necessary in a gateway.
- They are modular in physical architecture, allowing for the addition of analog or digital voice interfaces via expansion slots in their chassis.

One of the primary selling points for the router VoIP gateway platform is the embedded base of routers and the comfort level with which network managers (and the IT departments that support them) regard them. There is a great deal to be said for the difference in perception between deploying an entirely new server for VoIP, with all the complex configuration and troubleshooting associated with it, and simply adding a card to an existing router or upgrading the router itself. In actuality, the level of difficulty and potential for trouble are probably not that dissimilar, but perception is a compelling force—and networking staff who are comfortable with their router platforms may regard the provision of voice as "just another protocol to support" over a relatively simple routed network. This factor can work against the router, however, because of the perception by some that "routers are for packets" and that therefore equipping them with internal voice processing is stretching their capability too far—especially given the disdain with which some voice networking professionals view connectionless data networking.

The VoIP router market is directed primarily at converting voice from a relatively small number of physical ports, since routers in the categories mentioned earlier seldom are capable of supporting more than two T-1 interfaces in a single chassis.

Remote-Access Server Gateways

This platform is essentially a specialized segment of the VoIP router market, in the same way that remote-access concentrators have come to represent a specific type of high-port-density router for the purposes of aggregating traffic from a large number of sources over a single backbone. The physical components are similar: typically a chassis into which voice-specific aggregation modules fit. The primary difference between them is that access servers, by virtue of a high-speed backplane architecture and card management system, support much larger numbers of cards, which in turn support greater numbers of ports per card. This means that while the function of the two types of platforms are essentially the same, access servers fit better into a network designed to provide support for large numbers of users. A common type of remote-access server physical architecture is shown in Figure 8-6.

Remote-access servers also provide a convenient and scalable means by which to integrate voice capability into the relatively new voice-enabled Web-site model, also known as "click and talk" services. These services and the new kinds of customer service they are able to help provide are responsible for a considerable amount of the interest that the electronic commerce industry has shown in VoIP and were discussed in Chapter 6.

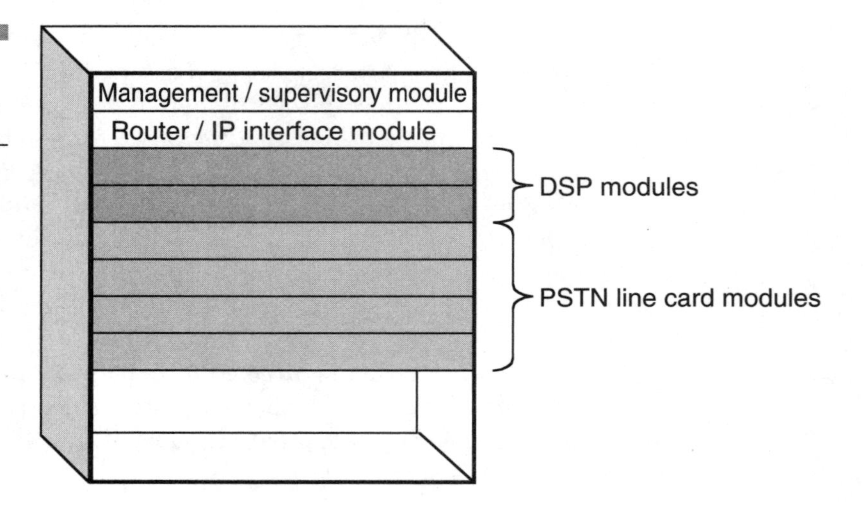

Figure 8-6 Remote-access server chassis.

PBX VoIP Cards

The preceding solutions have one attribute in common: They are all attempts to adapt devices traditionally associated in data networks to the world of voice networking, essentially bringing the data network to the edge of the voice cloud. This approach is perfectly suited to organizations in which data networks are seen as being as mission-critical as voice networks by management but places too much responsibility for and control over the voice stream for some tradition-bound voice managers' comfort. For enterprises looking for a way to deploy VoIP on the voice network's own terms, a number of traditional private branch exchange (PBX) vendors are producing cards that fit into their own devices' chassis to allow the compression and digitization to occur on the "voice side of the house."

By installing a VoIP card (more often called an *IP telephony trunk card* by PBX vendors) in a PBX, a voice service manager essentially gains the ability to use IP networks of virtually any type in the same way that traditional T-1 and ISDN PRI trunks have been used for decades. Regardless of the IP network that will be employed to carry the voice, the encoding and compression functions necessary for VoIP stay in the PBX, under the control and management of the voice service manager. In enterprises that must contend with "turf wars" (and budget battles) between the voice and data departments, PBX VoIP cards allow the voice department to take the initiative in reaping the benefits of connectionless networking.

Voice managers have a number of valid points when pointing out the advantages of such an approach. First among these is reliability: PBXs, as a component of the PSTN, have a long history of high reliability and availability, with features such as multiple redundant power supplies, integrated battery power inputs for uninterruptible power units, hot-swappable card buses, and robust physical environments. Keeping as much of the telephony processing inside the PBX as possible is one way to address concerns about entrusting voice to the vagaries of data networks.

Second, the management and administration procedures necessary to support VoIP are supported already by all PBXs, and the addition of an IP trunk card does nothing to change this. For instance, the manner in which CDRs are produced for billing and management—while a rather foreign concept to most router vendors—can proceed in many cases with no modification at all on the installation of a VoIP card into a PBX. Similarly, the integration of PSTN-type SS7 signaling into a PBX-based VoIP service is much less of an undertaking than all the new development that must be undertaken to make a traditional IP router understand it, because SS7 has been a part of PBXs for decades.

Finally, and perhaps most important, making the PBX itself a VoIP gateway allows for tighter integration between the voice and data networks, which makes the fact that their voice calls are now (at least in part) traveling over IP transparent to the users of the service. For instance, users of standalone gateways—regardless of exact architecture—often must dial a special code to manually access the VoIP network. While this simple step may seem to be a trivial task when compared with the cost savings and other benefits of VoIP, it has proven tremendously difficult to train users to step outside the bounds of "normal" telephone use to obtain a benefit that is not immediately apparent to them (most corporate workers display little concern for the cost of the calls they make). The fact that a VoIP-enabled PBX may choose between VoIP and traditional PSTN trunk transport dynamically, based on the called number, port availability, and other factors *without the knowledge of the users,* and with no special action required by them, is a huge benefit to managers who wish to roll out VoIP services without any disruption to the workforce they serve.

IP Telephony Switches

The last category of VoIP gateway is the newest on the scene and represents the ultimate integration of PSTN and IP telephony. These switches, also known as *carrier gateways* and *mediation devices*, are purpose-built telephony switches that attempt to make the boundary between the two types of network all but invisible by providing the best features of each to users and managers of voice networks. IP telephony switches are distinguished from other approaches to IP/PSTN interworking primarily by the following:

- Very high port density. These switches are intended by their manufacturers to be fully capable of replacing a class 5 central office (CO) switch, which is itself capable of supporting tens of thousands of input ports.
- High scalability. An IP telephony switch must be able to begin small and grow very large without the need for replacement of major parts of the architecture to do so.
- Comprehensive integration into the SS7 intelligent network (IN), with support for redundant connections to the IN, as well as built-in sup-

port for IN features—regardless of the type of trunk network a particular call is using.

- Transparency to the PSTN network in both the voice and the SS7 networks. An IP telephony switch should be capable of appearing identical to a carrier switch in terms of call routing, advanced services, etc.
- Transparency to callers. Users of the network should not be able to distinguish VoIP calls from PSTN calls.
- Management and billing features on par with and fully compatible with standard provider CDR/billing systems.
- A very high degree of fault tolerance and the ability to recover from failures of various components without total disruption of service.
- Compliance with the Bellcore (now Telcordia) Network Equipment Building Standard (NEBS) specification for continued operation (and survivability!) in various situations, from ground-fault current loops to earthquakes.

To the extent that a switch or gateway exhibits these features, it is able to claim "carrier class" status, a nebulous concept, but one that is defined by most as having the capability for large-scale deployment in the network service of a major service provider.

Only a few vendors have claimed all the preceding characteristics for their products, with most of those rolling out variants of all the preceding architectures, each with its own spin on exactly what makes a VoIP switch "carrier class." At some point we will undoubtedly see CO switches based on an IP architecture from the ground up, with "legacy telephony" or PSTN interfaces as an extra cost option. Major development of these products, while not that far off, awaits proof that IP telephony can work for large-scale carrier networks—something that will only come as a result of providers trialing, offering services, and eventually making money with their less robust brethren.

Gateway Components and Features

Regardless of the category into which a particular gateway fits, its hardware and software must share some characteristics in common with other gateways, if only because—regardless of network size—their functions are similar within the networks they serve. Listed below are some of the components that make up VoIP gateways, including the features that sometimes set gateways apart when comparing similar units.

Digital Signal Processors

The DSP is in many ways the heart of the gateway, being the hardware that performs the voice coding and decoding, and often other functions, that are so central to its function. Digital signal processing, as a procedure executed in hardware, has existed since the 1960s, but it is the breakthroughs in microelectronics design and manufacturing that have occurred in the last 10 years that have allowed the successful application of these techniques to VoIP. Early attempts at the computation of DSP algorithms resulted in multiple-chip boards or even whole computers dedicated to coding functions, which made them large in size and expensive to manufacture. In the late 1970s, however, advancements in VLSI chip manufacturing allowed all the functions of the DSP—addressing, control, arithmetic, input-output, and storage—to be accomplished on a single chip. First attempts at single-chip DSPs were made by AMI, Intel, NEC, and Bell Labs in the early 1980s.

DSPs are central to the operation of much of today's communications equipment and consumer electronics, such as central office switches, cellular telephones, CD players, and other gear. As such, DSP chips have reached a near-commodity status, with manufacturers offering both general-purpose and application-specific DSPs in their product lines. While many gateways are built with general-purpose DSPs performing the coding function, DSPs specific to the VoIP gateway market are beginning to emerge from a number of vendors and usually offer standard codings for voice such as G.723.1 and G.729a, as well as that vendor's proprietary codings. Depending on the architecture used in a particular gateway, capacity (number of simultaneous calls) sometimes can be increased simply by adding DSP cards. Advancements in DSP design mean that the number of DSPs does not always equal the number of simultaneous calls, since many newer DSPs can handle multiple calls per chip.

Physical Ports

Links to the telephone side of the network—usually a PBX—generally are made via a channelized T-1 or E-1 allowing the connection 24 or 30 lines of 64-kb/s PCM digital voice. Gateways with support for ISDN will make the telephone connection with basic rate interfaces (BRIs) or primary rate interfaces (PRIs). Some gateways are also available with analog interfaces, which allow for the connection of key systems and smaller PBXs on the telephone side. T-1 and E-1 connections are made physically via 75-Ω BNC coaxial or 120-Ω RJ-48C connectors, with one or more per board. ISDN interfaces are usually made via the RJ-48C.

On the IP side, network connections are usually via 10- or 100-Mb/s Ethernet, which allows easy interoperability with virtually any network connectivity hardware such as a frame-relay access device or router. Some gateway platforms, especially router-based models, have the ability to route IP directly out synchronous serial ports via a V.35 or RS-232 interface.

Interface and Hardware Slots

Physical expandability is most directly affected by the physical architecture of the box in which the gateway is housed. PC-based gateways are limited to the number of slots available in the ISA, EISA, or PCI slot schemes of their host computers, although limits imposed by the gateway design may come into play before the ultimate limit of the hardware itself.

The most flexible hardware configurations usually come from access-server-type gateways, which have purpose-built chassis able to accept (in the larger designs) many more cards than PC backplanes, each of which typically can provide a higher density of connectivity because of their larger size. For example, PC gateways usually provide only one T-1 or E-1 connector per slot, whereas some access server models can support as many as eight connectors per slot.

Means of Configuration

All gateways, from the simplest PC-based small-office unit to the class 5 replacements, require a certain amount of configuration. There are certain parameters, such as the IP address of the Internet interface and the up/down status of each of the individual ports, that all gateways must configure to operate in the context of the network in which they are installed. As features are added to the platform, so increases the number of options that must be chosen among and must be configured globally. For instance, a set of configuration options for a medium-sized gateway might include

- *IP communications parameters*: IP address, subnet mask, default router(s)
- *Coding algorithm*: proprietary or standard

- *Call data reporting (CDR) procedures*: data included in and format of CDRs
- *Diagnostic facilities*: error logging and other troubleshooting tools
- *Dialing support*: one- or two-stage dialing
- *Address resolution*: address and routing tables, PSTN/gateway address tables
- *Call security*: allowed and nonallowed numbers and area codes
- *Error handling*: fallback mechanisms, such as PSTN routing

Gateways that depend on a network device operating system, such as router- or access-server-based gateways, usually integrate commands necessary to configure all these options into the operating system (OS) configuration language, thus preserving the continuity of operation among their product line. Since remote management of the platform is a basic requirement of these devices, remote management of distant gateways is as simple as launching a telnet or WWW session to a port on the gateway and working as if the box were local.

Standalone gateways, particularly those built on a PC platform, must design and execute their own—usually proprietary—management software. The ease or difficulty of use of this software is of primary importance in the value of the platform, because gateways that are difficult to configure—regardless of their feature sets—are unlikely to be deployed in the most effective way. PC-based gateways, since they operate as applications under the PC's operating system, are usually configured via a graphic management program on the gateway box itself, which—in order to be remotely manageable—must be either standards-based (WWW or telnet) or have built-in remote accessibility as a feature. Some vendors, wishing to avoid the complications this entails, require the gateway to run a standalone remote-control program such as Symantec's pcAnywhere to access the gateway remotely.

Reporting System

Most gateways integrate some type of call reporting system, although the amount and nature of information gathered about each call vary widely. Service-provider-type large gateways have the largest number of options, including start and ending time of each call; incoming and outgoing port identification; IP address(es) used in the call; packets, bytes, and errors encountered during the call; as well as QoS parameters such as average delay and delay variation encountered during the call. Smaller "enterprise" gateways usually deliver less information about each call, reporting only call time and port information.

ITSPs are likely to seek out the more feature-rich of these gateways and consider this reporting function as important as the gateway capability itself because their ability to bill for their services depends on it. In addition, the traffic engineering and call control that are so necessary for the day-to-day functioning of a service provider benefit greatly from the larger amounts of information that these units can provide.

Operating System

Gateways that run on a PC or other general-purpose computer—whether entirely in software or with the help of special hardware for digitization and compression—usually run as an application under a standard operating system (OS). It is the job of the operating system to manage the resources of the computer, controlling the hardware and software access to the microprocessor and other components of the platform. As such, the OS becomes a de facto part of the gateway itself.

Most PC-based gateways run under one of the Microsoft operating systems that come as "standard equipment" on most Intel-based PCs. Many of the small-office, less capable (and therefore less expensive) systems are able to run under the desktop OSs Windows 95 and Windows 98. These OSs have the comparative advantage of being very inexpensive to purchase and run (indeed, it would be difficult to find an IS department worker who did not know how to install an application on a Windows 95 computer) but are less stable and flexible than other operating systems. Windows 95 and 98 are most often seen as "desktop" OSs, and most IS professionals feel that they are best left in that role—with the more "serious," centralized work left to other OSs.

The most popular OS choice for the PC-based gateway market is Microsoft Windows NT, which is Microsoft's entry into the "enterprise server and workstation" market. Windows NT does not suffer from the backward-compatibility issues that so complicated the design of Windows 95 and 98 and therefore takes better advantage of the hardware on which it runs, allowing for faster and more stable performance on par with what IS departments expect from a platform for their centralized data and services. While more expensive to purchase and maintain than Windows 95 or 98, Windows NT also benefits from a large user base that ensures the availability of qualified help for the platform (if not for the gateway software).

While Windows NT has a larger share of the market, many of the larger gateways based on general-purpose hardware run on variants of the UNIX operating system, most notably Sun Microsystem's Solaris (which is capable of running on both Intel-based and Sun SPARC–based platforms). UNIX has a number of advantages over other OSs in that

- It is the research and development operating environment of choice.
- It is a traditional IS department choice for "application servers" (hardware designed to support a single important application) because of its power and flexibility.
- It is among the most stable of operating systems and the least prone to "crashing."
- It has a long legacy in the telephony arena, being the operating system of the workhorse of the PSTN, the 5ESS switch.

There is therefore a perception by many IS staff that UNIX is the "right" operating system to run telephony on, and some feel as one "anti-VoIP" telephony manager did: "If I'm going to trust my voice service to a computer, at least it's going to be a UNIX computer."

Standard Protocol Support

It is the nature of data communications engineering that many gateway vendors have developed their own proprietary methods of signaling, encoding, and providing value-added services to calls handled by their gateways. Most vendors of these products can point to many ways in which their system of protocols and features is superior to the current ITU-T standard H.323 protocol, at least for certain applications. Yet most vendors also design their gateways to be able to switch between their proprietary methods and standards-based ones in order to allow their customers to retain control of exactly how their gateways function and to allow them to interoperate with other vendors' platforms.

QoS Support

Until recently, gateway vendors surprisingly have considered the quality of the IP network to be beyond their concern, and indeed, there are many who continue this line of thought, depending instead on the network architects to design an IP infrastructure that will provide the low delay and low loss that are necessary for successful telephony. With such a design, the deployer of VoIP services must know the characteristics of his or her network and manage them with voice traffic in mind—a task perhaps easier in networks purpose-built for VoIP but very difficult in converged voice/data networks. Prioritizing voice traffic over data traffic then becomes the job of the network itself, either by virtue of its ingress point (a particular physical port on a router being configured with priority queuing) or by classifying the voice traffic at the router using port numbers or another means of differentiating it from other traffic.

In an effort to ensure the best transport quality for their gateway's voice traffic, many vendors have embraced one or another of the growing field of IP QoS schemes. While there is no concensus on a "favorite" prioritization method for voice traffic, the market is largely split between RSVP and diffserv, with very few gateways supporting both. Since VoIP is perhaps the most obvious example of an IP application in need of QoS, network managers would do well to keep their gateway choice in mind when designing their network's QoS mechanism.

Security

In any telephone system, there is always the possibility of unauthorized use. Traditional phone systems are capable of per-port (per-phone) restrictions on incoming/outgoing use, allowing some phones to be used only for local calls or to call only among a particular set of numbers. Many corporate PBXs also require a personal identification number (PIN) to be entered before completing a call, thereby allowing authorization to be performed (e.g., only managers may be allowed to place long-distance calls) and providing a record of who made the call (regardless of which phone is used). These capabilities are slowly finding their way into VoIP gateways as well.

Some VoIP gateways can perform an IP security function as well by associating a particular IP address, or range of addresses, with a set of incoming call permissions. In this way, gateways can reject attempted connections from unauthorized gateways and PC phones (VoIP "crank calls").

Gatekeepers

Many current gateways integrate gatekeeper functions within them—largely because most gateway deployments are small enough to not require the services of a separate gatekeeper. According to the H.323

specification, a gatekeeper is an optional device whose services will be used if available but without which a VoIP network may be fully functional. Only in the largest networks (those expecting in the millions of calls per day) is it desirable to remove authentication, recordkeeping, call setup, and management away from the gateways themselves. By doing so, the system gains the ability to centralize these functions and thus provide for simpler management, but at the cost of maintaining an additional device or devices.

Most standalone gatekeepers are constructed around a PC or other microcomputer platform and are closely complementary to gateways produced by the same vendor. While the H.323 specification defines the roles of the gatekeeper quite clearly, it leaves unexamined many of the functions that are most important to service providers, such as billing and gateway status monitoring. Vendors therefore have integrated these services into their gatekeeper devices, which become more than simple gatekeepers as defined in H.323, becoming in addition full-featured management platforms that serve as the “control panel” for the entire VoIP network they serve.

For example, the most capable gateways function as capacity control devices, which will reroute a call to a PBX and out the PSTN if there is no VoIP capacity remaining in the system’s gateways to handle it. This “PSTN fallback” capability is necessary to allow providers to continue to offer service in the event of severe congestion or failure of the IP network. Gatekeeper products also can offer conference control, allowing and controlling multiparty conferences according to the role defined as a “multipoint controller” in H.323. Finally, gatekeepers serve as the VoIP system’s usual interface into external databases such as LDAP, RADIUS, and others, because the complicated procedures for accessing external data are best removed from the gateway devices that must process the actual voice.

The term *gatekeeper* in current VoIP networks must be viewed as meaning “call control and auxilliary service device,” and the requirement for a gatekeeper in a particular network depends as much on the unique services required by the network users and managers as it does on the need to move call control away from the gateways.

CHAPTER 9

LAN Telephony

In 1992, a consortium composed of Apple Computer, IBM, and National Semiconductor proposed a technology called *Isochronous Ethernet* (IsoEnet) that used an alternate coding scheme to allow 96 ISDN B-channels to coexist with a standard 10-Mb/s Ethernet channel over the same category 3 twisted-pair cable. Since ISDN allows for an isochronous (uniform in time) service, IsoEnet was intended to allow local-area network (LAN) managers to integrate their standard data networks with their voice, and later videoconferencing, networks. The result would be an integrated LAN capable of supporting standard (asynchronous) Ethernet data, multimedia conferencing, video services, and phone service—all over the same cable. The introduction of IsoEnet was nicely timed to coincide with the then-planned large-scale deployment of ISDN in the United States and around the world, so the stage seemed set—from the point of view of the vendors, at any rate—for IsoEnet to become a major force in LAN networking and drive the convergence of data and real-time services in the local enterprise. The future looked bright enough that the IEEE formalized the specification by standardizing it as 802.11 in the fall of 1995.

At about the same time, Asynchronous Transfer Mode (ATM) vendors developed their own means by which to offer both real-time services and non-real-time data over a similar type of twisted-pair cable by introducing 25-Mb/s twisted-pair adapters and low-cost "workgroup" switches. Paired with an innovative, complex means of offering LAN-like broadcast support called *LAN Emulation* (LANE), this configuration seemed ready to deliver on ATM's promise of shared data and multimedia networking from desktop to desktop. The ATM community foresaw a day when ATM would be used for all types of traffic, across all geographic spans, and its use for connecting desktop computers, multimedia terminals, and phones seemed only a matter of time.

LAN managers, however, had a different idea. No matter how fantastic the promise of IsoEnet, ATM, or a number of other new technologies, the people who design and maintain LANs in offices around the world had developed a taste for their own favorite technology: Ethernet. Developed in the early 1970s at Xerox's Palo Alto Research Center, Ethernet was designed to be precisely what ISDN, IsoEnet, and ATM were not: simple to plan, deploy, and maintain and inexpensive to implement. Not surprisingly, these were qualities that the rapidly developing LAN industry was happy to have in a product specification, which resulted in the IEEE's ratification of Ethernet as a standard—called IEEE 802.3—in 1985.

By 1990, Ethernet had largely developed into the form in which it is still best known today: a 10-Mb/s, best-effort, essentially unreliable LAN data transport operating over inexpensive twisted-pair cable. The 1990s saw even greater improvements in the system, including data rates of 100 and 1000 Mb/s, the capability for the use of optical fiber (vastly increasing the possible length of a single cable segment), prioritization functionality, and—perhaps most important—a rapid fall in the already relatively low cost of Ethernet equipment. These facts, taken together, help explain why Ethernet has become virtually the only viable LAN technology for most applications.

And what of the purpose-built competitors for the role of desktop multimedia network? Almost any analyst will agree that either is inherently better suited for the delivery of real-time data, if one considers all other things equal. Of course, all other things are *not* equal, and Ethernet has significant advantages of its own, particularly the enormous installed base and a tried-and-true reputation for working well in diverse environments—something that endears it to most LAN managers. In any case, the competing technologies' timing was off; they "foresaw" the need for LAN data/voice convergence before the hardware and applications were ready to support that type of traffic and before service providers believed that they could field products based on it.

Thus IsoEnet and desktop ATM installations survive in limited numbers, having been deployed by forward-thinking managers who did not mind spending the time and money it took to become early users of applications that "everyone knew" would someday become commonplace. Some of these networks undoubtedly were successful, having saved (or made) money for their owners by reaping the benefits of convergence outlined in this book. Others are maintained as legacy networks, being used as they depreciate, and are destined to be replaced at the next possible juncture. For the rest of the market, the voice and video applications would have to wait until Ethernet was ready for them. At the end of the 1990s, Ethernet is ready to provide these applications, and voice services are destined to be the first of such applications. This chapter explores how Ethernet can be used as the backbone of a LAN-based voice over IP (VoIP) system and investigates some of the approaches taken toward this goal by various vendors and standards bodies.

Convergence of the Local Area

By the late 1990s, the market forces that had driven the convergence of data and voice networks had extended their influence into the far reaches of the network. The reasoning is simple: If there are savings to be had by mixing traffic types in the carrier networks and over the wide area, then surely there must be something to be gained from doing the same in local networks as well. While there is not the toll-bypass justification, companies spend billions of dollars yearly on their private phone systems, purchasing PBXs or key systems and the special phones required by them and maintaining these units in parallel with LANs—which in many cases exactly duplicate phone system deployment. The investment necessary in phone systems is a cost center ripe for elimination or reduction.

Phone systems, however, are not an area to be trifled with. While most users of networked computing resources have come to expect the occasional outage of at least some portion of their networks from time to time, phone users have not. Redundancy and fault tolerance are built into every central office and every IXC switching point to make what is perhaps the most reliable networking system in the world. It is almost shocking to pick up a handset and not hear a dial tone, so accustomed have users become to the absolute dependability that phone systems exhibit.

Since they must not prove to be the weak link in the chain, office phone systems have come to resemble central office switches in miniature, with hot-swappable components, redundant cooling fans and power supplies, battery backup systems to prevent loss of service in the event of a power outage, and certifications attesting to their trustworthiness in these and many other areas of possible catastrophe. Users want more than dependability, however, and in addition to near perfection in that area, their phone systems must offer a wide array of services—ranging from music on hold to voicemail to call forwarding. Finally, these features must be as easy to use as possible, without requiring users to learn new commands to access them.

Clearly, those wishing to convince office managers to make their phone system more like their office computer system have their work cut out for them. The challenge for those seeking to develop VoIP products to replace PBXs and other office phone systems with LAN equipment is therefore to

- Continue the high dependability of the current system.
- Offer the same or better special features.
- Make users unaware of the change in technology.
- Reduce costs significantly.
- Finally, take advantage of the mixed data/voice environment to support applications that are impossible (or at least exceedingly rare and expensive) with current systems.

Office Telephony Environments

There are two types of telephone systems that traditionally have been used in offices: the key system (KS) and the private branch exchange (PBX). They are usually differentiated in the following ways:

- Key systems, usually employed in smaller offices, are distinguished by the presence of a number of separate incoming lines, each of which appears on every telephone unit. These lines are provided by the local exchange carrier (LEC) or other phone company on individual local loops and are arranged in a "hunt group" that allows them to appear to the outside world as a single phone number, with incoming calls rolling over to the first available line. When incoming calls are answered, the answerer simply directs the called party to "pick up" whichever line number is applicable. Users may place outgoing calls simply by selecting an unused line and dialing. Intraoffice calls are usually handled via an intercom system integrated into the key system telephones or control unit.
- PBX systems are defined by their control units, which are essentially small circuit telephony switches connected to the LEC via one or more T-1 or ISDN PRI trunk lines. Incoming calls are answered by an automated or human attendant, who switches the call to the appropriate extension, each of which is at the end of its own wire pair. Outgoing calls are placed by
  1. **1.** Lifting the handset and receiving dial tone generated by the PBX switch
  2. **2.** Dialing the outside access code (usually 9)
  3. **3.** Receiving dial tone generated by the LEC switch
  4. **4.** Dialing the desired number

  PBX systems allow for intraoffice communications by allowing users to directly dial the desired extension in the same way they would dial any other number but not requiring the outside access code. Rather than connecting directly to the called telephone (as would occur using the intercom feature of a key system), the call is switched by the PBX switch to the called extension—just as though the two telephones were in the same PSTN exchange. This relationship between switch and telephones is what gives rise to the name *private branch exchange.*

There is much room for argument in these definitions, and there is much debate between vendors who produce business telephone systems as to the correct classification of particular products. Many newer key systems are capable of behaving like PBX systems in terms of connectivity to the PSTN, and since the features and functions of each are so close, many of the more capable key systems are now known as *hybrid systems*, which seek to combine the best features of each type. For the remainder of this chapter, "traditional" office phone systems will be called PBXs, with the distinction between them and key systems being made where necessary. A typical PBX system is shown in Figure 9-1.

Regardless of the exact nature of the system, one thing all office telephone systems seek to provide is features beyond standard PSTN connectivity. Some of these features include

- *Conference calling.* Connecting one or more local telephones to one or more outgoing lines to allow for multiparty calls.
- *Call forwarding.* Redirecting incoming calls to one extension to another or to another phone located somewhere else in the PSTN. A common application is for users to forward calls to their mobile phones when they are out of the office so that their calls literally "follow" them wherever they are.
- *Call pickup.* Picking up a ringing extension from another extension.
- *Caller ID.* Identifying incoming calls with either the calling number or the caller's name on a liquid-crystal display (LCD) mounted on the phone.
- *Music on hold.* Playing a recording to incoming callers while they wait for the called extension to answer.
- *Night bell.* A form of call pickup in which all telephones, or centralized ringers, ring when the attendant's desk is unmanned. This allows calls to the main number to be answered by anyone in the office.

While not always integrated with the PBX itself, two other features are so commonplace that they deserve mention:

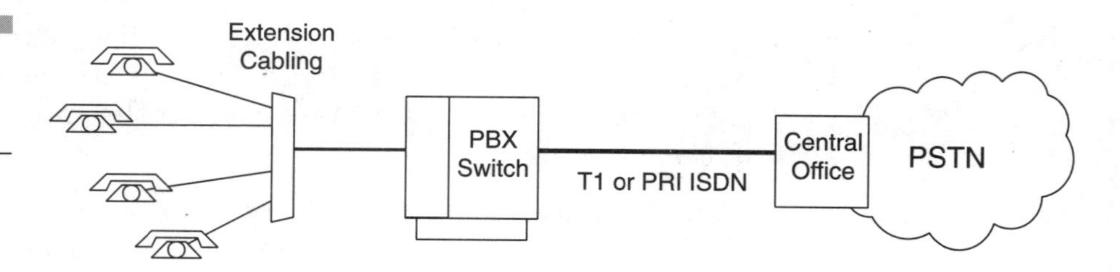

Figure 9-1
A typical PBX network.

- *Voice mail.* In its simplest form, a private answering service for each extension. This feature has developed over time to include its own wide array of services, such as message forwarding and notification.
- *Automated attendant.* A computer-based method for determining the desired extension and directing the PBX switch to complete the call. Usually, this involves a digital recording answering the initial call (in much the same way a switchboard operator does) and directing the caller to "Dial 1 for sales, 2 for service," etc. This system can range from the very simple to the extremely complex, with the more full-featured systems actually providing automatic call distribution (ACD) capability.

All these features must be configured and managed in accordance with the policies of the office, convenience, and security. For example, passwords should be necessary to restrict voice mail access and sometimes to place long-distance calls. Therefore, the PBX equipment must have a management console that allows for configuration in as straightforward a manner as possible. Ease of use is particularly important for smaller companies, because there is an assumed simplicity regarding the configuration of phone systems that belies its complex nature.

Companies that require the services of a PBX-like system but that do not have the expertise, staff, or desire to deploy their own can instead outsource the provision of a "private" phone system by purchasing Centrex, a phone company "virtual PBX" service that mimics the operation of a PBX yet provides it over standard or ISDN telephone lines. Centrex offerings typically provide free intraoffice calling and basic convenience features such as call forwarding and call hold. For an additional monthly per-line fee, other services may be added to Centrex service, including voice mail and automated attendant features, allowing Centrex service to compete effectively with full-featured key systems and PBXs.

LAN Telephony Components

In order to provide telephone service comparable with, or better than, the systems just outlined, the functions that a PBX system provides must be assigned to various LAN-based components, each of which will play its own role in the telephone system. A basic LAN telephony system is shown in Figure 9-2.

The following is a discussion of each of these components and the responsibilities each has in the LAN telephone system.

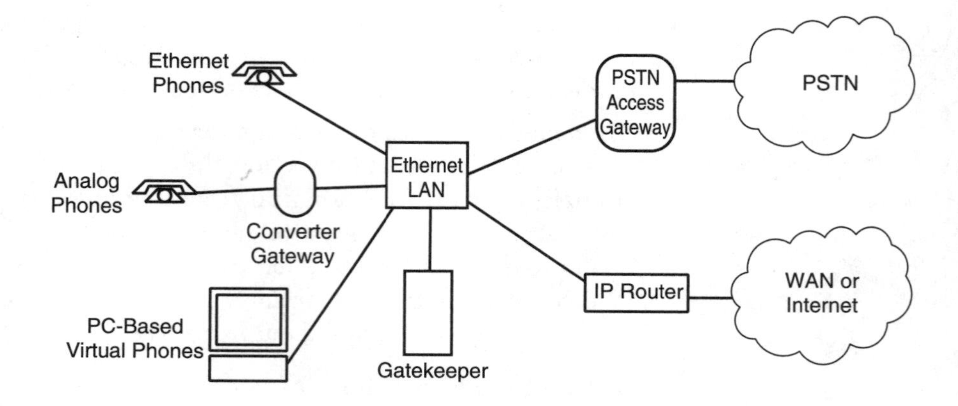

Figure 9-2
A LAN telephone system.

A Local-Area Network

As mentioned at the beginning of this chapter, there have been a number of attempts to develop LAN technologies specific to the provision of voice and other multimedia traffic, but none has succeeded in gaining enough of a market to play any significant role in the wide-scale deployment of LAN telephony. This fact is due to issues related to cost and marketing failures but also to the emergence of a de facto standard in LANs: Ethernet. By achieving a LAN market penetration of over 90 percent, Ethernet has become the one "given" in almost every office environment. The exceptions to this statistic tend to be the very largest offices, which still retain significant amounts of IBM's Token Ring technology. This fact does not weigh too heavily on LAN telephony vendors, however, since most of their systems target the small and medium-sized office environments.

In many ways, the development of Ethernet has followed a path similar to that of the Internet Protocol (IP): From humble beginnings as research and development experiments, both followed a slow path to respectability and wide-scale implementation, despite their essentially connectionless nature. The fact that both these technologies provide only "best effort" delivery of data and yet have been selected as being suitable for such applications as telephony simply shows how good the current notion of "best effort" is.

The perception that Ethernet LANs must be less reliable than PBX networks is a mistaken one in any case. A large part of the development that Ethernet has undergone in the last two decades has aimed at increasing its reliability. Amendments to the physical layer of the

Ethernet specification have resulted in Ethernets in which a problem with one computer or segment does not affect the entire network—an issue that plagued early, physical-bus configurations. Similarly, the quality of the twisted-pair cable used to interconnect Ethernet devices has improved considerably, and the optional use of fiber means that Ethernet traffic can be quite sure of an unimpeded path to its destination, given a well-engineered network.

While users, especially those in smaller offices without a full-time network management staff, often complain that "the network is down," it is seldom true that the network infrastructure itself is malfunctioning. It considerably more likely that there is a problem with the user's own computer or with the resource that he or she is trying to access. Problems with the network itself are usually limited to those caused by congestion—the result of too many Ethernet frames vying for the same network segment or device at the same time. The result of congestion for data applications is that file transfers take longer than they should, while the devices attempting to transmit "back off" and wait for congestion to subside. This annoys users and makes some applications difficult or impossible to use, but the network continues to function—albeit at a slower pace—until the congestion is resolved. For voice traffic, however, such congestion, if it did not prevent the connection of the call in the first place, would prove deadly.

Therefore, with most of the physical layer issues that affect availability of the network having been addressed, Ethernet development recently has focused primarily on means of reducing congestion. There are essentially three methods of doing so, and all three have seen remarkable advances in recent years:

Segmentation Ethernet was developed originally according to a model in which all devices shared a single cable. Since devices must transmit at a fixed speed, the ultimate capacity of a shared cable must be itself apportioned among the devices according to some agreed-on access control scheme. Ethernet uses the carrier sense multiple access with collision detection (CSMA/CD) method of access control, in which devices contend randomly for available bandwidth according to probability, ensuring that—over time—each device receives an equal share. This state of affairs persists even in most modern Ethernet systems, which, while more sophisticated from the cabling point of view, still require devices to contend for available bandwidth.

Under such a system, the nominal amount of bandwidth available to any one device is simple to calculate by dividing the speed of the shared network segment by the number of devices sharing it. For example, a 10-Mb/s segment shared among 10 devices results in each receiving something less than 1 Mb/s, "something less" because of the contention and framing overhead inherent in the Ethernet system itself.

One of the common-sense ways to increase the amount of available bandwidth is to reduce the number of devices per segment, thus decreasing the "contention domain" experienced by each device. The modern means for achieving this comes in the form of a LAN switch: a central connecting point for Ethernet segments that separates LAN segments for the purposes of contention while allowing intersegment traffic to pass when necessary. The ultimate in contention control is to provide each device with its own private segment connecting it to the switch, an option that was cost prohibitive until recently but one that falling equipment prices have made possible for even smaller networks. In such a network, each device receives the full "wire speed" of the medium to which it is attached. It also gains the ability to transmit at any time, without the need to contend for access to the medium. Figure 9-3 compares two types of switch deployments with a traditional, shared Ethernet network.

The most recent development in LAN switching is the capacity for virtual LAN connectivity, in which individual devices are assigned membership in a group that has connectivity only to each other and whose traffic does not "mix" with other traffic as it crosses backbone networks, distant switches, and shared hubs. Many of these groups, or vLANs, may coexist on the same network infrastructure. The implications for LAN telephony are considerable, in that a measure of separation between voice traffic and data traffic may be maintained by placing telephony equipment on a different vLAN from data equipment. Such a configuration is depicted in Figure 9-4.

While just described as a common-sense approach to the problem of

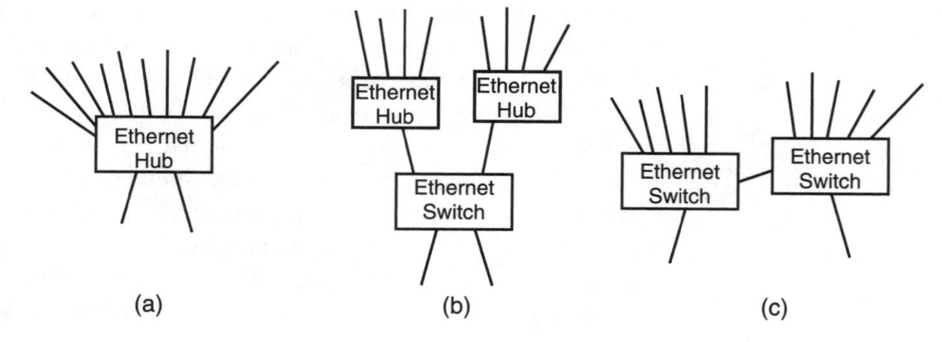

Figure 9-3 LAN switch deployments: *(a)* shared topology; *(b)* partially switched topology; *(c)* fully switched topology.

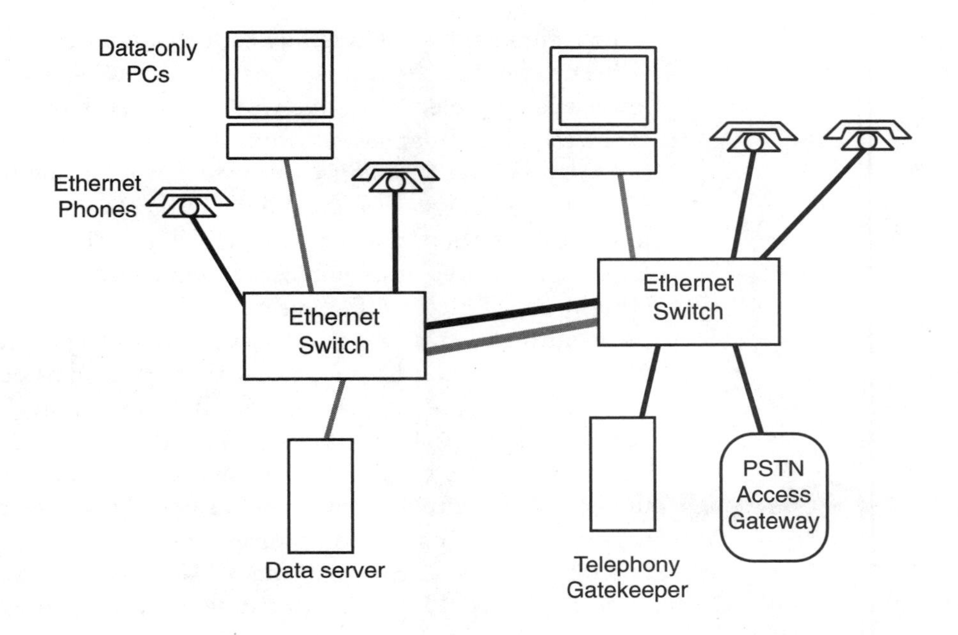

Figure 9-4 Virtual LANs supporting VoIP.

congestion, segmentation is not the simplest way of increasing LAN data capacity.

Increased Speed If available bandwidth can be estimated by dividing the total bandwidth by the number of devices contending for it, another, even simpler way to increase each device's portion is to provide more in the first place. While the earliest Ethernet ran at about 3 Mb/s, the technology took hold of the market in a form delivering 10 Mb/s—an astonishingly fast data rate in small networks but one that was saturated rapidly as networks grew to include tens and hundreds of devices. A consortium of vendors therefore extended the IEEE 10Base-T specification (essentially twisted-pair Ethernet, with devices wired to a central hub) to include data rates of 100 Mb/s, an effort that the IEEE ratified as 802.3u. In addition to higher data rates, this specification also provides for the automatic negotiation of speeds in networks that support both 10 and 100 Mb/s. 100Base-T requires some changes in the limits imposed on physical topologies but in virtually all other respects is "plain old Ethernet, just faster" and therefore interworks with 10 Mb/s flawlessly.

Finally, the late 1990s saw yet another set of vendors take the 802.3 specification to data rates that virtually no one had believed possible just a few years before: 1000 M/bs—called *Gigabit Ethernet* by most. Gigabit Ethernet retains the frame structure and other essential components of lower-speed implementations while making changes in areas not visible to the casual user: apparent frame size, maximum segment length, encoding method, and transmit clock rate. Initial implementations of Gigabit Ethernet ran only over optical fiber, but a standard for the use of twisted-pair cabling was added by the IEEE in mid-1999. For the moment, Gigabit Ethernet is primarily implemented for interswitch links and other areas of high traffic concentration.

In addition to the actual data rates of the medium themselves, another innovation has appeared that offers particular promise for the provision of real-time services such as telephony over Ethernet: full-duplex capability. From its inception, Ethernet has operated as a half-duplex technology, a constraint imposed by its shared-cable origins. An illusion of full-duplex operation can be achieved in traditional Ethernet only by virtue of its high data rates, and this illusion breaks down as contention increases. However, as LAN switches have come to form an important part of LAN architectures, they have in many cases been used to create segments that consist only of two devices: either two switches or an end station such as a PC and a switch. Such a segment, by virtue of its physical medium (two wire pairs or two optical fibers, one exclusively for the transmissions of each device), is therefore immune to contention, and each device can transmit at will, without regard for the other device on the segment—which has its own wire pair or fiber on which it is doing the same. Assuming that both devices are capable of and configured for full-duplex operation, this effectively doubles the amount of data that a given segment can carry in a period of time. The ramifications of this are particularly beneficial for real-time applications such as voice, because a fully switched full-duplex connection between two Ethernet devices can in almost all respects emulate a circuit-switched environment.

Prioritization As in the Internet, however, there is probably some limit beyond which throughput in Ethernet LANs cannot grow, even in the best-engineered networks. While network managers have continued to upgrade network equipment to support LAN switching, virtual LANs, and higher data rates, there is a point at which these paths become inadequate or simply too expensive to continue.

Efforts to allow for the prioritization of some LAN traffic over others—like similar efforts in the IP market—therefore have received much attention as data/voice/multimedia convergence looms over the LAN world. While vendors have equipped their switches with proprietary, and somewhat rudimentary, means of prioritization on a per-device level for years, the provision of a multivendor, multinetwork means of granting priority on a per-packet basis has been made available only recently in the form of the IEEE 802.1p/Q specification. While nominally useful with almost any LAN technology, the greatest focus for 802.1p/Q has been the provision of multimedia services over Ethernet. The use of 802.1p/Q is shown in Figure 9-5.

802.1p/Q defines a 32-bit header that, when appended to the Ethernet header, allows for the specification of priority levels, vLAN membership, and multicast and broadcast control. The "p" section of the specification (that concerned with priority) defines a 3-bit field that may be set by switches or end stations, including telephones or multimedia terminals, to allow up to eight levels of priority. The presence of this field with a high-priority field value can then be a cue to switches, routers, and other equipment to treat the packet with the designated quality of service (QoS). Since the tagging of frames with priority information is performed on a per-frame basis, this provides a means by which even a multimedia PC, which may be simultaneously sending both data and voice traffic, can dynamically indicate which frames should be given special handling and which are to be treated in the normal, "best effort" way.

While this system shows a great deal of promise, it provides as yet no specific mapping between priority levels as set in the frame and QoS parameters as requested by particular applications such as VoIP. The only requirement is that "higher is better" with regard to the value in the priority field. Therefore, while most switch manufacturers support the 802.1p/Q priority specification, they all do so in their own individual ways, leading to a very inconsistent approach to priority across multi-vendor networks. Furthermore, there are not yet multivendor specifications for the mapping of 802.1p/Q priority levels to the QoS parameters

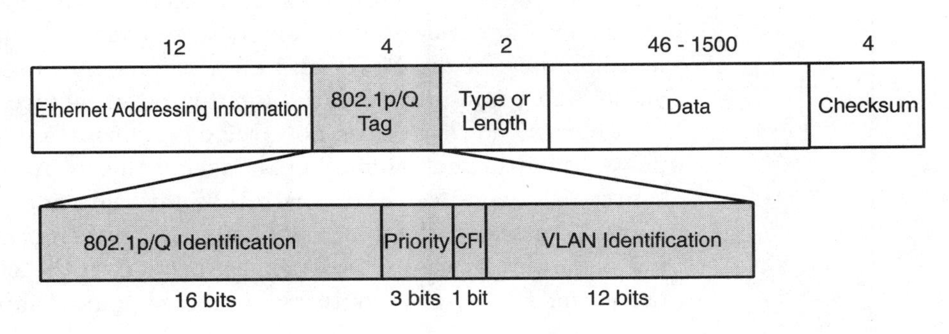

Figure 9-5 VoIP carried in Ethernet, with 802.1p/Q tagging.

(e.g., low delay, high throughput, etc.) becoming available in IP networking, meaning that these priority levels—whatever their final utility—become nearly meaningless once the packets that they contain leave the LAN and cross into the Internet or other IP WANs.

An Ethernet network suitable for voice—one in which switches, vLANs, high-speed links, and prioritization work together to provide a low-latency, high-bandwidth environment for VoIP—is clearly possible. The provision of simple VoIP over such a network is actually relatively simple, requiring only PCs with sound capabilities and client VoIP software to function. The challenge for network managers seeking to replace their PBXs with LAN telephony is in designing and managing their LAN using the technologies discussed earlier and in implementing features in a manner that best supports their business.

LAN Telephony Phones

The second component of a LAN telephony system is arguably the one on whose success the entire technology hinges: the telephone itself. These products come in a number of form factors, from normal-looking handsets that attach to cards mounted in a desktop PC to completely standalone units that are impossible to tell by appearance from a standard office phone. (Since the current discussion is focused on near-transparent PBX replacement, a discussion of the use of PCs as "phones" will come later in this chapter.) However, while Ethernet phones may not differ much from standard PBX phones in outward appearance, their approach to providing phone service is considerably different.

First, the electricity necessary for a standard PBX phone to operate is usually furnished by the PBX switch, which becomes in essence a power supply for its entire network of phones. Providing the switch with a failsafe source of power, such as a battery backup unit, ensures that the entire phone system will continue to function in the event of an electrical outage or other, more serious emergency. This "bulletproof" survivability is a hallmark of the PSTN and an attribute that users understandably value highly and that potential liability often requires.

Ethernet, however, does not provide any constant source of electrical power, so Ethernet phones must derive their power from some other source. The simplest option is to use an ac/dc adapter connected to a standard power outlet near the phone, which then provides a plug carrying dc power (typically 48 V) for connection to the phone itself. Unless the power supply or its outlet is protected against failure, however, this approach does nothing to provide high availability. The second option, one that is being supported by an increasing number of LAN telephone vendors, is called *fourth-pair power*. This system uses one or more shared power supplies to distribute 48-V power over the unused fourth pair of wires in a typical Ethernet wiring plan (in a convenient bit of happenstance, Ethernet premises wiring is typically furnished via four-pair cable, only two of which are actually used for Ethernet signals), as seen in Figure 9-6. It must be said that this relatively new use for those previously unused pairs has yet to be tested with all makes and models of networking equipment, and some hubs and other connectivity equipment surely will not appreciate having 48 V suddenly appear on their interfaces—particularly those hubs which autonegotiate between two- and four-pair (TX and T-4) operation. In addition, how this method of power provision will affect the interference characteristics of marginal segments remains to be seen.

Having been furnished with power, an Ethernet phone becomes a completely legitimate network node, requiring a set of addresses: a six-octet MAC address for Ethernet's use, as well as a four-octet IP address. The LAN adapter comes coded with a unique address from the factory, but the IP address must be assigned by the network administrator to fit into the network scheme. IP address management traditionally has been one of the constant headaches of a LAN manager's life, as the number of network devices grows and he or she is forced to provide an addressing system that handles them all in an organized manner.

Figure 9-6 Fourth-pair Ethernet power.

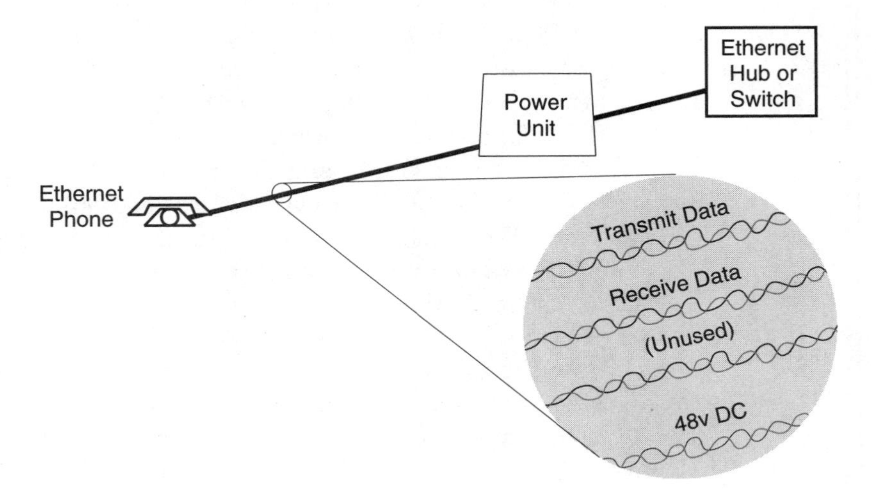

The manual configuration of IP address information into every node of the LAN can be obviated by using the Dynamic Host Configuration Protocol (DHCP), which uses a centralized server or series of servers to provide address and other IP information to hosts on an as-needed basis. The DHCP standard is supported by virtually all Ethernet phone manufacturers, so furnishing an address to a new phone is as simple as plugging it in—provided that there is a DHCP server configured to give it an address. While the use of DHCP is common, it is by no means a given, particularly in the small to middle-sized office environments that LAN telephony primarily targets. Most Ethernet phones are able to be configured manually with an IP address, although doing so eliminates the ability to move the phone from network to network with no configuration, since the address will need to be changed each time. Therefore, network support for DHCP, in the form of a DHCP server, should be considered a prerequisite for the deployment of Ethernet phones.

Once configured with an IP address, an Ethernet VoIP phone is a fully functioning member of the LAN and (assuming a suitable Internet or WAN connection) theoretically can send IP packets to any other compatible phone or gateway, provided it has some means of resolving the dialed phone number into the appropriate IP address. This capability is provided by the call manager, which is discussed below.

The outgoing VoIP packet stream, as well as incoming packets, contains voice data encoded according to one of the various specifications for VoIP encoding, either proprietary or standards-based. Some phones are capable of multiple encoding, schemes; for example, calls traveling over LAN and high-bandwidth WAN links may be encoded according to the 64-kB/s G.711 PCM scheme, whereas calls that must travel the Internet or other networks where bandwidth is a concern may use the 6-kB/s G.723 codec. This particular dual capacity is within the specifications set by H.323.

If the phone and network support it, outgoing voice data packets should be tagged for preferential treatment by the network, according to either (or both) the 802.1p/Q specification and one of the various IP QoS methods. While LAN telephony manufacturers have agreed that priority and QoS are of concern even in LAN environments, actual implementations of phones that conform to newer specifications for providing it are as yet rare.

How the phone operates beyond the standard ability to place and receive calls, and receiving and decoding the audio signal, is limited only by the imagination of the manufacturer, and Ethernet phones are as capable as standard PBX phones at providing all manner of "convenience" features, such as speakerphone capability, programmable buttons, distinctive ringing tones, line-status and message-waiting lights, and LCD displays.

Yet another feature that some Ethernet phones offer is a two-port Ethernet switch integrated into the phone unit. This simplifies physical connectivity by allowing the phone and an office PC to share the single Ethernet cable run that most one-desk offices have, thus obviating the need for additional cabling to support the phones. The disadvantage of such a setup is the interdependence thus created—disconnecting the phone to move it to another jack must interrupt the connectivity of the PC.

In addition to purpose-built Ethernet VoIP telephones, several manufacturers have announced converters that allow the use of one or more standard dial phones (the ubiquitous "2500 set") on an Ethernet LAN by providing interfaces for both the phone(s) and the network on a single small device, which appears like a phone network to the phone(s) and an Ethernet phone to the Ethernet. While such devices allow for the simple and relatively inexpensive connection of a number of phones, they are essentially gateways in miniature and therefore complicate the management of security and configuration.

Finally, there are also "virtual phone" software packages distributed by LAN telephony vendors to allow implementers to deploy gatekeepers and other components of the system without the need to purchase Ethernet phone hardware for all users. These packages can run on almost any PC with Ethernet and sound capability and provide all the functionality of a standalone phone via a "virtual keypad" on the user's screen. While most users will insist on a traditional handset of some type, there are third-party handsets that attach to a PC via a universal serial bus or standard serial (com) port, providing a reasonable approximation of a regular phone's "feel."

Along with the disadvantages of having one's phone (an extremely reliable device) dependent on one's PC (a much less reliable device) comes the need to learn a new paradigm of phone use—a simple matter for some people that may prove to be an insurmountable obstacle for others who are inherently less adaptable or less willing. These difficulties for virtual phone deployment are somewhat offset by the ability for phone and data service to share the same Ethernet drop and the ability for users to have multiple "phones" using a mixture of virtual and hardware devices. These considerations mean that virtual phone packages are probably best seen as complementary to, rather than replacements for, hardware phones in all but call centers and other vertical telephony markets.

Gatekeeper or Call Manager

H.323, Session Initiation Protocol (SIP), and other proprietary solutions define *multimedia terminals* that are fully capable of operating in pairs or groups to connect and complete VoIP calls. Ethernet phones, because of their required low cost, simple interface, and need to operate according to standard telephone operating skills, are not capable of this except for a small percentage of their potential calls. Only when the user (or network manager) has programmed a particular speed-dial button with the IP address of another gateway or phone can a user accurately specify the device he or she intends to dial. Simple dynamic dialing (i.e., being able to dial any phone number and have it connect), as well as any features that depend on the cooperation of other devices (e.g., conference calling), requires the services of a call processing unit. This device goes by many names. H.323 includes some of the required responsibilities under its gatekeeper definition; vendors use the terms *call manager* and *communication center*.

Chief among the responsibilities of a PBX gatekeeper is call establishment, the first step of which is address resolution. When a user picks up an Ethernet phone and dials a number (or when a call establishment message is received from across the wide area), the called number information is passed to the gatekeeper. By comparing the called number with a directory database (either internal to it or provided via a directory service such as LDAP), the gatekeeper can determine how the call should best be handled. A local extension lookup may return an IP address on the same LAN, whereas a phone number of a phone in a different city may return the address of a gateway in that city, which may be reached via a private WAN, a managed IP network, or the Internet. For numbers that have no gateway or phone associated with them, the gatekeeper may return no address at all and therefore must use the PSTN to route the call. PSTN-routed calls require the services of a PSTN access gateway, a kind of "VoIP gateway in reverse," discussed below.

The gatekeeper then proceeds to initiate the call, by using H.225, SIP, or proprietary signaling to contact the called party. Upon a successful connection, the gatekeeper "hands off" the call to the two connected phones or phone-gateway pair. If the call cannot go through, because the called phone was in use or otherwise unavailable, the gatekeeper will then inform the caller with a busy signal or redirect the call to an automated attendant or voice mail system.

As the "brains" of the LAN PBX, a LAN gatekeeper is also charged with providing all the services that the center of the traditional PBX—

the PBX switch—typically has provided. Most of the basic ones, such as call hold, call transfer, call forwarding, and call waiting, are relatively simple to provide at the gatekeeper, since all call establishment and control must occur through it. Still, there are an estimated 500 different functions that traditional PBX systems are capable of providing, and while no implementation uses them all, one or more of them may be extremely important in vertical markets or to the employees of a particular office. Managers wishing to replace their PBX systems would do well to ensure that the features their users currently depend on are supported by the LAN PBX under consideration.

Common PBX features are necessary, but somewhat pedestrian, and offer little interest to those seeking an answer to common business problems from their phone system. However, one of the more exciting developments that LAN telephony, and in particular the gatekeeper function within it, makes possible is the final emergence of a wide variety of computer-telephony integration (CTI) applications. The arrival of applications such as unified messaging and data/voice conferencing has been one of the most anticipated events in the telephony universe, but the reality to date has been a disappointing mixture of complex products with limited applicability and expensive custom development and programming. By hastening the arrival of data/voice convergence on the LAN, the LAN PBX may be just the infrastructure that CTI applications need to be developed and deployed in earnest. These features are discussed in detail below.

PSTN Access Devices

While the day may arrive when all voice traffic is carried in IP packets, for the foreseeable future, LAN telephony systems must provide the capability to connect to the PSTN in order to connect those calls which cannot be routed over the IP network cloud. The device that makes this possible is known as a *PSTN access gateway* (also called an *access switch*), and it is in all respects a gateway, albeit one used in a manner entirely backward from the more common gateway concept as a means to reduce carrier costs.

Figure 9-7 shows the opposite relationship between the LANs and WANs that the gateway spans when used for PSTN access or PSTN avoidance. In a LAN telephony system, the PSTN access gateway is the device to which outgoing calls are directed by the gatekeeper when

- No association between the called number and a remote gateway (one that is presumably closer to the callee than the caller) is found.

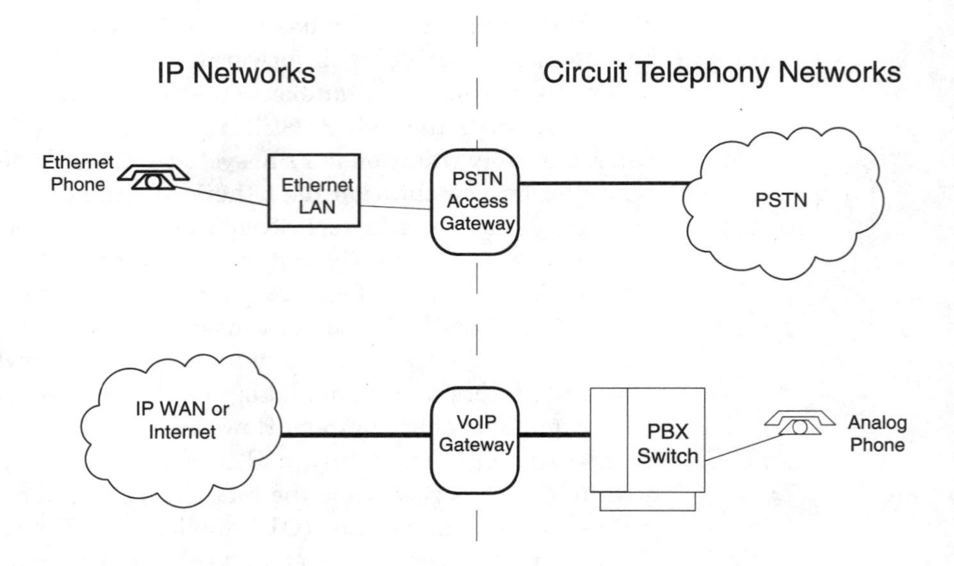

Figure 9-7 Gateways in both directions.

- The destination gateway is unable to accept an incoming IP call for any reason, such as having filled all its ports or because it is shut down for maintenance.
- An IP route to the destination cannot be found, because of failure of a network link, a routing protocol error, or failure of a node such as a router.

Some errors of these types, particularly those caused by a busy or inoperative remote gateway, can be corrected by connecting to a gateway that, while not ideal, is at least closer to the called number than the caller. However, there are other errors, such as the failure of an office's main link to the WAN or extreme congestion on the Internet, that require a capability for *PSTN fallback*—the ability to always get a connection, regardless of the means. (The fact that this capability is perceived as necessary by virtually all vendors of LAN telephony equipment shows that—regardless of statements to the contrary even in this book—the packet-switched world still has much work to do to become as reliable as the PSTN.)

In addition to providing a means for placing traditional telephone calls when necessary, PSTN access gateways also provide the essential capacity to connect incoming PSTN calls to the LAN PBX. The access switch must both pass signaling information to the gatekeeper when receiving a call from the PSTN and—once connected—encode and decode the audio stream on behalf of the remote caller. Most office phone systems will continue to see the bulk of their incoming calls delivered by their LEC, which (at present, anyway) will only do so via the same T-1 and ISDN PRI lines that have been used for years to connect traditional PBX switches to the PSTN. The time may come when LECs offer to deliver incoming calls in packetized form, but there are no public plans to do so yet.

The use of PSTN access gateways provides room for some irony in LAN telephony systems, since an inoperative router may cause a call that normally would be routed end to end over IP networks instead to be switched to the PSTN on leaving one office, only to incur long-distance charges as it travels an IXC network, and finally be repacketized by the distant access gateway for receipt by an IP phone. The pessimist might observe that those long-distance tolls are precisely what VoIP promises to avoid; the optimist might remark that at least the call did go through, proving the essential robustness of an integrated phone system.

All in all, the job of the PSTN access gateway is perhaps the least exciting yet most important in the entire LAN telephony system. In addition to providing the system's connectivity to the great majority of people (who have no VoIP capacity yet), the reliability features that it provides, such as PSTN fallback, are an essential part of making LAN telephony systems robust enough to serve mission-critical telephony requirements.

Benefits of LAN Telephony

As more than one networking pundit has observed, it will take more than the promise of eventual savings to convince network managers to give up their tried PBX technology in favor of the new world of LAN telephony. Cost savings aside, a large part of the attractiveness of LAN telephony is the potential for new applications that it may enable. Additionally, the inherent advantages in ease of deployment, management, and reconfiguration that a packet-based technology has over a circuit-switched one may help convince managers to consider the switch—at least at a time when they have outgrown their current systems or their current contract for PBX (or Centrex service) expires. Listed below are some of the most compelling reasons to make the change.

Simplified Adds and Changes

The relationship between a gatekeeper or call manager and the Ethernet phones or terminals that it serves is based on a logical network diagram instead of a physical one. Traditional PBX switches must associate a physical port with a particular extension, requiring exacting cable management and physical reconfiguration in order to accommodate the physical relocation of a user's phone when workers change offices. Since membership in an IP PBX is based on logical information—the IP addresses of the phones—which may be changed from a central management console, physical reconfiguration of the switch is no longer required. Indeed, in many LAN telephony networks, a user could walk around the office plugging his or her phone into any available Ethernet port, maintaining the identity of, and connectivity to, "his" or "her" extension all the while. The addresses associated with the phone allow it to be identified regardless of its location in the network topology.

The logical nature of LAN connectivity also allows for much simpler addition of users and phones to the network, since there is no requirement for a particular number of physical ports on a PBX switch. With traditional PBX switches, the only way to extend the capacity of a switch is to add an extra port module—an expensive proposition if it is necessary to accommodate one or two additional telephones. Gatekeepers, while they have recommended limits beyond which the number of phones they serve should not grow, usually can be "convinced" to process the calls of that "one more phone."

Multimedia Terminal Capability

Although this chapter has focused on the use of Ethernet phones as a way to replace a current PBX system with no change in the behavior of its users, an Ethernet phone provides only a small part of the functionality available under the protocols that LAN telephony uses. The H.323 specification, for example, while it *requires* the support of audio signals and therefore is a natural for telephony use, also provides for the existence of multiple streams of multimedia traffic between two or more terminals. Desktop videoconferencing, when seen from the point of view of a LAN telephony user, is simply the existence of another stream of media traffic between the two units. Since H.323 also supports T.120 data conferencing over one or more additional data streams, allowing the dynamic sharing of data files, shared whiteboard collaboration, and other innovative uses.

The fundamental infrastructure of LAN telephony, therefore, provides one of the components for new applications such as distance learning, broadcast or closed-circuit television, and "virtual meetings."

To take advantage of this infrastructure, the phones must be replaced by devices capable of using the media and data streams provided. Many offices, however, are already in possession of such devices in the form of their desktop PCs—many of which are equipped with the essential components for these new applications. At least two organizations interested in encouraging the spread of desktop multimedia (the Multimedia PC Marketing Council and the Information Technology Training Initiative) have produced specifications for the minimum requirements necessary in such a system. While minimum requirements vary depending on the exact application, a reasonable approximation of such a specification might be

- A reasonably fast processor, on the order of a Pentium- or PowerPC-class chip running at 133 MHz or more
- A minimum of 8 MB of RAM
- A 540-MB hard disk with 1.5 MB/s sustained throughput
- A CD-ROM drive capable of a 600-kB/s transfer rate
- A video card capable of displaying MPEG1 video
- A 16-bit sound card capable of high-quality audio
- A microphone and speakers or, alternatively, a handset developed expressly for LAN telephony.

Most PCs that meet these rather modest minimums are capable of participating in the majority of multimedia sessions that are likely to be implemented in the next few years. Moreover, the short life cycle of the typical corporate PC (many are replaced every few years) means that the installed base of multimedia-capable desktops will increase at a rapid rate. Indeed, it is nearly impossible to find new PCs being sold in 1999 (even among discount resellers) that do not far exceed the specifications in the preceding list, so commonplace have these components become.

One item that remains a relative rarity in the deployed PC base is a video camera. While not necessary for all types of video sessions (one-way video with two-way audio is suitable for much distance learning and some other types of conferencing), a camera allows for the origination as well as receipt of video signals. Although most PC manufacturers do not currently have plans to include cameras in their typical configurations, they can be purchased as third-party items beginning at around $100.

Much of the software available for desktop conferencing allows for the connection of a number of different types of video cameras.

In addition to the hardware requirements just listed, a PC operating as a multimedia terminal must run software capable of supporting the terminal functions and communications protocols supported by the LAN telephony system. A great number of software packages are capable of supporting standards-based multimedia conferencing such as H.323. The Microsoft NetMeeting package, now distributed as part of Microsoft's desktop operating systems, is an H.323-compliant multimedia client that is already present on the vast majority of new PCs currently shipping. In addition to this notable example, there are a number of multimedia packages, for all operating systems, distributed for free or as shareware—a fact that results from the education community's early adoption of multimedia teleconferencing as a research endeavor.

Computer-Telephony Integration

Call centers and other vertical applications that depend heavily on advanced features have employed some means of computer-telephony integration (CTI), such as Automatic Call Distribution (ACD), for years. However, the difficulty and expense of the integration have limited the availability of most of these features to a relative small set of potential customers. By delivering call control and voice data information via the same protocol that is already used for data, and thereby making it inherently accessible to computers, IP telephony is seen by many as the technology that will finally allow CTI to deliver on its promise for the market at large.

At the core of CTI is the computer's ability to record, store, analyze, process, and otherwise act on telephony signals, including DTMF tones and, of course, digitized voice itself, as interpreted by the main building block of CTI, the voice response unit (VRU). While telephony signaling information is usually thought of as the capture of DTMF digits, interactive voice response (IVR) software may be used to capture a caller's voice response to a prompt and analyze that response for a match with a digital sample of the word *yes* or *no*, the spoken words *zero* through *nine*, etc. More complex forms of voice processing are in development, but actual implementations of them are as yet rare.

Once interpreted by the VRU, control information can be forwarded to the computer itself and used to do password authentication and menu-selected functions such as the intelligent routing of calls and message store and playback. One very common menu-driven CTI application is the automated delivery of fax pages (such as sales literature or technical support information), which may be stored in digital form on the computer and simply "printed" to a fax driver on being selected from a menu by a caller at a fax machine. In an IP telephony situation, this same document may be attached in its original form to an e-mail message, which can then be sent directly via IP.

CTI-enabled LAN telephony terminals, on receiving calling number information from the gatekeeper, may use this information to do a database lookup for information on that caller, querying a customer database housed on the network or that user's own personal information manager (PIM) software. This capability is already integrated into a number of PIMs for the PC. In fact, a wide range of telephony services application programming interfaces (APIs) have appeared, allowing programmers to key on any identifiable portion of an incoming call to perform computing tasks.

Finally, the capability for unified messaging (UM) (the ability to retrieve, and send, e-mail, voice mail, and sometimes fax transmissions from both telephones and computers) has been the most eagerly sought-after CTI application for many years, albeit one that had remained out of reach for most organizations because of complexity and expense. In a UM environment, users are able to access all their messages from either the IP or PSTN network. Most UM systems work by providing a message center, which interfaces into all applicable networks and serves as a central repository for all incoming messages. In traditional environments, this results in a server that is comprised of various combinations of telephony switch components, voice mail system parts, e-mail system parts, and other interfaces—all tied together with an operating system that manages their interaction.

IP telephony greatly simplifies the implementation of UM by providing a common, IP-based format for all messages, regardless of source. As shown in Figure 9-8, incoming e-mail messages may be stored by the

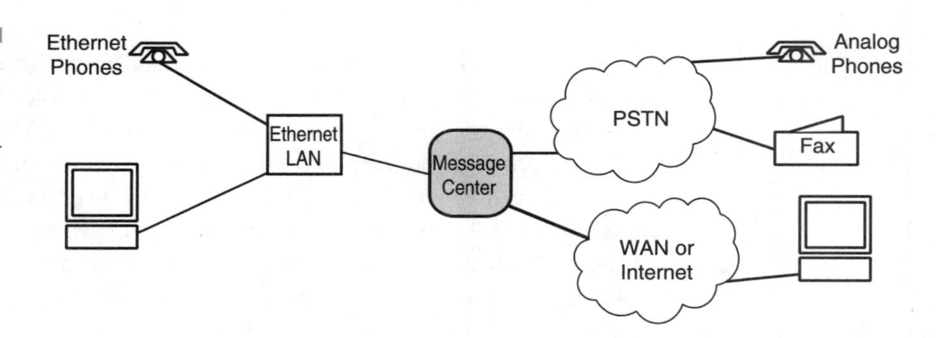

Figure 9-8 Unified messaging via LAN telephony.

server in exactly the same way that they are now, whereas incoming voice mail messages can be stored in the same mailbox, having been received by the message center in predigitized format over the Ethernet interface. A number of LAN telephony gatekeeper packages already implement this capability, formatting incoming audio messages as .WAV audio files and attaching them to e-mail messages that then can be stored by the e-mail system in the standard way. To retrieve his or her messages, a user may download his or her e-mail and click on the .WAV file icon to hear the message contained therein. This same user also could call the message center and, on being authenticated via passcode, retrieve both kinds of messages: the stored .WAV files being played over the audio signal in a manner similar to standard voice mail systems and the e-mails being "read" to him or her via a speech synthesizer. Integration of fax messages is somewhat more complex, because while an image of a fax easily can be digitally scanned for inclusion in an e-mail, the files are large, and the quality of most faxes is so poor that they become unreadable after scanning. Phone retrieval of faxes requires either a human operator to read and interpret the fax to the caller or optical character recognition (OCR) software that can "read" the fax (assuming it is mostly text) and place the message in an e-mail.

Although the implementation of UM is easier in LAN telephony environments than it has been traditionally, it is still a new technology that requires some growth before it emerges as a standard application. In order to reap the full benefits that being "constantly connected" will bring, UM systems must be simple and intuitive, reliable, and inexpensive enough for wide-scale implementation.

Challenges and Future of LAN Telephony

It is clear that there is much to be gained from the coming wide-scale deployment of telephony services over LANs. As telephony APIs become more widespread and begin to accommodate the LAN telephony model, many standard personal productivity applications such as e-mail, word processing, and spreadsheet applications will be capable of using the telephone network for dynamic help, content interpretation, and marketing.

Future applications will tie even more functionality into unified mailboxes. Executives can have their unified mailbox become a personal agent, sending personalized information and notification preset by the subscriber to those who leave messages. This capability has the potential to help vendors provide better customer service by allowing them to automate much more of the support service. It is conceivable that the telephone "customer support" personnel of many organizations may well find that in a few years their jobs have become less providing personal support and more maintaining the telephony support system.

In order for LAN telephony to become a contender for deployment in the PBX market, a number of objections regarding LAN/PSTN integration and packet-based voice communications in general must be overcome. Issues such as the preservation of QoS from Ethernet to WAN (or Internet) and back to Ethernet must be met with standards-based solutions. Experience in designing and installing LAN telephony systems must be developed in research and other environments. Interworking of LAN telephony with entrenched systems must be made possible to protect the significant investment that businesses have in their current systems. And a credible business case for the migration must be made by vendors and IT managers.

Finally, reliability must become a given, as it is for users of traditional PBX systems today. Toward this end, some vendors have introduced fail-safe features into their switches that are almost impossible to defeat, such as using backup physical relays to provide secondary connectivity. When the power fails, these switches are able to connect certain preconfigured phones directly to outside lines. While such an approach still depends on the PSTN to provide the robustness that business users demand, it is an important first step in the eventual development of "bulletproof" VoIP.

The ultimate success of LAN telephony depends on market perception, since managers who make deployment decisions must become comfortable with the idea of the LAN, and the PCs that inhabit it, as a viable alternative to the circuit-switched PBX. The eventual emergence of the advantages of packet-based telephony, however, will slowly yet surely tip the balance. LAN telephony, playing its role as the "new customer premises equipment" of the VoIP world, will play a key part in the ultimate consolidation of IP telephony.

CHAPTER 10

Internet Fax

While most people think of the facsimile machine as a relatively new invention, the principle for the electrically assisted transmission of images and text has existed since the nineteenth century. A Scottish psychologist named Alexander Bain proposed a method for using "electric current...in electric printing and signal telegraphs" in a patent application of 1843, only 7 years after the invention of the much more famous electric telegraph by Samuel Morse.

Bain's invention used two distant synchronized metal pendulums, each of which swung between the ends of identical metal rollers, over which paper was passed slowly. The sending side's paper was marked in electrically conductive ink, which allowed current to pass from the roller to the pendulum when it passed over words or images. The electrical impulses thus produced traveled up the pendulum, across the wire connecting the two machines, and down the pendulum on the second machine, where it marked the photosensitive paper passing over the second roller and completed the circuit. The first actual commercial use of this principle was developed by an Italian abbot, Giovanni Caselli, who with the support of Napoleon III began offering his "pantelegraph" service between major French cities in 1865.

Ironically, the spread of this remarkable technology was suppressed by the simultaneous success of the much less flexible telegraph, which nevertheless was considered by most to be a superior method of sending text over long distances. The "digital" nature of telegraphy, in which the Roman alphabet and Arabic numbers are encoded in Morse code, offered a much more accurate representation of the original message when sent over long distances and was much faster than the pantelegraph. The commercial success of the first facsimile transport therefore suffered, and the Morse telegraph became the first wide-spread telecommunications service.

While facsimile technology lived on in research laboratories in Germany and China (where its graphic nature made it much more suitable than Morse code for the reproduction of handwritten Chinese characters), it was not until the 1920s that commercial facsimile products again appeared. While expensive, and therefore available only to companies with a profitable use for the service, these found limited use in news service bureaus and banks. It remained for the electronics revolution of the 1960s to make the technology cheap enough for small offices and eventually even home use. The Japanese phone company NTT was the first service provider to allow the connection of fax equipment to its ordinary dialup phone lines, which it did in 1973 for the same reasons that so interested the Chinese a century before. From the 50,000 fax machines worldwide in 1970, the real revolution in facsimile has occurred in the last 20 years, resulting in a world-wide base of tens of millions today.

Why Fax Persists

The overwhelming success of electronic mail and the Internet, which allows the distance-insensitive transmission of text and images (through message attachments), would seem to spell the twilight of the fax era. Yet fax machine sales continue to climb, even as the Internet finds increasing utility in the daily life of more and more people. From a data communications point of view, the use of fax machines to send most types of messages is almost maddeningly slow and inefficient. For example, using current state-of-the-art fax technology, a fax machine takes roughly 1 minute to transmit a typical single-page text memo about 150 kB in size. The same text can be transmitted in an e-mail message in less than a few seconds, due in part to the modem's faster transmission rates but mostly due to the representation of the text as digital data from start to finish, requiring no digital-to-analog conversion or mechanical scanning and printing. Today's fax machines are nothing more than faster, somewhat more accurate versions of the same technology Bain developed in 1843, with many of the same drawbacks—and yet their popularity continues to grow. Why is this so? There are, in fact, a number of excellent reasons:

- *They are easy to use.* Put in the paper, dial a phone number, and press the send button—that's all the training most people need to operate a fax machine. There are units with somewhat more complex features, and some fax-based technologies (such as interactive fax response and fax broadcast) are more like computer systems than fax machines. But most fax transmissions are simpler than using the postal service.
- *They are always the same.* While there are dozens of fax machine manufacturers, the user interface remains remarkably consistent between makes and models—another benefit of the inherent simplicity of the interface.
- *They are inexpensive to purchase.* Full-featured "home fax" systems, which are built for less constant use than business models, are available for less than $200, while the average business finds fax machines inexpensive enough to consider them office commodities.

- *They are good enough.* Given a good-quality original and a decent phone connection, the quality of text and graphics reproduction, while far from reproduction quality, is acceptable for most business uses.
- *They are everywhere.* Virtually every office in the developed world has one or can purchase one inexpensively if business warrants. Hotels, business centers, airport lounges, and many waiting rooms have "public use" fax machines for use at no cost or for a small fee. Airlines have even begun to equip their business- and first-class seating areas with wireless fax machines for use in flight. Add the proliferation of PC modems with fax capability (included by default in almost all modems produced after 1996), and it is difficult to see how anyone could claim not to have at least temporary fax access if they need it.
- *They always work.* One of the benefits of the underlying simplicity of fax technology is its robustness—the machines tend to be very reliable, even under constant use. The electronics have been perfected to the point where mass production of fax chipsets is a commodity process, resulting in very dependable products. Paper jams and other failures related to the paper-handling process are the most common errors. The Group 3 fax specification that almost all machines adhere to (and which is discussed below) ensures compatibility across a wide range of manufacturers and models.
- *They are legal.* Despite an increasing trend toward the acceptance of digitally signed electronic mail messages as contracts and other legal documents, there is already a significant amount of legal precedent upholding the status of facsimile transmissions as legally valid. This status is due in large part to the fact that they can easily represent a handwritten signature.
- *There is nothing virtual about them.* Finally, in a world increasingly comprised of intangible methods of delivering information, many people feel reassured by the tangible presence of a fax message when it arrives. Faxes may be written on with a pencil and filed in a cabinet—not to mention slammed on a desk for emphasis, torn up in a fit of rage, or kissed when they bring good news. They are physical entities that are simply more traditional and comfortable for some people to work with, particularly those who have not fully embraced the digital age.

Thus, while the inefficiency and slow speed of fax perhaps make it technologically less interesting and less "beautiful" (if such a word can be used to describe any data communications technology) than some other solutions, its benefits are very real to the hundreds of millions of people who use it. In fact, some users have suggested without irony that before electronic mail, the World Wide Web, and other Internet technologies can become truly ubiquitous, they must become “as easy to use as a fax machine.” This is a sentiment that deserves serious consideration by software and hardware developers.

Why Fax Must Change

Despite all these excellent attributes, fax does have its drawbacks:

- *Transmission costs.* While the machines themselves are inexpensive, current facsimile technology is inexorably tied to the circuit-switched public switched telephone network (PSTN), and is therefore subject to the same call pricing model as telephone calls. While local use is inexpensive or even free—depending on the telephone service contract—long-distance and international faxes can be staggeringly expensive when the document to be transmitted is more than a few pages long. Since many calling plans charge more for the initial minute of connection time than for subsequent minutes, even the shortest trans-Pacific fax message, for example, can cost close to $10 if sent during business hours. There are many estimates of the amount of money business spends each year on the transmission of faxes, and while they are quite divergent, for a typical Fortune 500 company, the amount is in the millions of dollars. This fact alone creates a significant incentive for the exploration of alternatives and the improvement of the current technology.
- *Lack of intelligence.* This is the unfortunate corollary to their ease of use. There is a limited number of intelligent applications for current fax technology, beyond call-in fax-on-demand servers for document distribution and fax broadcast servers, whose dependence on PSTN pricing makes them relatively difficult to scale to large populations.
- *Weak business integration.* Nothing about faxes besides their ease of use makes them integrate well into typical office procedures. Electronic documents are usually printed to paper before they are faxed, leaving the original paper to then be disposed of unless it is to be filed. Fax modems do allow the direct transmission of faxes from computers to remote fax machines but have yet to become commonly used, largely because the differences in operating procedure between them and normal fax machines neutralizes many of the benefits of fax in the first place.

- *Poor security and privacy.* The common use of centralized fax machines, usually located in a relatively public place such as an office corridor or secretarial area, means that unless attended by the sender or receiver, documents in fax machine input and output trays may be perused at leisure by anyone who happens by. Unsolicited or poorly coordinated transmissions are particularly at risk of snooping. Finally, the use of paper in the fax process leaves two copies of each transmission (one at the sender's side and one at the receiver's) to be managed or securely disposed of.

While these and other problems persist, they have not proved a significant enough disincentive for most businesses to stop, or even reduce, their use of fax machines. Taking into account all its positive and negative attributes, while facsimile service may one day succumb to the practices of an increasingly digitized world, it is clear that its use will continue to be a part of common business procedure for the foreseeable future.

This does not mean that users are not aware of the problems. In particular, the high price of long-distance PSTN fax is a target for cost reduction in many offices. Since it persists, fax must change in whatever ways it can to take advantage of the inexpensive worldwide transmission capability of IP networks, particularly the Internet. It must do so without, at least in the short term, losing the characteristics that make it so attractive.

How Fax Technology Works

The principal elements of fax technology have not changed that much over the years, the main developments coming incrementally, in the form of faster transmission rates and better image quality. Currently, by far the most widely implemented procedures for fax operation are defined in two IUT-T recommendations, T.30, *Procedures for Document Facsimile Transmission in the General Switched Telephone Network*, and T.4, *Standardization of Group 3 Facsimile Terminals for Document Transmission*. In summary, T.30 defines the methods of communications session management, whereas T.4 defines the page definition and coding schemes used to describe the image being transmitted. These two recommendations taken together describe the Group 3 (G3) method of operation, which is the standard adhered to by virtually all machines manufactured since the early 1990s.

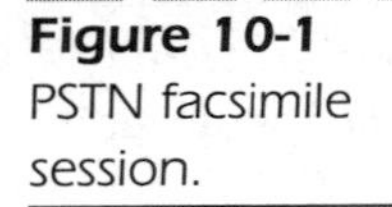

Figure 10-1
PSTN facsimile session.

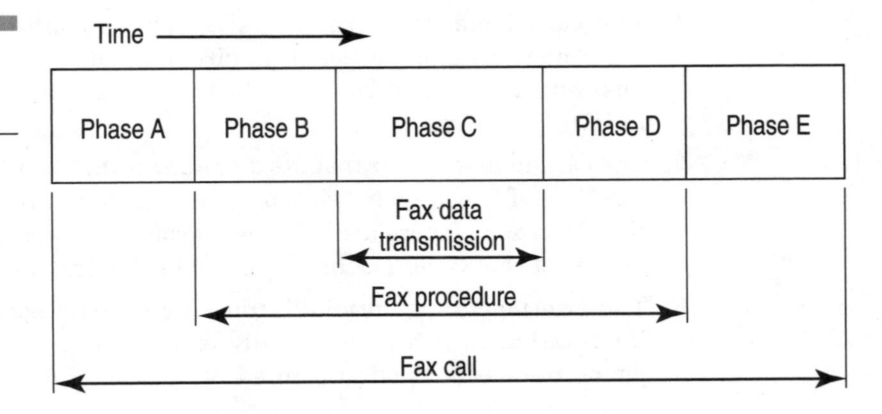

T.30 defines a number of different manual and automatic terminal modes to allow for different combinations of attended and automated operation. All modes share similar basic phases of operation as described in the following example, in which it is assumed that the calling machine is the sender of the fax (Figure 10-1).

Phase A: Call Establishment

This phase is simply the dialing and answering of the phone call used to provide the circuit over which the fax will travel, each of which may be accomplished either manually by the person operating the machine or automatically (e.g., according to a timer or signal). In automatic (unattended) operation, the calling machine sends a series of 1100-Hz tones called *calling tones* (CNG) immediately after dialing the phone to allow a human mistakenly answering the call to identify it as a fax call and take appropriate action. In any mode of operation, the answering fax machine is required to transmit a *called station identifier* (CED), a 2100-Hz tone to inform the calling machine that it has in fact connected to a like station.

Phase B: Premessage Procedure

Included are both an identification section and a command section. Immediately following the connection of the call in phase A, the two machines exchange capability information to determine what types of features each is able to offer, as well as optional identification information. This exchange occurs according to the following procedure:

1. The called machine transmits a frame to inform the calling machine of its capabilities (e.g., size, resolution, supported transmission speeds, etc.). This and all other messages are encapsulated in HDLC frames.
2. The calling machine transmits a similar frame to inform the called machine of the transmission parameters that will be in effect for this document. These are selected, of course, according to the capabilities of the called machine learned in the first step.
3. The two machines optionally may now exchange identification information, which is essentially expressed as corporate ID and phone number and often printed at the top of a printed fax message or displayed on an LCD screen.
4. Following this, the two machines determine the maximum transmission speed that the link between them can reliably sustain, through a procedure known as *training*.

Phase C: Message Transmission

The fax message itself is now sent, according to the formats and procedures stated in the T.4 recommendation. T.4 standardizes the possible paper sizes, as well as the orientation of the paper, in order that the two may be the same at both ends of the transmission. The sending machine then begins to scan the document left to right, proceeding from the top of the page to the bottom. Resolution of the scan (in both the horizontal and vertical dimensions) may be either 200, 300, or 400 pixels per inch, depending on the capabilities exchange and the parameters chosen in phase B.

The actual fax data would be enormous if it were encoded as a simple string of ones and zeros. A piece of letter-size paper, scanned in its entirety, would require almost half a megabyte if transmitted in this manner—even in the lowest resolution. Instead, since faxes often contain large repeating runs of white or black pixels (think of the margins of a letter or the white space in between lines of text), they are encoded according to a run-length encoding (RLE) scheme, where alternating "runs" (consecutive pixel groups) of either white or black are encoded as unique binary words. For example, a line that is composed entirely of white pixels can be represented with the string 010011011 rather than the 1728 zeros it would take if each pixel were given its own bit value. In practice, this disparity is not so great, because the "average" line is a mix of black and white pixels, but in the long run—so to speak—RLE saves an immense amount of transmission.

Each coded line begins with a white word (if the line's beginning is in fact black, the white zero run word is used) in order to maintain synchronization. There is also a special codeword for the end of a line (EOL)—a string of 11 zeros followed by a single one—to notify the receiver that the next codeword begins a new line. The end of the last line in the page is marked by a special return to control (RTC) code word, comprised of 6 consecutive EOLs, which indicates the start of phase D.

Phase D: Postmessage Procedure

On the completion of the page, the postmessage procedure allows the calling machine to indicate its desires to the called one, including specifying the end of the fax or the transmission of another page. Depending on the content, the called machine simply may confirm receipt of the page or may indicate its readiness to receive a second page. It also may indicate that it must abort the connection or continue only following a retraining. The two machines then proceed to either phase B or C or—if the end of the message has in fact been reached—to phase E.

Phase E: Call Release

This contains the procedures to hang up the phone connection, thus ending the session.

Internet Fax Standards

As with the rationale for voice over IP (VoIP), the main interest in Fax over the Internet Protocol (IP) is related to cost savings. Therefore, the typical IP fax network topology closely mimics the topology of "toll bypass" VoIP, with standard fax machines substituting for the standard PSTN phones at both ends of the call, which is itself completed by local or corporate gateways (Figure 10-2).

There are a number of other possibilities, ranging from purpose-built "IP fax machines" that connect directly to the IP network via an Ethernet interface to "IP fax servers" that are capable of acting as gateways for print jobs, thus avoiding a paper stage at the sending end. The

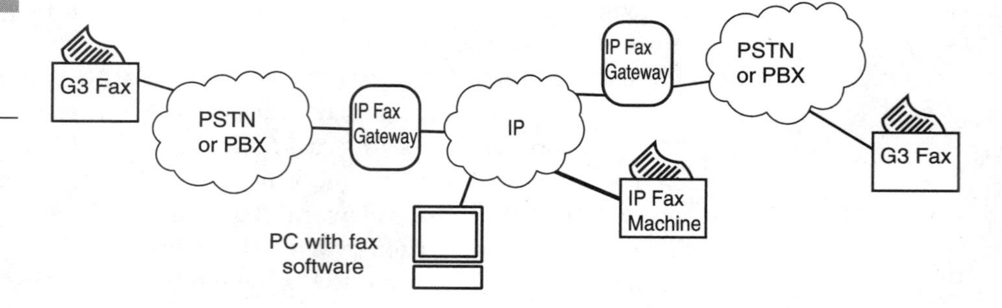

Figure 10-2 Typical fax over IP network.

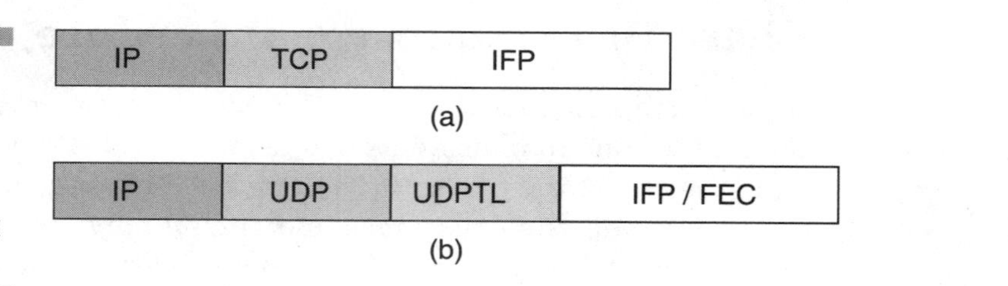

Figure 10-3 Encapsulation methods for fax over IP: *(a)* TCP/IP and *(b)* UDP/IP fax packet structure.

following discussion will address differences in these approaches where necessary.

The most watched specification for the construction of IP fax products is the ITU-T T.38 specification, entitled *Procedures for Real-Time Group 3 Facsimile Communication over IP Networks*. The "real time" designation is important because it is this quality that preserves one of the most important characteristics of traditional facsimile technology: the ability of the sender to know that the transmission is complete when the sending machine hangs up. Another ITU specification, T.37, addresses the possibility of store-and-forward fax but has not received nearly the amount of industry attention that T.38 has.

T.38 defines two packetization methods for fax data, one using the reliable Transmission Control Protocol (TCP) and one using the essentially unreliable User Datagram Protocol (UDP). There are benefits and disadvantages to each approach, the general encapsulation for which is shown in Figure 10-3. TCP has the advantage of being error-free, with lost or errored packets being retransmitted automatically at the transport layer so that the fax application is never given "bad" data for which it must request retransmission. Unfortunately, this error recovery process can add unacceptable delay to the data stream. TCP also has the disadvantage of being the first protocol to be discarded by most routers when congested (excluding routers implementing random early discard or other relatively advanced traffic-shaping algorithms), making the need for the error recovery more likely with TCP than with UDP.

UDP, while not error-free, contributes less overhead and requires no call establishment messages, thus contributing less to end-to-end delay than TCP. The lack of error correction means that the fax application must provide this function, correcting both errors introduced by the fax processes (which is what the HDLC checksum is supposed to do) and errors introduced by the packet network (which it was not intended to do). Most manufacturers of fax over IP (FoIP) software and hardware seem to be betting that the general low error rate of today's IP networks will allow their products to perform error correction at the fax layer only and are therefore supporting primarily fax services using the UDP transport method. The error correction and other TCP services therefore are provided by the fax layer or by proprietary protocols.

Carrying the fax data within the TCP or UDP messages are the Internet Facsimile Protocol (IFP) packets, of which there are two types: T30_INDICATOR and T30_DATA. As their names imply, both these packets are simply formalizations of the standard T.30 protocol messages used in standard PSTN fax communications. Both types of packets are employed in systems using UDP, whereas implementations using TCP may use only the DATA type and employ proprietary means of indicating CNG, CED, and other T.30-type control messages.

The IFP message flow between the gateways follows the same general procedure as that between two PSTN fax machines, with the exception of the training procedure used to determine the transmission speed (part of phase B of T.30). Since successful (or at least efficient) communications depends on both "PSTN ends" of the fax call operating at the same rate, training information must be communicated over the packet network to allow end-to-end synchronization. This procedure, called *data-rate management* in the T.38 specification, may be dictated by either the receiving gateway or between the two fax machines.

Once the data rate has been established, T.30 data packets may begin to flow from the transmitter to the receiver. When employing UDP, each packet consists of a UDPTL (UDP transport layer) packet containing a sequence number determining its place in the data stream, followed by one or more IFP packets—themselves containing the T.30 data. Depending on session information negotiated at the start of the call, the data may be protected from errors either by redundancy (in which multiple copies of the T.30 data are transmitted in each UDP packet) or by forward error correction (or FEC, in which some number of previously sent packets are included along with the current message to guard against the eventuality of some packet loss). Since TCP itself guards against data loss, the redundancy and FEC schemes are not necessary when using it.

IETF Fax Efforts

Despite the recent high profile of the T.38 recommendation, the ITU-T is not the only standards body working on an IP fax specification. The IETF has a working group chartered to produce a store-and-forward-based fax messaging specification, with later, optional extensions for "immediate delivery" faxes (a service similar to that provided by T.38). These specifications both use the standard for Internet electronic mail, the Simple Message Transport Protocol (SMTP), as a transport mechanism and add new protocols and procedures to handle addressing and other functions necessary to complete fax transmissions. Interestingly, it is this framework that resulted in the ITU T.37 specification.

The IETF chose to develop a fax framework around the electronic mail model with a particular agenda in mind. While the ITU-T was most interested in developing a series of standards for the interoperation of fax gateways, the IETF (as might be expected given its place as the primary Internet standards body) concentrates its efforts on allowing seamless communications between Internet e-mail and fax users. This capability would be useful, for example, to a user wishing to send a message to a group, some of whose members were reachable by fax and others by e-mail. Toward this end, the IETF approach (as outlined in its RFC 2305 document) is to define a number of simple communications cases, e.g., e-mail to fax machine or network scanner to e-mail. By defining these cases, the framework allows for the construction of complete fax/e-mail interworking solutions.

The IETF defines an Internet fax (IFax) device, which is simply an IP node capable of sending or receiving Internet faxes. IFax devices may be fax machines with local-area network (LAN) interfaces or gateways for the connection of standard G3 fax machines to an IP network. Whatever the actual IFax device, it must encapsulate its fax messages according to the standard RFC 822 Internet mail header format. The fax message itself is identified according to Multipart Internet Mail Encapsulations (MIME) standards and is encoded as a Tagged Image File Format—Facsimile (TIFF-F) file. The operation of the Internet mail subsystem, dealing as it does with protocols that have existed for some time and are implemented in virtually all networks, means that standard methods of encryption, authentication, and other aspects of data security may be employed with no complications.

Gateways or hosts sending fax data to IP hosts use standard Internet mail addresses, in the format *user@host.domain*. When the destination is a PSTN fax machine connected to a so-called offramp gateway, the message header must contain the telephone number of the fax machine in order that the offramp may dial that number to complete the transfer.

The significant differences between the ITU T.38 and IETF approaches to IP fax procedures mean that interworking between the two systems is not yet possible using a standards-based approach. It remains to be seen whether commercial implementations of the IETF IFax standards becomes widespread enough to generate interest in an interworking specification.

VoIP and FoIP

Before examining the IP fax protocol architecture in detail, it makes sense to first address what seems an obvious question: Why can't Internet fax calls use the same components and protocols as Internet voice calls? Since PSTN fax uses the same infrastructure as PSTN voice, it would stand to reason that fax data simply could "ride for free" over VoIP networks.

The trouble with this approach is that G3 fax data are encoded in a way that assumes the presence of the relatively high-bandwidth PSTN, which operates as an essentially analog carrier as far as the end nodes can tell. While digital circuits may indeed carry a large portion of the call, the digitization scheme used on them is the standard 64-kb/s PCM. This encoding method allows the near-transparent translation of almost any analog signal, including signals that exhibit characteristics never seen in human speech. Standard fax machines and modems, designed as they are for use on the PSTN, capitalize on this capability by using encoding methods that use phase shifting and other signaling components—secure in the knowledge that the 64-kb/s PCM network will recreate them faithfully.

In order to produce a bit stream smaller than the 64-kb/s PCM standard, codec designers for VoIP therefore have concentrated on accurately reproducing only those portions of the spectrum (and those analog "events") which most characterize speech—leaving a problem for fax and modem traffic that goes beyond those signals. The result is that while the G.723 codec may allow good voice quality in one-tenth the bandwidth demanded by PCM, it will wreak havoc with a stream of G3 fax data. Therefore, Internet fax developers must attack the problem of fax encoding differently from that of voice encoding.

There are essentially two solutions to this problem, both of which allow the integration of VoIP and FoIP traffic at the gateway, albeit in different ways. The first approach depends on the calling gateway to discriminate between voice and fax calls being placed through it (something it can do by listening for the 1100-Hz calling tones that are transmitted at the beginning of many fax calls). After making the determination, the gateway can then choose the codec that will best support the type of call being placed, e.g., G.723.1 for voice and standard G.711 (64-kb/s PCM) for a fax call. While an approach like this will work, it does obviate one of the primary advantages of FoIP in the first place: the use of only the amount of bandwidth necessary to transmit a certain amount of data. There is nothing inherent about a fax document that requires the use of 64 kb/s; it is simply the characteristics of the encoding method that do so. This approach, therefore, can be seen as a very simple, yet inefficient, way of providing limited fax capability over an essentially VoIP network.

The second approach, and the one that is used by the T.38 standard and an increasing number of proprietary solutions, is to essentially end the "standard" portion of the fax call at the calling gateway, which itself demodulates the analog signal back into the original digital data stream, packetizes it in IP, and sends it to the called gateway. The called gateway then remodulates the data back into the fax signal, which is transmitted over the PSTN or PBX to the called machine. This process also requires the calling gateway to discriminate between voice and fax calls, because a gateway attempting to demodulate a voice signal would wind up with garbage as a result. The advantage of this approach is that it can be very efficient, because only the scanned fax data must be contained in IP, with none of the overhead inherent in the digitization process.

This second approach does, however, create a rather ironic situation in which the scanned fax page passes through a number of very different analog and digital states before making it to the terminating fax machine. Figure 10-4 shows this manifold set of transformations and illustrates the complexity of the typical FoIP architecture.

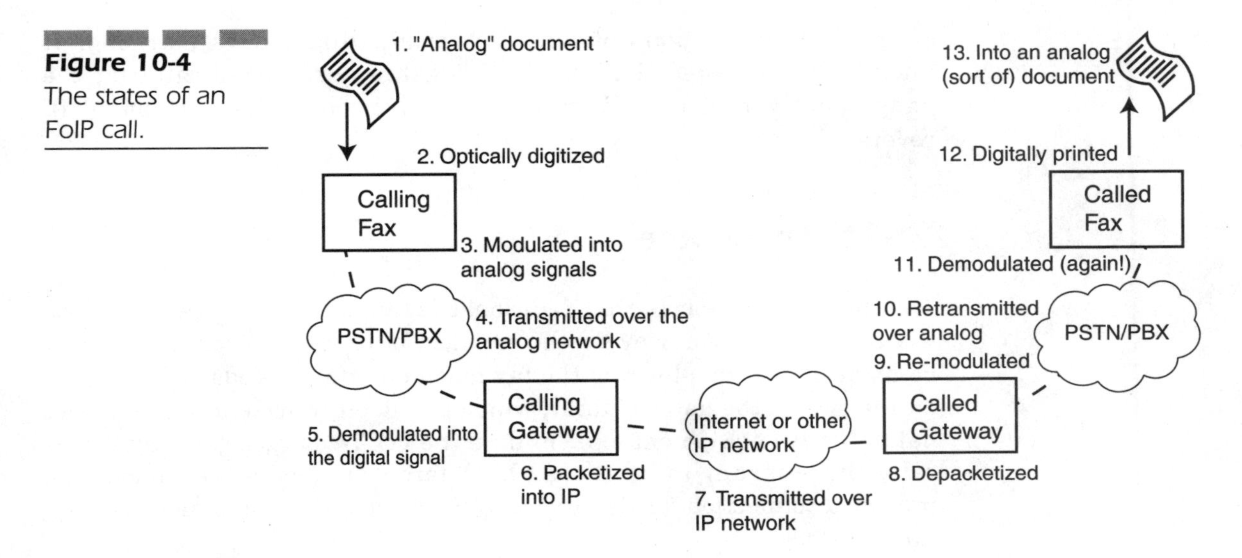

Figure 10-4 The states of an FoIP call.

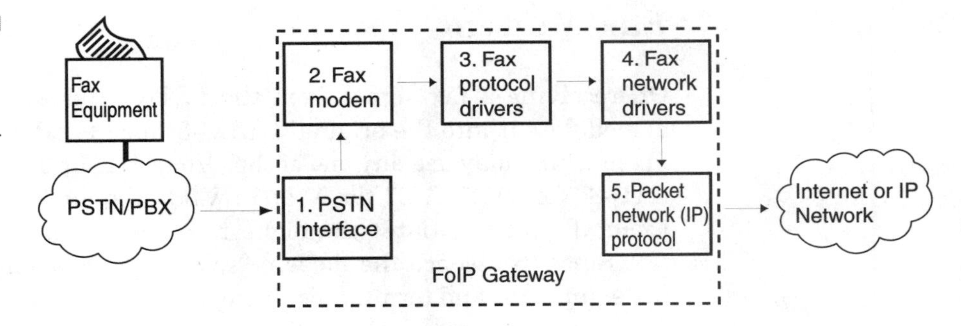

Figure 10-5 Fax gateway components.

Architectures for Internet Fax

Most vendors define, as the fundamental component of their FoIP systems, a T.38 fax gateway (also referred to as a *fax relay, fax interface unit*, or—when it forms a component of an integrated VoIP/FoIP system—*fax module*). The components of such a gateway can be seen in Figure 10-5.

An FoIP gateway has a minimum set of requirements (which may be extended by the manufacturer with proprietary or other functions) that can be divided into the following modular structure. This structure is seen here from the point of view of an originating gateway, although it must be remembered that should this gateway receive a call over the packet network, the following procedures must be performed in reverse.

PSTN Interface

This interface connects the traditional fax equipment to the FoIP network interfacing gateway. In the smallest of units, this connection may be made simply by plugging the fax machine into a standard RJ-11 jack on the side of the gateway unit, which can itself provide a dial tone signal to the fax equipment and emulate the PSTN. Larger gateways, typically those meant to be shared by a large number of fax machines, provide a connection for the premises' PBX, via either an analog or a digital interface.

Fax Modem

On receiving a fax signal from the PSTN interface, the modem must demodulate it into the original HDLC-framed fax data. Since the calling fax machine may use any one of the Group 1, 2, or 3 modulation protocols (ITU-T V.21, V.17, V.27 ter, V.29, or V.33), the modem typically supports them all and negotiates the highest possible data rate with the calling fax. Since in essence the modem "spoofs" the existence of the receiving fax equipment and terminates the modulated portion of the call, it therefore also must support the calling tones and calling station ID (CNG and CED) detection and generation.

Fax Protocol Drivers

This module provides the true Group 3 T.30 fax protocol processing, performing signaling on behalf of one end to the other (via the modem to the calling machine and via the fax packet interface to the called one). It therefore compensates for the inherent delay of the packet network to the local machine by sending empty HDLC frames as "keepalive" messages and also (when receiving data from the packet network) requests retransmission of lost or errored frames with the CRP (command repeat) frame.

Fax Network Drivers

As the primary component of the T.38 specification, the fax network driver constructs outgoing Internet Facsimile Protocol (IFP) packets and hands them to the IP layer. When handling incoming IFP packets, this layer must buffer incoming data and present them to the fax protocol in such a way as to remove jitter from the data stream. Some gateway architectures use either proprietary or standards-based (e.g., RTP) signaling to do this; others simply "even out" the incoming data by presenting them to the fax protocol according to a schedule related to the negotiated fax data rate.

Packet Network (IP) Protocol

Finally, the IP layer must encapsulate the IFP packet in an IP datagram and transmit it to the distant gateway.

These architectural elements of an FoIP service may be implemented in a standalone manner or as a part of a comprehensive VoIP/FoIP solution. As shown in Figure 10-6, when integrated into a system that also provides standards-based VoIP, the T.38 protocol occupies a position within the H.323 model similar to that of the G.7xx series voice codecs, simply offering an alternative method of encoding from the point of view of H.323. This means that the addressing, capabilities exchange, and other session procedures used in VoIP may be used to establish, control, and break down fax calls. Standalone solutions benefit from the use of these same procedures, without the need for the development of fax-specific ones.

Figure 10-6 VoIP/FoIP integration.

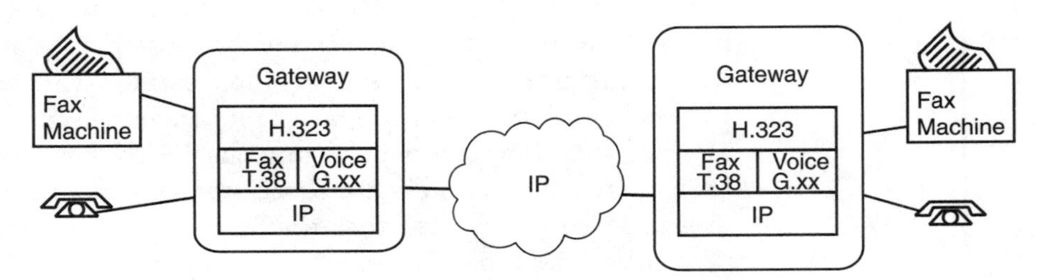

Internet Fax Product Models

While both T.38 and the IETF define fax gateway devices and IP-enabled fax equipment, neither series of specifications addresses how these devices should be used to provide fax services. Hardware manufacturers, as well as service providers, have focused on a number of different network implementations, each of which solves a particular need.

Fax Servers

While they are simple to use, fax machines typically have a drawback when using in a shared office setting—they must serve a number of users. Besides the privacy concerns that sharing communications equipment raises, a great deal of the difficulty and inconvenience of using fax machines are due to the shared nature of the hardware itself. Fax servers allow users to fax documents directly from their desktop, obviating the need to print a paper document, retrieve it from the printer, queue up at the shared fax machine, and manually feed the paper into the machine.

Fax servers are implemented as combinations of software and hardware, comprising a client package that is loaded on each of the desktop PCs that wishes to use the service, as well as a centralized fax server that works in many ways like an Internet fax gateway in reverse. The fax server appears to the client PCs as a standard printer, and the user simply selects it instead of a printer in the PC's printer configuration facility and "prints" the document as he or she would if it was being printed to paper. The document is then encoded according to the fax protocol in use and transmitted to the fax server. The fax server receives the fax "print job" and dials the PSTN phone number indicated by the user when he or she gave the print command.

Fax servers of one type or another have been in use at least since the mid-1980s, but only recently have they begun to use IP as the transport mechanism. While this has little or no consequence in the local area, the increasing use of IP means that fax servers can be located over the Internet or other IP network far distant from the sender. Faxes sent in this manner therefore can avoid using the PSTN for significant portions of their routes and not incur the associated long-distance or overseas tolls.

Fax servers also ease the receipt of fax transmissions by allowing a centralized fax inbox for an office's incoming faxes. Once received by the server, faxes may be routed to a shared printer or even to individual mailboxes as e-mail attachments if the proper incoming PSTN service is configured. In order to perform this more intelligent routing, direct inward dialing (DID) allows the server to receive faxes on shared PSTN lines yet discriminate between faxes meant for different numbers.

IP-Enabled Fax Machines

Among the variations in the high-end fax equipment market are a new class of standalone fax machines with integrated Ethernet interfaces. This type of solution is all-hardware-based and presents a user interface that can be indistinguishable from a standard G3 fax. On inserting the document to be faxed, the user enters the remote PSTN fax number just as he or she would when sending a traditional fax. At this point, the machine consults an internal database, previously programmed by office staff, that relates phone numbers to other IP fax machines and their IP addresses or DNS names. When it finds that the dialed number has an IP equivalent, the machine can then contact the remote fax equipment via the T.37, T.38, or another Internet fax protocol and transmit the fax via IP. If the sending machine finds that the dialed number has no Internet equivalent, it can use its PSTN interface to send the fax over the PSTN in the usual manner.

This same idea is also being implemented in PSTN/IP "portal" devices, which are essentially the IP interface and database function integrated in a single device, into which a standard G3 fax machine is plugged. When using the Ethernet interface to send a fax, the portal emulates the PSTN to the local fax machine and "spoofs" the existence of the remote fax to it.

This approach has the enormous advantage of being transparent to the user, in that he or she need not even know that his or her fax is sometimes traveling over the Internet rather than the PSTN. Users therefore need not learn new procedures for sending faxes. The disadvantages to the IP fax machine approach, however, are related to the advantages. Since it works "just like a regular fax machine," it is therefore subject to the same privacy concerns, queues of users waiting to use it, and paper handling concerns. It also must be configured, usually manually, with the lookup database of other machines that it can contact via IP. In particular, this limitation means that these devices are best suited for offices that spend most of their fax money calling a limited number of other offices, such as branches of a larger organization.

Variations on this scheme allow the sending user to enter an Internet mail address at the machine's keyboard and thereby route the fax as part of an e-mail message to any Internet inbox.

Dedicated IP Fax Gateways

For offices or organizations that wish to implement IP fax capability on an enterprise level, many vendors offer dedicated gateways that are meant to serve the IP connectivity needs of a large number of local fax machines, which connect to the gateway via the corporate phone system—usually a PBX. Such gateways typically offer services similar to those described earlier, but extend those capabilities to include the ability to handle tens or even hundreds of fax calls simultaneously. With this increased size, the ability to effectively scale the network of fax gateways becomes paramount. Dedicated fax gateways therefore usually allow the maintenance of a single gateway directory for fax number/IP address resolution, which is often hosted on a server dedicated to this purpose. Fax gateways usually are capable of producing detailed call records, which not only provide a record that can serve as proof of receipt but also allow the accounting of fax calls.

It is not just the number of simultaneous calls that differentiates the enterprise-class server from the standalone IP fax. In order to gain a competitive advantage, gateway vendors continue to push the number of services offered by their products upward, resulting in lengthy feature lists that include e-mail-to-fax translation, broadcast capability (the ability to send the same fax to many different IP destinations simultaneously), and even the use of optical character recognition (OCR) to identify the recipient indicated on an incoming fax.

Dedicated gateways are usually implemented as hardware and software products and as such involve installation of PSTN and Ethernet interface boards, as well as a fax protocol/processing board, into the backplane of a server PC. A gateway software application generally runs on top of the PC's general-purpose operating system, such as Microsoft Windows or a UNIX variant. This open architecture allows relatively easy access into the fax process from a software point of view, and many gateway vendors provide Telephony Application Programming Interface (TAPI) options to enable deployers of their gateways to write custom solutions to their own needs.

Integrated VoIP/FoIP Gateways

As the VoIP market continues to grow, gateway vendors are striving to make their products "as good as the PSTN." One of the things that the PSTN offers, however, is fax capability, so gateway vendors have fixed on

T.38 as the means for offering a seamless fax service over a network of VoIP gateways. An integrated gateway must be able to distinguish voice calls from fax calls and implement the appropriate method of encoding once it has determined the status of the call. In order to do this, most gateways rely on the CNG tones that are transmitted by G3 fax machines when operating in unattended mode. The codec in use is then identified to the remote end during the H.323 capabilities exchange that takes place at the beginning of the call, causing the remote gateway to complete the connection in the same manner that it was established.

Prior to introduction of the T.38 standard, many vendors who have had VoIP gateway products available for some time either have been unable to support fax calls or have found it necessary to implement proprietary protocols to do so, resulting in incompatibilities between different vendors' gateways. Successful pre-T.38 fax capability is usually supported via the G.711 codec, which trades low efficiency (not that important in a free packet network such as the Internet) for the ability to handle the phase shift and other signals on which traditional T.4 fax encoding depends. The introduction of T.38 allows support for fax to rise to the same level of efficiency as that for voice.

IP Fax Service Providers

Finally, network service providers have capitalized on the corporate world's appetite for inexpensive fax services by constructing managed networks of IP fax gateways, the use of which they charge for on a contract or per-use basis. These services are particularly attractive for those companies without the means or need to build fax networks of their own yet still wishing to take advantage of the cost savings inherent in IP fax. Even many larger organizations find it easier to simply purchase an IP fax service rather than build one by purchasing and configuring gateways and designing the network from scratch. Typically, service providers design their networks by placing gateways within major metropolitan areas, connecting them over the Internet or a dedicated IP network cloud.

These services can provide, in addition to basic low-cost fax capability

- Multiple methods of fax message submission, via desktop fax modem, e-mail (SMTP), or simple PSTN fax calling. Some even provide a custom API for integration into in-house vertical applications, which permits the sending of faxes from the user to the ingress gateway via IP.
- The rendering and delivery of faxes as graphics files attached to e-mail messages.

- Store-and-forward service, which will cache a sent fax message if the destination machine is busy or off-line and attempt to deliver it for a period of time.
- Fax broadcast services, which use the IP network to simultaneously deliver multiple copies of fax messages to multiple gateways for PSTN transmission to many fax machines.
- Faxback services, which allow the provision of just-in-time customer support or sales information via the IP fax network.

IP fax service providers succeed by leveraging the tremendous savings of IP fax against the willingness of users to pay in order to save the considerably higher expense of PSTN faxing.

The Future of IP Fax

IP fax, and the facsimile machine in general, are two technologies whose futures are inextricably linked to the future of the Internet. In a very real sense, the Internet—whose promise of free faxing regardless of distance provides the main attraction of IP fax—will also support services that eventually will replace the fax machine as we know it today. For all the reasons discussed at the beginning of this chapter, however, that day will not come soon.

In the meantime, the success of IP fax seems assured, although for reasons that have as much to do with services it will make available as with the cost savings it makes possible. In fact, the cost savings argument begins to weaken as PSTN per-minute long-distance charges continue to decline. As vendors and implementers discover ways to exploit the full promise of IP fax—by using APIs that allow customer service representatives to immediately fax information to interested parties, for example—the full benefit of faxing over IP will become clear. And as the PSTN integrates with the Internet and other IP networks, IP fax itself will become one of a wide range of services that the new global network supports.

CHAPTER 11

International Internet Bypass

Any voice call involves some cost to the service provider. These costs, and more, must be passed along to the caller. But how much more? What cost elements beyond the purchase price and operating cost of equipment go into the price of a call charged to the caller? If there are charges above and beyond simple costs, how high should they be set before profits become price gouging? Is it possible to *bypass* some of these charges so that telephone call prices more closely reflect the actual cost of providing the service? This chapter discusses these issues and answers these questions as they apply to Internet telephony.

It is important to have a firm idea of what bypass is. If a local carrier charges a long-distance carrier 5 cents a minute to complete voice calls on its local network, then if the local carrier can be "bypassed" for less than 5 cents a minute and the call still can be completed, a savings will accrue to the long-distance company. Any telephone carrier can be bypassed in a variety of ways to save money. This chapter explores bypassing the charges involved in international telephony.

International calling is big business. The current patterns of global commerce, family mobility, and air travel mean that the telecommunications industry has had to adjust to an environment where more and more telephone calls do not just go down the block but rather go around the world. With few exceptions, usually due more to political than technical considerations, anyone in the United States can pick up a telephone and dial another telephone anywhere in the world directly, without human intervention.

Yet in the United States, international public switched telephone network (PSTN) calling remains quite pricey compared with local and long-distance services. Some of this price-per-minute disparity is due to the longer distances that international trunks must span. However, this is only part of the answer. A 2500-mile call from east to west within the United States might cost 10 cents a minute. So why does a 2500-mile call from west to east across the Atlantic to Europe sometimes cost 10 times or more the cost of the domestic call for the same distance?

Undersea cables are expensive, but so are underground and aerial cables. In some ways it is much easier and more cost-effective to drop a cable off a ship than to span mountains and valleys. There must be more to it than this. As it turns out, there is, and this is the main topic of this chapter. International calling is an attractive target for Internet bypass using Internet Protocol (IP) telephony as part of the total voice over IP (VoIP) package. The high costs of international calling are due in part to the regulatory environment set up in the PSTN, or GSTN—global switched telephone network—a term often used when international calling is the main topic. Regulation traditionally has limited the number of international telephony carriers to a handful. And scarcity contributes to higher pricing configurations. Also, many countries still have tightly regulated and controlled telephony environments, despite recent efforts at deregulation and competition around the world. In many cases, there is only one domestic carrier available to complete calls in-country.

The most exciting thing about international Internet bypass is that using the Internet for telephony circumvents two of these obstacles to lower international calling rates. First, there are many more Internet service providers (ISPs) with international links than there are international telephony carriers. Second, ISPs can be more flexible in handling in-country connections and routing than more traditional telephone carriers. In many countries, the national telephone company is a bureau or branch of the government, but the ISPs are not affiliated with the central government at all.

Taking these two factors together, the ISPs, or Internet telephony service providers (ITSPs), can get away with many activities that would draw the attention of authorities in other service environments. Everybody likes the Internet and wants to encourage its use. As long as ITSPs do not try to do too much, too soon, the efforts of the ITSPs can be placed neatly under the deregulation umbrella. After all, in many cases, revenues lost by the national PSTN carrier to the ITSPs in the form of international Internet bypass not only affects the PSTN carrier but also the revenues of the national government and therefore the whole economy of the country.

This chapter examines international calling from mainly a U.S.-centric point of view. That is, the international calling scenarios detailed here are mostly of the to/from the United States and some other country variety. This provides a focus for the chapter, but in many ways international calling to and from the United States has unique aspects absent in other scenarios. For example, deregulation of domestic long-distance calling in the United States has been a reality since 1984 and remains just a dream in other places around the world. In order to fully appreciate the impact of IP telephony on the structured world of international calling, inclusion of the United States is absolutely crucial.

The International Telephony Environment

In many countries, telecommunications is structured in a kind of three-tiered arrangement. Services are local, long distance, or international in scope. In the United States, for example, different companies provide the different services, although there is some "mixing" of services where the same company is both a local carrier and long-distance carrier (sometimes not in the same geographic area). Any one (or more) of the levels potentially can be bypassed by a call, of course, eliminating some revenues from the service company being bypassed.

Within the United States, competition has more or less made long-distance bypass unattractive except when the local carriers are bypassed as well. In fact, the lowest per-minute voice costs in the world are effectively on the domestic backbones of the long-distance carriers in the United States. It makes little sense to even attempt to bypass U.S. long-distance carriers except when higher-cost local per-minute charges are bypassed as well. This is not true in other countries around the world, however.

The United States has always been pretty much a "free market" society. That is, companies enter and leave industries all the time, whether their purpose is to sell can openers or telephony services. Regulations provide some uniformity and can be substantial barriers to competition, but a determined enough group can get permission in the United States to do almost anything, even in the realm of telecommunications. In other countries, however, the telecommunications industry is absolutely critical to the economic, industrial, social, and even political structure of the nation. Some form of regulation or government control is therefore sure to exist, and this is true even in the United States. The question is how much or how little regulation and government control exist.

Regulation and government control over telecommunications usually are enforced by five main methods. Some of the methods apply more to domestic situations, and others apply more to international situations, but all are affected by the rise of IP telephony. Table 11-1 lists the methods traditionally used to control telecommunications in a given country and how Internet bypass can change the status quo virtually overnight.

A few words about each control method are in order, if only because many of these methods will come up again and again in this chapter. *Tariffs* are documents that define the nature of a telecommunications service, the geographic area where the service is offered, and the price of the service. Tariffs must be approved before they go into effect and are

TABLE 11-1

How Internet Bypass Challenges International PSTN Regulation and Control

Control Method	Internet Challenge
Tariffs required for service provision	Internet versatility makes definition difficult.
Service quality requirements	Users determine quality levels.
Information policy on transport and content	Transport/content mixing is a given.
Transborder flow limitations	Emphasis on global client-server interactions.
Landing rights for connections, settlements	One termination does it all, freer atmosphere.

public documents that anyone can see on request. Naturally, all pricing must be in accordance with the tariff. Tariffs can be quite restrictive and stifling in some environments. For example, a new competitor with an approved tariff 10 percent below another carrier's basic price for services can siphon off a lot of revenue unless the incumbent carrier's tariff is quickly revised to match the newer rate. However, if the procedure to revise a tariff takes a year of filings, public hearings, review, and so on, then the tariff process easily can be used to control the entire industry.

Now try to define the service that the Internet provides on a piece of paper. Include the Web, file transfer, remote logins, and so on. What about the geographic area? There is no limitation on where clients and servers on the Internet can reside. And pricing for each "service"? Impossible to compute, let alone detail enough to form the basis for a rate of return. And when voice is added to the mix, it is clear that the very versatility of the Internet makes it next to impossible to control through a tariff system.

Regulating agencies and governments also control telecommunications by requiring a minimal quality for the service. This is obviously true of voice and also applies to data services. If a government or bureau sets a minimal limit to network delay and maximizes network voice quality (for example), this in and of itself will restrict entry into the marketplace. The higher the quality of service requirements, the fewer the potential entrants that will be able to enter the field. Traditionally, voice quality requirements have been very high.

On the Internet, however, service quality is mainly a user, not a network, concern. The Internet emphasizes universal client-to-server connectivity, but that's all. If it is a guaranteed delay, or bandwidth, or anything else that the user needs, the Internet is spectacularly indifferent to these requests (at least the way that the Internet functions today). All users, meaning the people that control the client or the server, determine their own service quality levels. The situation is simple: If an Internet connection will not support the voice quality a user expects, he or she will get a new ITSP or use the PSTN. As long as there are alternatives, less control over service quality leads to more innovation and more widespread service availability.

Then there is the method that imposes some sort of information policy on transport network and content. For instance, in the United States, telecommunications services are strictly divided for the most part into *basic* and *enhanced* services. Basic transport service providers, usually the local exchange carriers (LECs) that enjoy a local monopoly on access lines, can do nothing except transport bits from place to place. The bits cannot be stored, converted, or processed in any other way. If the LEC also acts as an ISP, this line of business must be separated carefully from the regulated access line monopoly, and the ISP traffic must be treated in exactly the same way as any other LEC customer's traffic.

In the United States, the ISPs are enhanced service providers (sometimes seen as ESPs) that are allowed to store, convert, and otherwise freely manipulate bits as they make their way across the network. It could hardly be otherwise. What ISP does not provide content through its own Web site? What ISP does not offer e-mail boxes for customers? On the Internet, the mixing of transport and content (routers and servers) is a given, and there is no easy way to separate the two functions.

So far, all the control methods outlined apply as much to domestic telecommunications as to international communications. The last two control methods relate directly to international issues. The PSTN (or GSTN) regulators and governments can and do impose limits on the flow of traffic across national boundaries. These transborder flow limits are the most direct form of control exercised in international telecommunications. For example, there are limits to the number of other countries (called *transit countries*) that a telephone call can pass through before reaching its destination. In some cases, bits that originate and terminate in a given country can never leave the country, even if a neighboring country's telecommunications system is much more cost-effective to use. In extreme cases, some forms of traffic (such as VoIP) can be outlawed from entering or leaving a country altogether.

The Internet, however, with its emphasis on and ease of global client-server interactions is a much more difficult network on which to impose transborder limits. This problem has already been an issue in international export considerations. Suppose it is illegal to export a certain program (such as an encryption technique, to use a real Internet example) without a government license, which has not been granted. If the program is on a server in one country, and a client from another country accesses it, has the program been exported? If so, by whom? Is the client's ISP and/or the server's ISP equally guilty or innocent? What about "transit" ISPs, especially if they are the ones who actually transported the program across the border? There are no easy answers, and there will be none for some time to come.

The last control method considered is the need for *landing rights* for connections and the *settlements* process. The settlements process is important enough to consider in a section of its own. However, the concepts behind landing rights (the same term is used in the airline industry) are treated here. All countries tightly control the connections used to cross national borders. A service provider cannot just arbitrarily "land" an international cable used for telephony or other services by dragging a cable onto a beach and linking up to the domestic network. Either permission is needed on the part of the service provider itself, or a path must be purchased (really a long-term leasing arrangement) from a service provider that had links to the country already. Permission from the regulator and/or government can be yet another expensive, elaborate, and drawn-out process that controls what services are available.

With the Internet, however, one termination does it all. In many cases, the Internet is not interfacing a complex circuit-switched signaling environment with metered services but a simple packet-switched connectionless environment. If signaling for voice calls is needed, this is usually provided by another server on the network and need not be an integral part of the network itself, since signaling is on the PSTN/GSTN. PSTN links for telephony are concerned first and foremost with interconnection of networks for billing purposes.

The Internet, by definition, is already an *internetwork*. Linking a new country onto the Internet is as easy as running a link to a router. There is no real concern for billing information, and the Internet is more of a "I'll deliver your packets if you deliver mine" environment. There is a freer atmosphere in general on the global Internet. Few countries try to keep the Internet out, and the packet-switching nature of the Internet means that a lot can be accomplished with a simple 64-kb/s link to a router. The physical link can come from an "approved" international carrier with landing rights, but now the Internet—the whole Internet, not just a part—is part of the entire telecommunications fabric of the entire country. The whole Internet is now open to those with access to the router. VoIP connections may be slow and choppy, but as long as people will use them, there is little incentive to make them better.

Thus the Internet is a different world, and it functions by different rules when it comes to international networks. Should the same rules that apply to the PSTN apply to the Internet when it comes to international telecommunications? No one seems to really want this. However, there are strong arguments for going slow, introducing gradual changes, in controlled stages, rather than risking haphazard chaos. The end result would be a telecommunications environment much more closely modeled on the Internet than on the current PSTN.

Should Anyone Pay for IP Telephony?

If some service has a value, it must be paid for. The service provider must pay for the cost of the hardware and software used to furnish the service, and the customer pays the price established for using the service, which produces revenue for the service provider. If costs exceed revenues, a loss results. If revenues exceed costs, then the difference is the service provider's profit. Unless there is a chance for profit, few service providers would offer the service.

Telephony services follow a complex model where different entities all contribute to complete and carry a call. There are many reasons that the telecommunications industry has evolved to support such a complex model. Telecommunications was, and to some extent still is, a capital-intensive industry where large amounts of money are needed to gather and support even a small number of customers. Telecommunications traditionally has been considered just another form of messaging, just like the post office, an activity always firmly under the control of the government and related regulators. There are even other reasons beyond the scope of the present discussion.

Internet services follow a different model. There is a low-cost, often flat rate for all services. Singling out IP telephony on the Internet for separate handling would be as difficult as singling out file transfers for separate billing arrangements. What makes sense for telephony does not always make sense for the Internet, and the same is true the other way around.

However, it does not take much to change an industry to a form where the telephony model makes sense. Suppose taxicabs could operate only in given geographic areas because of restrictions of the type common in the telecommunications industry. Local taxis on public streets are limited to small numbers of passengers, low speeds, and so on. On the highways, congestion mandates that taxis be treated like small buses, with a minimum number of passengers, higher speeds, and the like. At the airport, small electric taxis are used to zip people to their terminals in highly congested environments.

But how does a person take a taxi to the airport? One carrier takes them from home to the highway, another cab company takes them to the airport, and a third delivers them in the electric cab to the terminal. This situation is shown in Figure 11-1.

This system may seem inefficient and somewhat contrived, but it actually addresses some of the problems and limitations with the door-to-door taxi systems used in many countries. The highway taxis can use the high-occupancy-vehicle lanes on the highway and not get stuck in commuter traffic. Small, local cabs cause less wear and tear on local roads. Electric cabs cut down on air pollution near airports, which suffer enough from jet fumes. Thus the situation is not as far-fetched as it might first appear. And any "why would anyone do this" reaction also points out the role of tradition in a given industry: The taxi industry does things this way and not that way. However, the same logic could easily be applied to telecommunications.

The biggest issue in this "tiered" taxicab world is not how people switch from one system to another (connections between rail and bus and air flights have not crippled any of these industries). The biggest issue is how the service providers get paid for the services they render. One possibility is to have each passenger individually pay the local, highway, and

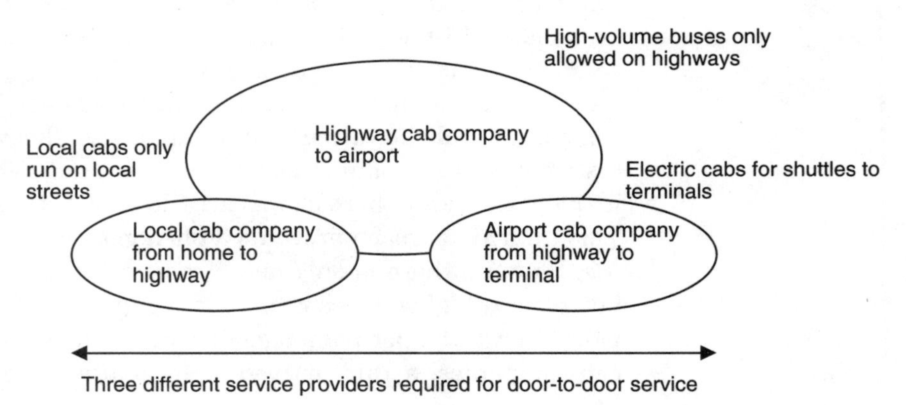

Figure 11-1 Taxicab companies as regulated service providers.

airport cab drivers as they enter or leave the vehicle. Not only is this inefficient, but it slows down the operation and makes it harder to figure out exactly how much the trip costs. Better to allow riders to pay once. But pay whom?

If payment in the taxicabs' world is along the lines of the telephony world, each rider would only pay one company once for the whole ride from start to finish. Perhaps this is the highway taxi company. But then how does the local cab company, which picks up the passenger in the first place, get its share? And what about the airport cab company? And should all three companies split the fare equally? Even if the highway ride is 50 miles and the local ride 5 miles and the airport portion is less than a mile?

The payment schemes that would evolve in such a taxicab scenario are probably no different from those which have evolved in the telephony world. Three major payment schemes have evolved to address this billing issue in the telephony world. These are

1. *Bilateral or reciprocal payment or traffic schemes.* In this scenario, all payment agreements are between two individual service providers. There are no real rules, other than that the arrangement is agreed on by both parties. In the taxi example, the payment agreement between the local and highway cab companies is totally independent of the agreement between the highway cab company and the airport cab company. Thus there are potentially many negotiated agreements, some of which are purely reciprocal ("you complete my calls and I'll complete yours") and some of which are also monetary ("I'll bill you 5 cents per call"). In the United States, reciprocal arrangements are common between competing LECs and between ISPs.
2. *The unified clearinghouse approach.* This method is much more systematic. In the taxi example, all local cab companies have the same payment rates established with all highway cab companies, and the same applies to the relationship between the highway cab companies and the airport cab companies. Whoever gets the fare distributes the amount according to these rules. The clearinghouse is the place where these claims are filed and settlements are made. This unified approach prevents a phenomenon known as *whipsawing*. Suppose there is only one airport cab company at an airport but several highway cab companies. Suppose the airport cab company charges $1 per passenger to each of the established highway cab companies, a rate arrived at by bilateral agreement. A new

highway cab company can be charged $2 per passenger because the new company is eager for business and basically has to agree to what the airport cab company charges. Then the airport cab company goes back to each of the highway cab companies and raises their rates to $2. They have little choice but to accept the new rate. In the United Sates, the Federal Communication Commission (FCC) sets international clearinghouse rates paid to other countries on behalf of all U.S. carriers just to prevent this whipsawing.

3. *The universal fund approach.* In this scenario, each cab company would pay a fixed amount per passenger into a service fund. This charge could be passed along to the customer, of course. Then any cab company that could show that it was losing money carrying passengers from another cab company could draw funds to cover operational loses on that segment of the service. In the United States, a universal fund has been set up to make sure that low-cost, low-revenue local telephone service is still offered when local competition would favor high-cost, high-revenue customers exclusively.

What the Internet would do to any of these "taxicab world" payment methods is simple. The Internet cab company can deliver passengers door to door, or at least use the same cabs for one or more of the "hops" involved in the trip to the airport. If there is a provision that says the airport cab has to deliver passengers that do not exit a highway cab to the terminal for half price (perhaps to encourage people near the airport to just walk on over), then the people arriving in an Internet cab get a real bargain. However, what if *more* people now jump into electric airport cabs from Internet cabs than arrive by highway cab? Then the airport cab company loses a lot of revenue but still has to carry the same number of passengers. And raising the rates charged to the highway cab company, which now carriers fewer passengers than ever, is hardly an option. Perhaps for the sake of the airport cab companies, the Internet cab companies should be outlawed.

By now it should be obvious that there are no easy answers to any of these questions, at least none that will satisfy all parties equally. This is the reason that Internet telephony bypass is so attractive at present and also the reason that governments and regulators are paying so much attention to the situation. However, it is equally obvious that if no one pays for Internet telephony, then traditional PSTN revenues will be eroded substantially. Then when the time comes to put *all* telephone calls on the Internet, there will not be enough capacity to carry them all and no revenues from IP telephony to increase ITSP capacity.

Fortunately, most ITSPs recognize the need to charge for services rendered. ITSP clearinghouses already exist, and their operation will be detailed at the end of this chapter.

How Should You Pay for IP Telephone Calls?

Unless IP telephone calls are to be entirely free, a situation that no Internet telephony service provider (ITSP) seriously contemplates, someone must pay for the calls. All telephone calls have value, and someone must pay both the cost to provide the value and the price to use the value. The differential between cost and price is the profit or loss that the service provider incurs. There must be profits if a service is to prosper and become widespread.

The financials of IP telephony are muddied by the fact that the ISPs are not primarily in the business of providing voice services but data services. Companies can accept some losses in peripheral activities as long as the core business remains solid. And things get even more complex when it is realized that packetized voice is virtually indistinguishable from data packets on the network. However, when all things are considered, IP telephony calls handled by ITSPs must be paid for separately from data services, especially when one or both ends of the call involve the PSTN. Charges are always assessed by PSTN service providers for handling calls, naturally. The charges may be flat rate instead of metered, but there are charges nonetheless.

This does not mean that IP telephone calls will be paid for in the same manner as traditional voice carrier calls are billed. When it comes to PSTN billing, there is about 125 years of tradition and regulation that determine how telephone carriers inform their customers about their charges. Customers usually must be sent a bill, typically every 30 days. Risky customers may be asked to supply a deposit before service commences, but these funds are not used routinely to pay for services that have not been rendered.

It certainly makes sense that a voice carrier cannot charge customers ahead of time for services not yet rendered. However, this makes it quite expensive to bill PSTN customers in this 30-day cycle over and over again, even when the invoice is for flat-rate local service and the same month in and month out. It has been estimated that it costs a large local telephone company in the United States $3.05 per customer *per month* to send their customers a bill. Considering that large local carriers have millions of customers, this adds up to quite an expense. Some LECs have experimented with putting customer invoices on a Web site (actually, many Web sites) and just letting the customer decide on payment options, whether a printout is needed, and so on.

The ITSPs are now experimenting with this concept. A Web site with up-to-the-minute information on IP telephony customer account activity is not a luxury—it is a necessity for these new carriers. The operating margins that appear in low-cost IP telephony would not allow for traditional voice paper-mailed-in-an-envelope billing systems anyway. Thus IP telephony charges are posted routinely on the ITSP's Web site. The customer typically is assigned a simple password and/or account number to secure access, but more elaborate protection and privacy schemes are possible, of course.

ITSP Web billing is in some cases even better than the billing supplied by traditional PSTN accounting systems. For example, calls usually are posted within 15 minutes of the time they are made. A local carrier usually only processes calling activity once every 24 hours, although there are exceptions. Thus it is easy for an IP telephony customer to check his or her account for up-to-the-minute activity with any Web browser. It is much more difficult for a PSTN customer to get the same information, and checking account status in the middle of the 30-day billing cycle is somewhat of a chore. The information is usually supplied only verbally, and getting a printout is next to impossible with a PSTN carrier. This is one reason that fraudulent PSTN account use is only discovered when the bill arrives in the mail.

A typical Web site with account information for an IP telephony customer is shown in Figure 11-2. Note the presence of the "remaining balance" information.

In contrast to PSTN billing practices, IP telephony service providers almost uniformly charge for services before they are rendered. ITSPs are not burdened by years of tradition and hampered by regulations. Many charge beforehand simply because they can. This avoids much of the risk of bad debts, although there is some risk of customer dissatisfaction if the service is abandoned with a remaining balance (there is no easy way to *return* used balances). Customer credit checking is now done up-front, as part of the check-clearing or credit card validation process, and service is not provided until background checking is completed or the check clears.

Prepayment may be done by way of a personal check or credit card, with credit card charging being the preferred method of payment. Credit card validation and charging can be done on-line, as part of the sub-

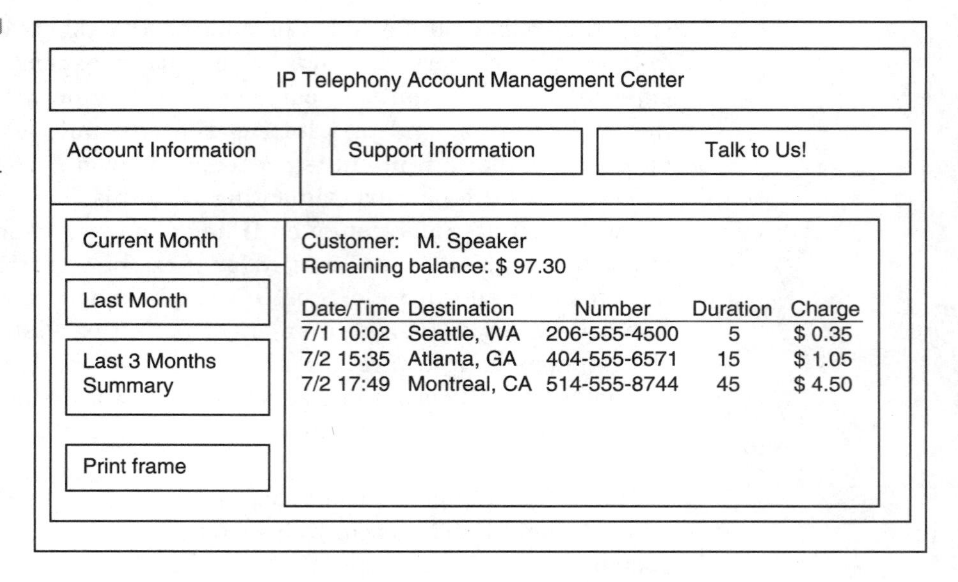

Figure 11-2 Billing for IP telephony calls through a Web site.

scription process, but check payment requires a longer delay and use of another system entirely (postal mail delivery). The customer's credit card account is charged for a bulk amount and is then debited as IP calls are made. When the balance reaches zero, another bulk charge is made regardless of date or time. The customer must approve of this use of the credit card account, of course, but few complain about the rules.

The fact that customers accepted prepaid calling came as somewhat of a surprise to PSTN carriers. The pay-as-you-go concept is so embedded in the PSTN that conventional wisdom held that customers would simply not pay for telephone service ahead of time. One of the reasons that the IP telephony prepaid plan went over so well was that by the 1990s, customers were used to paying ahead of time for many services, especially in the transportation field. Buses, subways, and trains routinely issued electromagnetic fare cards that were debited with each use and then discarded or (more rarely) "recharged." Since recharging carried the risk of fraudulent recharging (some could be recharged any time with a simple 3½-in diskette drive), the disposable cards became more common.

Pay telephones lead the way for prepaid telecommunications services in the PSTN. Coin-operated telephones were the frequent target of vandalism with the intent of theft, which often rendered the telephone useless. Not only that, but the rise of the credit card economy meant that fewer people carried the required coins, even for local calls. Prepaid calling cards changed all this pay telephone coin fumbling and crowbar bashing. A calling card fit in the wallet or purse along with a credit card. Some were electomagnetic and so just bore an account code. When the balance was exhausted, the card was useless.

The prepaid calling card made consumers realize that prepayment was not necessarily a bad thing. There was still the risk of losing a card or having it stolen, but the risk was at most only up to the face value of the card. From pay telephones, the concept of prepayment made the leap to wireless services and then to IP telephony.

Today, calling cards are marketed and sold for international calls to specific countries at a predetermined cost per minute. Some just offer lower rates by taking advantage of bulk-rate discounts offered to resellers on international trunks. Others are offered by true IP telephony service providers. The Internet takes the place of the international voice carrier, and this is where the savings comes in. This is how IP telephony is used for international Internet bypass.

International Carriers and the Internet

One of the reasons that international Internet bypass can be accomplished so inexpensively is that the global PSTN (GSTN) has a different structure from the global Internet. The Internet is much "flatter" in terms of infrastructure and lacks the tiered local, long-distance, international carrier infrastructure found with the PSTN. The structure of the PSTN is shown in Figure 11-3.

Thus domestic carriers handle all local and long-distance calling within a country, but special international carriers handle calls between countries. Now, the international carriers are usually owned and operated by one or more of the large domestic telephone companies, but they typically function independently of their parent organizations. For a while, satellite usage for international telephone calls changed this structure somewhat, since satellite international switching centers (called ISCs) can be located almost anywhere within a country. However, the delays through satellite circuits were very high, and coupled with the rise of undersea fiber-optic cables with many times the capacity of even the largest telecommunications satellite (which were limited in total number due to orbital considerations), today international calls are back to using terrestrial links in many cases.

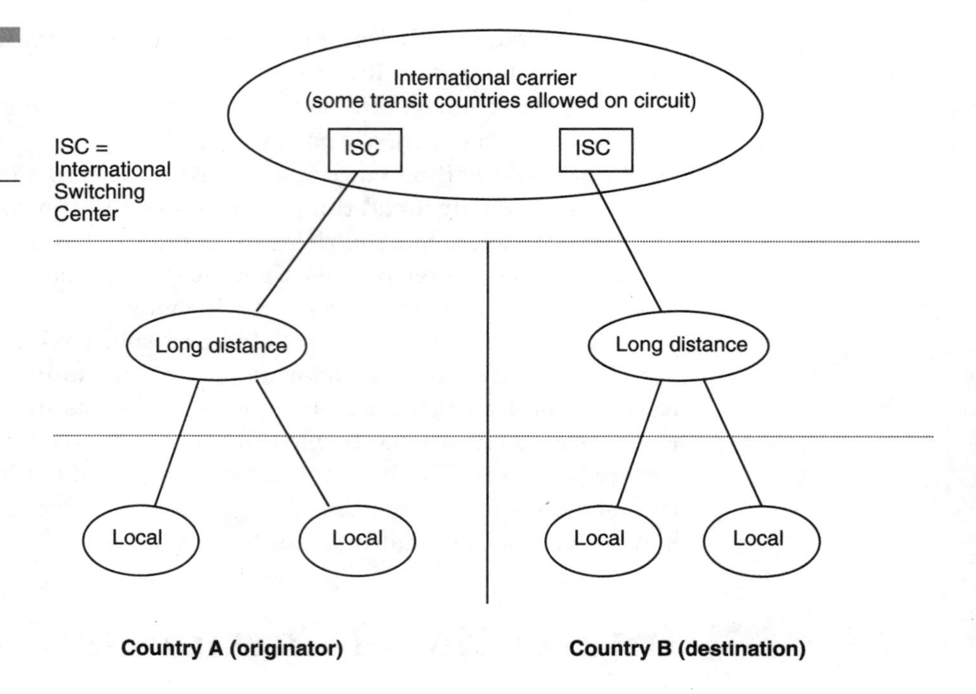

Figure 11-3 The global PSTN and international calls.

Now consider the global Internet. Today, it is even somewhat misleading to insist that there is even such a thing as the Internet. The term *Internet* is just shorthand for *Internetwork*, and this describes the situation exactly. Every ISP has its own access network for its users and backbone network for connecting its routers. It is the fact that each ISP has one or more links to other ISPs that makes the whole thing the Internet as people know it today. Thus the Internet is in fact a loose agglomeration of thousands of ISPs all haphazardly connected together that collectively compose the entity known as the *Internet*. This is shown in Figure 11-4.

The ISPs that make up the Internet have essentially just a framework of individual bilateral and reciprocal agreements between each other. There is no universal fund or clearinghouse for compensating ISPs for packets handled on behalf of other ISPs. This is one reason that ISPs come and go, but it is also one reason that Internet telephony is so inexpensive.

Most ISPs have loose "you deliver my packets and I'll deliver yours" arrangements. In contrast to the PSTN, where the international carriers have no real individual customers or users to bill directly, every national ISP has a large customer base to draw on for day-to-day operational

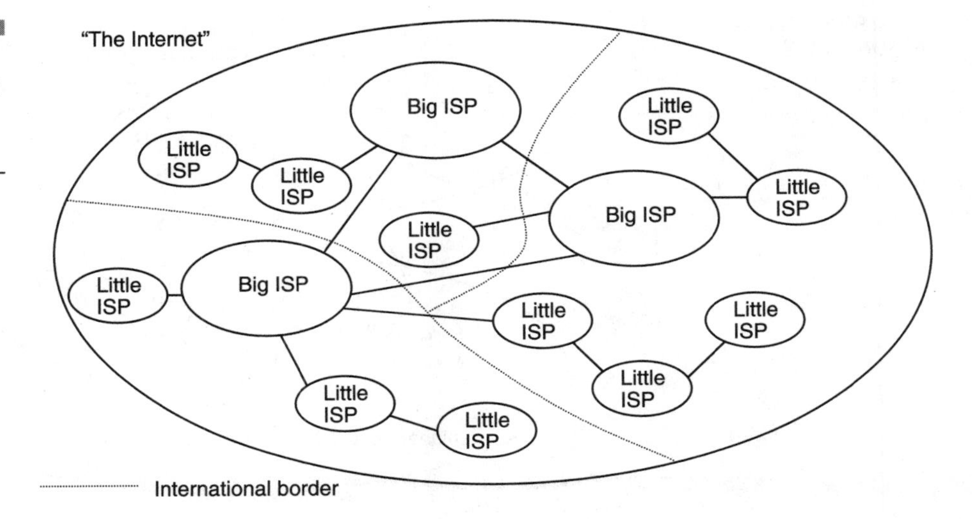

Figure 11-4 The global Internet and international calls.

costs. Thus funds derived from other ISPs are usually for physical connections alone and not for packet load on the backbone.

This is not an ideal situation. Two other ISPs can deliver vastly different packet loads onto a *transit* ISP with a fast national backbone. The transit ISP must route every packet but delivers none of the packets to a client or server owned and operated directly by one of the backbone ISP's customers, which makes the backbone ISP a transit ISP in the first place.

Not only that, each individual ISP tends to develop a strategy of passing packets as soon as possible to another ISP to minimize the total work expended to deliver a packet. However, since the other ISP has the same strategy, the path through each ISP's network in each direction from client to server and back is not the same. If there are more routers in one ISP "cloud" than the other, which is pretty much a given, then the network with the greater number of routers more than likely has a higher delay, a higher risk of congestion, greater jitter, and so on. Thus there can be wildly varying delays in each direction, a situation just not encountered in circuit-switched networks such as the PSTN, where outbound and inbound paths are always paired by the call setup process. This "get off of my cloud" packet-handling policy and the result are shown in Figure 11-5.

Despite this apparent limitation, Internet routing is much more dynamic and more resistant to link failures than the global PSTN.

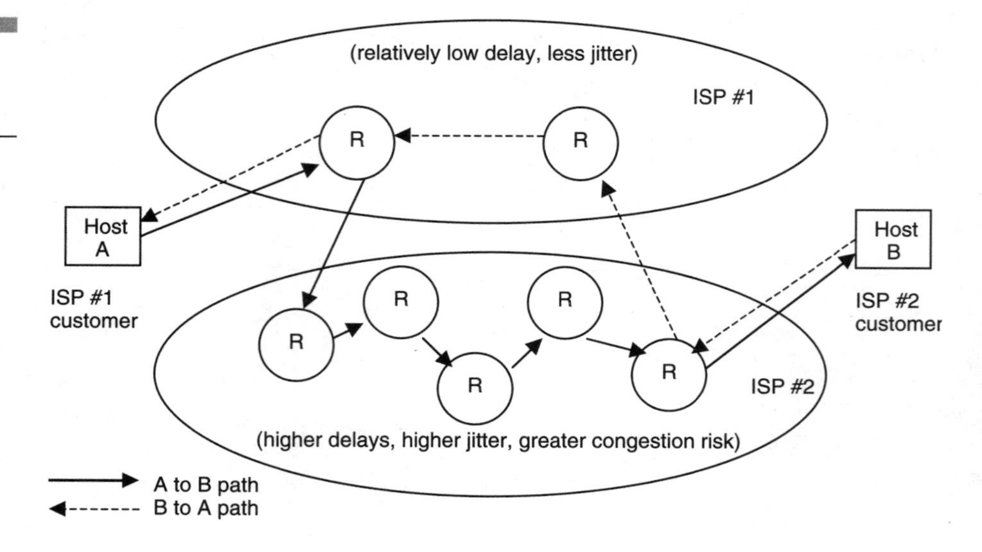

Figure 11-5 ISPs playing "get off of my cloud."

Therefore, why doesn't the global PSTN just mimic the flatter structure of the Internet? Because the global PSTN cannot. In the name of stable and low delays, the ITU-T has limited the number of circuits that can be used on an international call. The maximum number of circuits between PSTN switches that can be used for an international call is 12, with an exception of 14 made for special, low-volume cases. The number of international circuits used in tandem is limited to 4, leaving a maximum of 4 circuits for the national networks at each end of the call. This limitation more or less mandates the existence of international carriers, as shown in Figure 11-6.

In practice, this limitation is quite restrictive. International calls cannot just wander from country to country until the call terminates in a particular nation. Internet traffic delivered by an ISP, however, can wander from ISP to ISP until it pops out at the desired destination. For example, all an ISP needs to handle IP calls to Finland, conceivably, is a link to an ISP in Spain, as long as the ISP in Spain is somehow linked to an ISP in Finland that can complete the call. There may be 4, 8, or more ISPs in the total chain from the United States to Europe, but this is how the Internet works. However, the net result is variable delays and very high delays for IP telephone calls, unless efforts are made to keep the voice traffic under the control of one particular ISP as long as possible. Such ISP control over IP telephone calls is not a given, however.

So, is the best reason to use international Internet bypass simply cost

Figure 11-6 Limitations on global PSTN circuits.

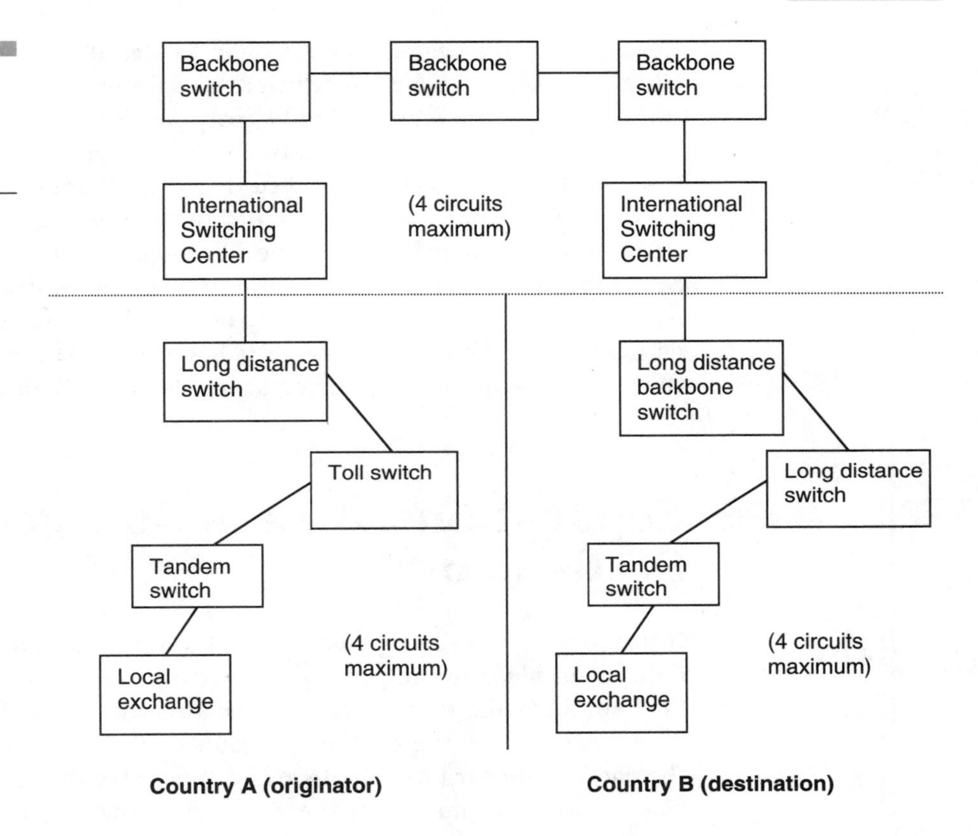

TABLE 11-2 International Telephony Prices

Per-Minute Cost, U.S. to Israel	Comment
$2.98	1990 rate, before high-volume international links, competition, and Internet bypass
$1.00	1995 rate, with high-volume international links
$0.30	1997 rate, with competition
$0.17	1999 rate, with Internet bypass

savings? In a word, yes. Table 11-2 shows a lot of effects that combined over the years to produce the cost savings. Each one continued to operate throughout the time period. Thus at least some of the causes for the drop in price from $1.00 to $0.30 per minute also was due to high-speed international link efficiency as well as the introduction of competition, of course.

Nevertheless, the overall trend is clear. In fact, if Internet bypass were as effective as the other factors, which caused about a 67 percent reduction in rates, Internet telephony should cost only $0.10 per minute from the United States to Israel. Obviously, there is a limit to the impact of these changes that seems to have reached the point of diminishing returns.

None of this implies that Internet telephony is anywhere as good as PSTN circuit-switching voice either. For example, an Internet telephone call made at a major conference from the United States to Hong Kong in April of 1998 over the general-purpose global Internet suffered from delays ranging from 1 to 2 s and was hobbled by 10 percent packet losses that made the voice barely understandable. Nevertheless, everyone loved it anyway.

Separations, Access Charges, and Settlements

International Internet bypass avoids the international settlements system put in place to compensate transit nations and destination nations for handling calls they do not bill for directly. If the ISP has a point of presence (POP) in the destination country, for all intents and purposes, the packets appear to originate at the in-country POP instead of overseas at the actual origination of the voice. Thus the only charges that apply are local, domestic charges.

This hardly seems fair, since the destination country expends a lot of effort to complete the call and cannot use the local facilities to carry other, revenue-generating domestic calls while the international call is in progress. In fact, this problem has been around since the early days of telephony in the United States, when the issue revolved around the proper separation of the charges on an access line between local and long-distance calling. First, the issue was addressed through the use of *separations*. Then, after divestiture of local service by AT&T [more properly, with the passing of the Modification of Final Judgment (MFJ) rendered in an earlier antitrust agreement between AT&T and the U.S. Department of Justice], the issue was addressed by the creation of local access transport areas (LATAs) and the assessment of *access charges*. When applied to international calling, this concept of access charges becomes the *settlements* process for PSTN interconnections.

There is no reason that Internet telephony cannot be used in the United States to bypass long-distance telephone companies such as

AT&T, MCI, and Sprint, just to name the "big 3." However, competition among these three companies has been so fierce that long-distance calling is quite inexpensive already. And most of the remaining charges are not direct charges by the long-distance carriers but local access charges that the FCC has ruled that the ISPs are technically subject to in any case.

Thus long-distance access charges and international settlements are related concepts, but they apply to different situations. Since both access charges and settlements are important in IP telephony environments, a few words about both are in order. It is probably best to start at the beginning.

AT&T Long Lines and Separations

AT&T Long Lines was the long-distance company of the Bell System. AT&T Long Lines began as a company called simply The Long Distance Company to handle calls beyond the scope of the local telephone companies, such as from state to state. However, all long-distance calls included two local calls, handled by the local service providers at each end of the long-distance call. Obviously, all long-distance calls used the same local access line as local calls. The local switch also handled both local and long-distance calls. The question was how much of the local line and switching facilities costs should be charged against long-distance calling.

This question was not a trivial one. On the PSTN, the number of calls falls off rapidly as the distance between the end points increases. Thus long-distance facilities were not used as much as local facilities, but they had to be there anyway. People would have to pay more for long-distance calls, but long-distance calls were still made, as long as they were necessary and kept as short as possible. Local calls had to be inexpensive enough to make sure enough people had telephones. Thus the separations process was invented to take some of the revenues generated by long-distance calling, where economies of scale were in effect, and contribute to the cost of providing local service.

What this boiled down to is that long-distance calls were priced higher than they naturally would be in order to keep the price of local service down. Since in most cases the Bell System controlled both the long-distance call (through AT&T Long Lines) and the local calls at each end (through the local Bell telephone company), the separations process was more or less an accounting procedure.

As time went on, especially in the economic boom following World War II in the United States, more and more long-distance calls were made. Thus more and more money was available to subsidize local service in the Bell System. As long as the local telephone companies had to use AT&T Long Lines to complete long-distance calls, no one was overly concerned that long-distance revenues were being used to keep the price of local service down.

The Internet was never burdened with a need for separations. The Internet never distinguished between packets delivered to a local server and packets delivered to a server halfway around the world. One major reason for this lack of concern with packet delivery distance is that on the Internet, interactions between a client and a server do not fall off rapidly with the distance between the client and server. The contrast between distances with regard to PSTN voice calls and Internet packet delivery is shown in Figure 11-7. No absolutes are used, but the contrast is the important point.

This means that a Web browser, for instance, is as likely to fetch a Web page from the local automobile dealer as from an art museum in another country. Distinguishing between "local packets" and "long-distance packets," therefore, was never a critical factor on the Internet.

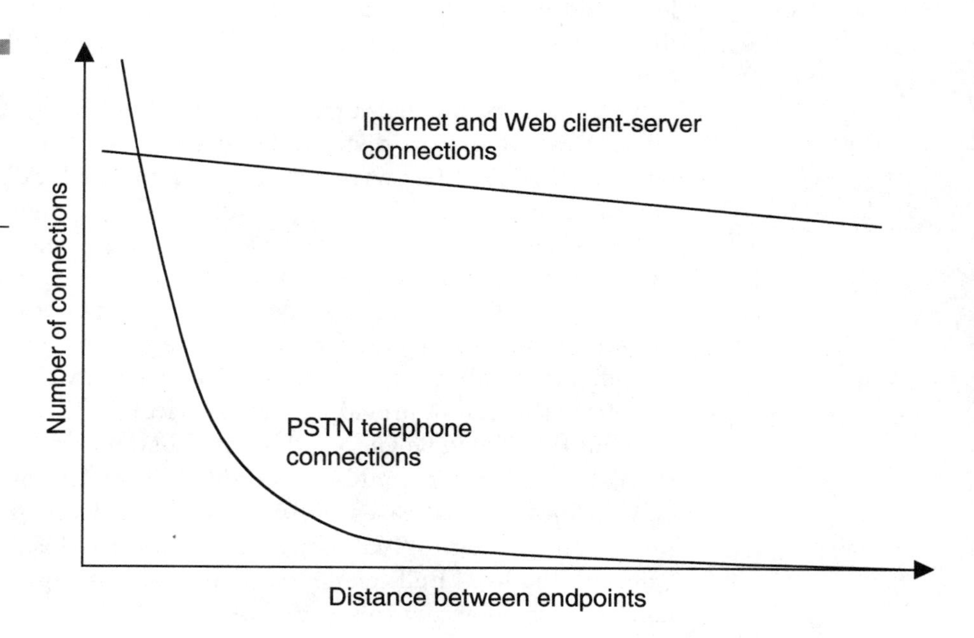

Figure 11-7 Connections and distance in the PSTN and on the Internet.

The MFJ, LATAs, and Access Charges

Separations kept long-distance charges high and local calling charges low for many years. However, things changed with the Modified Final Judgment (MFJ) of 1984 and AT&T's divestiture of the local portions of the Bell System. The Bell System local companies were reorganized as seven Regional Bell Operating Companies (RBOCs), and each customer had a choice of long-distance company, a concept known as *equal access*.

Under equal access, an RBOC customer did not have to use AT&T for long-distance calling. The customer might choose to use MCI or Sprint, among others. But what if the MCI or Sprint switching office was located far from the nearest RBOC switching office? Was the RBOC obliged to ship calls a long way to the nearest MCI or Sprint switch, calls that generated no direct revenue for the RBOC?

To solve this problem, divestiture established 200 or so local access transport areas (LATAs) of varying size around the United States. All LATAs were given names and numbers. The first number indicated one of the seven original RBOCs, which have merged since 1984 in a variety of ways. LATAs in the 800s were used for areas outside the continental United States, and LATAs in the 900s were used for independent companies such as GTE that were never part of the Bell System.

All calls where both telephones (or modems) were within the same LATA could be completed without using a long-distance carrier. These were *intraLATA* calls, although since some LATAs were quite large, especially in the western United States, a toll (premium) was charged on many of these intraLATA calls. However, all revenues went to the local telephone companies.

All calls where one telephone (or modem) was within one LATA and the other was in another LATA had to be completed using a long-distance carrier. These were *interLATA* calls, and although most interLATA calls also were interstate calls, many were not. All that a long-distance carrier needed to handle interLATA calls from a particular LATA was a switching office called a *point of presence* (POP) within the originating LATA. All revenues generated by interLATA calls went to the long-distance telephone company, although these were still usually billed by the local originating company, which also collected the entire billed amount. The long-distance charges, however, were paid to the long-distance company.

A lot of thought was given to the LATA structure. If LATAs were too large, there would be too few long-distance calls, and the restricted revenue would not be enough to support more than a handful of long-dis-

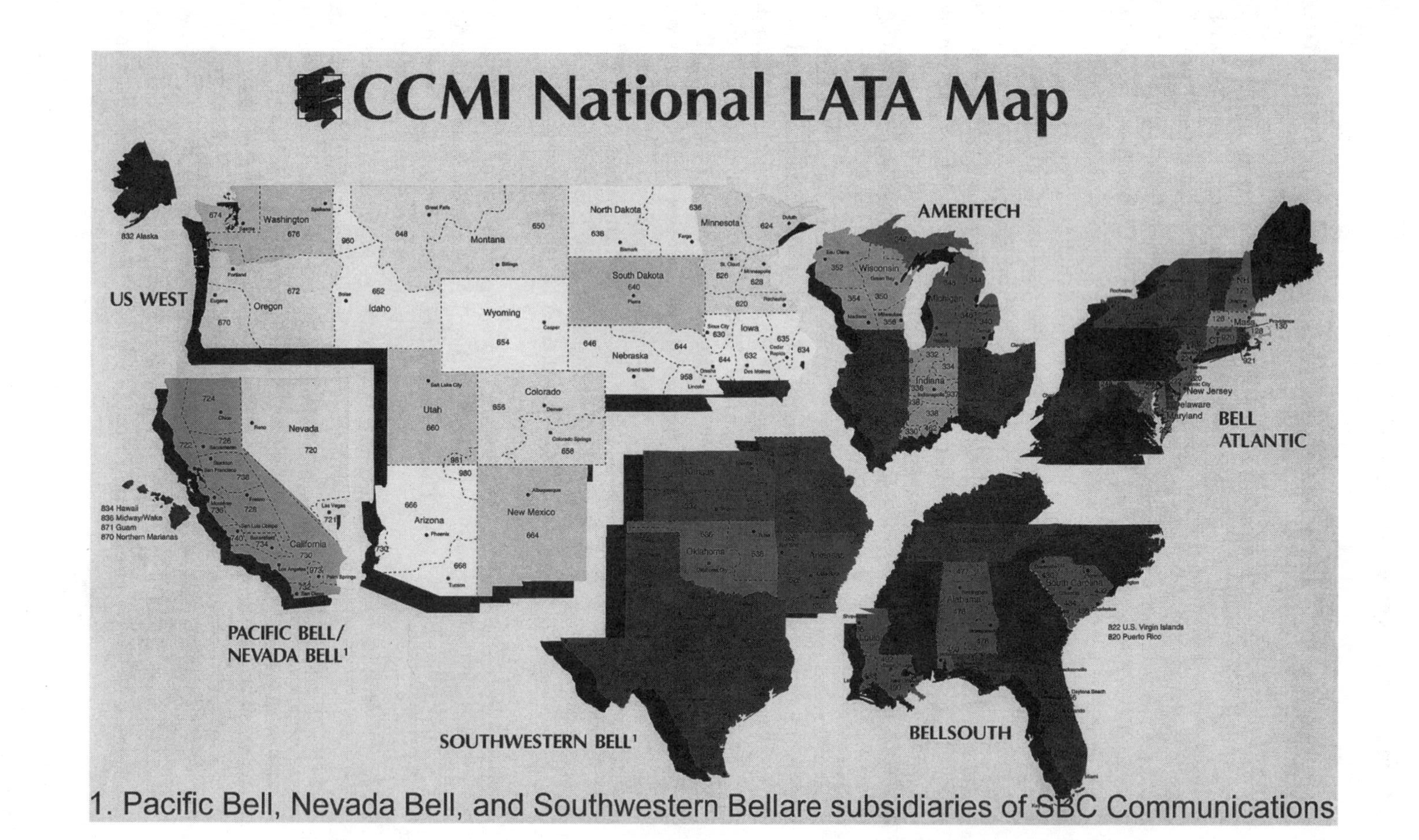

1. Pacific Bell, Nevada Bell, and Southwestern Bellare subsidiaries of SBC Communications

LATA #,LATA Name (State)

120 Maine (ME)
122 New Hampshire (NH)
124 Vermont (VT)
126 West Massachusetts (MA)
128 East Massachusetts (MA)
130 Rhode Island (RI)
132 New York Metro (NY)
133 Poughkeepsie (NY)
134 Albany (NY)
136 Syracuse (NY)
138 Binghamton (NY)
140 Buffalo (NY)
220 Atlantic Coastal (NJ)
222 Delaware Valley (NJ)
224 North Jersey (NJ)
226 Capital (PA)
228 Philadelphia (PA)
230 Altoona (PA)
232 Northeast (PA)
234 Pittsburgh (PA)
236 Washington (DC)
238 Baltimore (MID)
240 Hagerstown (MD)
242 Salisbury (MID)
244 Roanoke (VA)
246 Culpeper (VA)
248 Richmond (VA)
250 Lynchburg (VA)
252 Norfolk (VA)
254 Charleston (WV)
256 Clarksburg (WV)
320 Cleveland (OH)
322 Youngstown (OH)
324 Columbus (OH)
325 Akron (OH)
326 Toledo (OH)
328 Dayton (OH)
330 Evansville (IN)
332 South Bend (IN)
334 Auburn/Huntington (IN)
336 Indianapolis (IN)
338 Bloomington (IN)
340 Detroit (MI)
342 Upper Peninsula (MI)
344 Saginaw (MI)
346 Lansing (MI)
348 Grand Rapids (MI)
350 Northeast (WI)
352 Northwest (WI)
354 Southwest (WI)
356 Southeast (WI)
358 Chicago (IL)
360 Rockford (IL)
362 Cairo (IL)
364 Sterling (IL)
366 Forrest (IL)
368 Peoria (IL)
370 Champaign (IL)
374 Springfield (IL)
376 Quincy (IL)
420 Asheville (NC)
422 Charlotte (NC)
424 Greensboro (NC)
426 Raleigh (NC)
428 Wilmington (NC)
430 Greenville (SC)
432 Florence (SC)
434 Columbia (SC)
436 Charleston (SC)
438 Atlanta (GA)
440 Savannah (GA)
442 Augusta (GA)
444 Albany (GA)
446 Macon (GA)
448 Pensacola (FL)
450 Panama City (FL)
452 Jacksonville (FL)
454 Gainesville (FL)
456 Daytona Beach (FL)
458 Orlando (FL)
460 Southeast (FL)
462 Louisville (KY)
464 Owensboro (KY)
466 Winchester (KY)
468 Memphis (TN)
470 Nashville (TIN)
472 Chattanooga (TN)
474 Knoxville (TN)
476 Birmingham (AL)
477 Huntsville (AL)
478 Montgomery (AL)
480 Mobile (AL)
482 Jackson (MS)
484 Biloxi (MS)
486 Shreveport (LA)
488 Lafayette (LA)
490 New Orleans (LA)
492 Baton Rouge (LA)
520 St. Louis (MO)
521 Westphalia (MO)
522 Springfield (MO)
524 Kansas City (MO)
526 Fort Smith (AR)
528 Little Rock (AR)
530 Pine Bluff (AR)
532 Wichita (KS)
534 Topeka (KS)
536 Oklahoma City (OK)
538 Tulsa (OK)
540 El Paso (TX)
542 Midland (TX)
544 Lubbock (TX)
546 Amarillo (TX)
548 Wichita Falls (TX)
550 Abilene (TX)
552 Dallas (TX)
554 Longview (TX)
556 Waco (TX)
668 Austin (TX)
680 Houston (TX)
562 Beaumont (TX)
564 Corpus Christi (TX)
566 San Antonio (TX)
668 Brownsville (TX)
570 Hearne (TX)
620 Rochester (MN)
624 Duluth (MN)
826 St. Cloud (MN)
628 Minneapolis (MN)
630 Sioux City (IA)
632 Des Moines (IA)
634 Davenport (IA)
635 Cedar Rapids (IA)
636 Fargo (ND)
638 Bismarck (ND)
640 South Dakota (SD)
644 Omaha (NE)
646 Grand Island (NE)
648 Great Falls (MT)
650 Billings (MT)
652 Idaho (ID)
664 Wyoming (WY)
656 Denver (CO)
658 Colorado Springs (CO)
660 Utah (UT)
664 New Mexico (NM)
666 Phoenix (AZ)
668 Tucson (AZ)
670 Eugene (OR)
672 Portland (OR)
674 Seattle (WA)
676 Spokane (WA)
720 Reno (NV)
721 Pahrump (NV)
722 San Francisco (CA)
724 Chico (CA)
726 Sacramento (CA)
728 Fresno (CA)
730 Los Angeles (CA)
732 San Diego (CA)
734 Bakersfield (CA)
736 Monterey (CA)
738 Stockton (CA)
740 San Luis Obispo (CA)
822 US Virgin Islands (VI)
832 Alaska (AK)
834 Hawaii (H I)
836 Midway/Wake (US)
846 Puerto Rico (PR)
920 Connecticut (CT)
921 Fishers Island (NY)
922 Cincinnati (OH)
923 Mansfield (OH)
924 Erie (PA)
927 Harrisonburg (VA)
928 Charlottesville (VA)
929 Edinburg (VA)
932 Bluefield (WV)
937 Richmond (IN)
938 Terre Haute (IN)
939 Fort Myers (FL)
949 Fayetteville (NC)
951 Rocky Mount (NC)
952 Gulf Coast (FL)
953 Tallahassee (FL)
956 Bristol (TN)
958 Lincoln (NE)
960 Coeur D'Alene (ID)
961 San Angelo (TX)
963 Kalispell (MT)
973 Palm Springs (CA)
974 Rochester (NY)
976 Mattoon (IL)
977 Galesburg (IL)
978 Olney (IL)
980 Navajo (AZ)
981 Navajo (UT)

Figure 11-8 LATA map.© 1999 by CCMI (1-888-ASK-CCMI). Reproduced by permission.

tance companies. If LATAs were too small, then requiring each long-distance company to maintain a POP in each LATA would limit the number of long-distance companies by raising expenses. All in all, LATAs have fulfilled their function admirably in supporting an active and competitive long-distance industry in the United States. Figure 11-8 shows the current structure of the LATAs in the United States today.

LATAs are a totally artificial construct invented solely to address the needs of divestiture. Although efforts were made to preserve "communities of interest" and group things naturally, in most cases LATAs did not cross state boundaries. Thus almost all interstate calls were long-distance calls or, more properly, interLATA calls, although there were notable exceptions. This is important to keep in mind because cellular (wireless) telephony is often deployed based on a U.S. Census and Department of Commerce concept known as the *basic trading area* (BTA). BTAs are thought to more closely reflect actual patterns of living and working than LATAs. For instance, BTAs frequent cross state lines. People quickly discovered that land-line calls between office and home were interLATA long-distance calls, but wireless calls made from their cell phones were local calls. Perhaps, as wireless telephony becomes more and more common, LATAs will become less and less important. However, for the time being, LATA considerations are important in the United States.

The MFJ also established a system of access charges to compensate the local telephone companies for carrying calls to the POP and completing calls from the POP on each end of the long-distance call. The local companies bill the long-distance carrier for the local call at each end. This is usually done monthly through a process known as the *carrier access billing system* (CABS), but access charges are most easily understood if an example of the access charges assessed on a single call is given. Figure 11-9 shows the access charges on a long-distance call. It is assumed that no special billing arrangements, such as collect calling to reverse the charges to the recipient, are in use on this call.

In this example, the caller is billed a total of $1.00 for the 10-minute interLATA call, a not unreasonable 10 cents per minute. This charge appears on the long-distance portion of the caller's telephone bill (some long-distance carriers bill customers directly), and the entire amount of the bill typically is paid to the local carrier. The local carrier forwards the $1.00 to the long-distance carrier. The CABS process is used so that the local companies can recover the cost of maintaining the local access lines and switches so that customers can make long-distance calls in the first place. In this simple example, both local companies charge the long-distance company 3.5 cents per minute. This leaves only 3 cents per minute

Figure 11-9 Access charges and long-distance calls.

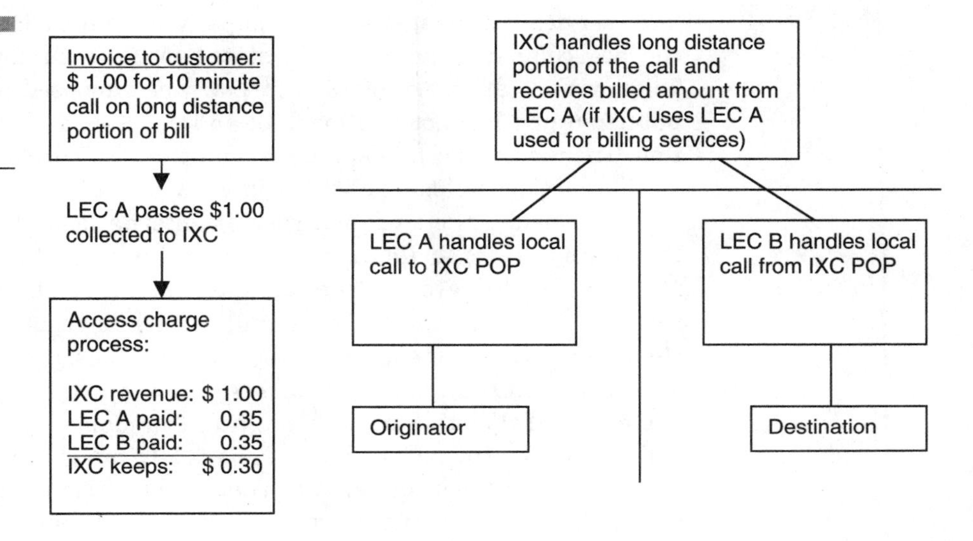
Invoice to customer:
$ 1.00 for 10 minute call on long distance portion of bill
LEC A passes $1.00 collected to IXC
Access charge process:
IXC revenue: $ 1.00
LEC A paid: 0.35
LEC B paid: 0.35
IXC keeps: $ 0.30
IXC handles long distance portion of the call and receives billed amount from LEC A (if IXC uses LEC A used for billing services)
LEC A handles local call to IXC POP
LEC B handles local call from IXC POP
Originator
Destination

for the long-distance carrier. Obviously, the long-distance carrier must be able to carry voice calls between POPs for less than 3 cents per minute to make any money at all. Most long-distance companies carry calls between POPs for 1 or 2 cents per minute, but most would agree that prices for long-distance calls cannot drop much as long as local access charges remain where they are.

This long discussion of access charges shows why Internet telephony bypass is seldom used in the United States. If the ISP is just another long-distance POP, there is little chance that carrying voice packets between ISP POPs will be less than 1 or 2 cents per minute. And even if it were, the limiting factor is not so much backbone costs as local access costs. Since the FCC has ruled that ISPs are just as liable to access charges as voice carrier POPs, there is little incentive for ISPs to make long-distance Internet telephony a major service offering. Consolidation of voice and data remains an incentive to use an ISP for everything, but voice cost savings is not a powerful attraction for Internet telephony in the United States.

International Carriers and Settlements

Why are access charges so high in relation to long-distance backbone charges in the United States? One major reason is that the local telephone company in the vast majority of the cases still has a monopoly on local service. The long-distance company has no choice but to pay what the local company wants (and the state has approved) for local access. The Telecommunication Act of 1996 (TA96) was supposed to change all this, but TA96 just has not had much of an impact in fostering local competition. The reasons for this apparent failing of TA96 are beyond the scope of this discussion, but competition will always result in lower prices.

Now consider the situation with an international call, this time from the United States to another country. In this case, the international carrier receives the revenue from the international call. Both domestic carriers must charge the international carrier for completing the call on their networks. Instead of access charges, this process is most often known as *international settlements*. However, the idea is exactly the same. And if the other country maintains a monopoly on completing the call in-country, the national carrier basically can charge almost anything it wants for this call completion. There is no alternative. Some countries charge the international carrier as much as 20 cents per minute to terminate a call on their local network.

This is where international Internet bypass comes in, of course. ISPs are not subject to international settlements put in place for telephony. All that is needed is a simple interconnection with the other country's ISP network to handle IP telephony calls. This interconnection cost can be as low as 1 cent per minute when computed over time (there is no easy way to meter connection times exclusively for packetized voice).

Mixing Politics and Technology

The FCC, in ruling that the ISPs were subject to access charges, effectively defined dial-up calls to an ISP for Internet access as interstate (interLATA) calls. After all, the server and clients could be anywhere. Prior to this ruling, Internet access was considered to be an *enhanced service* or *information service* (the FCC often uses these terms interchangeably) and specifically exempt from access charges. This was purely a political decision made to allow technology to prosper without worrying about additional charges that might limit service deployment.

The FCC has not defined what these access charges should be. Officially, this power belongs to the state public utility commissions, but the FCC has expressed a desire that Internet access should not become inordinately expensive if and when systematic access charges (or *Internet taxes*) are introduced. However, some confusion is sure to result.

Part of the confusion will come from the simple fact that the FCC's decision is 27 pages long and includes 110 footnotes—all this to allow each state to decide on its own what the relationship between LEC and ISP is to be. However, there is still some mixing of politics and technology in the FCC's decision. Paragraph 29 says that the compensation due the LEC from the ISP is "not likely to be based entirely on minute-of-use pricing structures" as are voice access charges. But what exactly do the words *likely* and *entirely* mean here? Is such a system *likely* to be based on something other than per-minute use? Such as what? Number of packets? If such a system is not *entirely* based on per-minute charges, does this not mean that at least some of the compensation scheme is based on per-minute charges? What else is there?

IP Telephony Settlements

Just because Internet telephony can bypass long-distance access charges and international settlements (if legally allowed by the nations involved) and that loose reciprocal "you complete my calls and I'll complete yours" does not mean that no money changes hands when an ISP becomes an ITSP and carries voice traffic along with Web pages, there has emerged a new breed of clearinghouse for Internet telephony traffic known as *IP or Internet telephony settlements providers* (an acronym unfortunately would collide with Internet telephony service providers, so one is not normally used). The business model used by the IP telephony settlements providers is exactly the same in form and function as an international settlements clearinghouse used to determine originating nation, transit nation, and destination nation compensation for handling calls they do not bill for directly. In the case of the ITSP, the settlements process is between IP telephony gateway owners and operators. The clearinghouse is a conduit that allows ITSPs to offer their services wider calling areas than they could support on their own by systematically allowing links to other ITSP gateways. The clearinghouses track which ITSPs terminate what voice traffic on whose IP telephony gateway or other network facility, such as the PSTN. The clearinghouses then provide settlements among the carriers based on the information they have gathered on these interconnections. The clearinghouses will take a cut of each transaction they handle, of course, but this is how they stay in business and attempt to make money. Some IP telephony settlements clearinghouses do even more, but they all offer this basic settlement tracking service.

As in PSTN settlements, the clearinghouse concept avoids the need for each ITSP to have bilateral agreements with every other IP telephony gateway owner and operator in the world. The settlements provider essentially buys terminating minutes on carrier gateways, whether other ITSP gateways or PSTN gateways, and resells those minutes (after adding the charge for the service) to the ITSP gateway owners that are originating the calls (and presumably collecting the fee for the call). To the terminating ITSP, the call appears to come from the clearinghouse carrier directly, and the terminating ITSP has no knowledge about the originating ITSP, nor does the terminating ITSP need to. Once all the minutes are used up, the settlements provider simply buys more.

The IP telephony settlements providers also make sure that all the calls are routed over Internet links from end to end, so calls do not get switched onto the PSTN only to reappear on another ITSP later on as the call is set up. Also, the clearinghouse makes sure that the gateway is available to handle VoIP calls. Thus network monitoring is part of the settlements provider's service. Naturally, the PSTN is always available to handle connectivity when Internet links are not.

The clearinghouse buys minutes from terminating ITSPs and usually also signs up the ITSP to be an originator of traffic. In this way, ITSPs get a more or less global footprint for Internet telephony calls. An originating ITSP gateway queries the clearinghouse gateway as to which terminating ITSP gateway to use. The clearinghouse provides the IP address of the gateway and usually an encrypted token to keep the transaction private and prevent fraud. The token is a type of IOU from the clearinghouse and is sent from the originating ITSP gateway to the terminating ITSP gateway. When the terminating gateway receives the token, it knows that it will be compensated for handling the call directly from the clearinghouse. Since the token is issued by the clearinghouse and not by the originating ITSP, which might be an otherwise unknown entity in a distant country to the terminating ITSP, the credit history of the originating ITSP is not an issue.

This Internet telephony clearinghouse system is not intended to replace the ITSP's existing network of routers and IP telephony gateways. If a given ITSP can handle an IP telephony call entirely on its own backbone and among its own gateways and collect the entire sum the user pays for the call, then the clearinghouse is not needed. Rather, the system is intended to make it easier for ITSPs to interoperate in a structured fashion using the same types of agreements used by PSTN carriers. An ISP can even sign up for the IP telephony settlements provider services if the ISP does not have an IP telephony gateway. The clearinghouse can offer the gateway service, and there are even variations on this theme. Sometimes the clearinghouse provides billing systems and services to participating ITSPs.

Standards for Internet telephony settlements hardware and software do not yet exist and may not for some time to come. The scramble right now is to get into the field, grow very large, and then make sure the now-dominant method used is adopted as the standard down the road. Thus variations abound. Some clearinghouses use a Web site, whereas others do not. Some clearinghouses have a Web site where ITSPs can register to post and review rates charged for Internet telephony minutes. Some even add wholesale and retail pricing information and benchmark the buy and sell rates against the average rates charged on a particular route. The whole field remains wide open.

IP Tele-*phony?*

This chapter has treated international Internet bypass use of IP telephony in a mostly sympathetic light. IP telephony voice quality is not great, especially over the public Internet, but will improve as IP networks gain quality of service (QoS) guarantees and digital signal processor (DSP) chipset design continues to improve. Links between the Internet and PSTN carriers will become more common and plentiful. The cost savings derived from IP telephony bypass will ensure the common use of IP telephony in the near future. And so on.

Perhaps there is another side to this rosy IP telephone future, however. Maybe it really should be called IP tele-*phony*, because each of the claims made by IP telephony is as phony as the rosy IP telephony future itself. From this point of view, IP voice quality will never be as good as people expect, so cost savings will remain the only reason the talk over IP. IP networks will never be able to mimic the QoS levels taken for granted on the PSTN (more of this point in the next chapter). DSP design breakthroughs are over, there is no place to go from 4-kb/s voice, and DSP chipsets are overengineered already. The PSTNs will view Internet connections as a threat to their revenues and so keep them to an absolute minimum and of poor quality. Any cost savings derived from the use of IP telephony are short term and a delusion, and fairly soon everyone will go back to using the global PSTN for telephone calls once the fascination with IP telephony wears off.

This section explores the IP telephony objects from the cost savings and bypass perspectives. This does not mean that the other issues are not important. They are. But issues such as the future evolution of DSP chipsets are best left to engineers, and the arguments are well beyond the scope of this book in any case. Periodically, similar technical barriers are predicted for PC memory, processors, and disk drives. Thus firm answers in these areas should not be expected anyway. But the bypass issue is not a technical issue.

Two cost elements that contribute to customer invoice charges can be bypassed using IP telephony. The first are the long-distance service provider's backbone charges bypassed when calls are carried by an ITSP instead of an IXC. The second are the international settlements bypassed when calls into a country are handled by an ITSP interconnection line instead of by an international carrier's termination agreement (landing rights). Access charges potentially could be bypassed from an ITSP POP to a subscriber's home, but since the vast majority of Internet access is still furnished by the LEC's access line, there is no easy way to accomplish this today (although cable modems are a very attractive alternative in some areas).

In the United States, fierce competition among long-distance service providers has kept the cost of voice minutes on the carrier backbone well below the access charges assessed by the LECs at each end of the long-distance call. There is some incentive to migrate from 10 cents a minute voice to 7 cents a minute voice, but not as much as there would be if IP telephony reduced voice to 1 or 2 cents a minute. A 3 cents a minute differential might not even cover the cost of the increased bandwidth and equipment needed for years to come.

Thus the focus of Internet telephony bypass has been on international calling. Restrictive and highly regulated telecommunications practices in some countries have kept global PSTN settlements very high. In many cases, the money charged other carriers to terminate calls goes directly to the nation's central government, since the telecommunications carrier is a branch or bureau of the government. Over the years, telephony in some countries has been seen as a cash cow to be milked as much and as often as possible, all in the name of contributing to the gross national product and thus raising the standard of living enjoyed by the citizens.

Countries where the telecommunications climate is more permissive, and there is competition and less government control (it is never entirely absent), have become increasingly dissatisfied with high settlement rates. High termination costs to other countries keep the price of international calling in the originating country high, since these costs must be passed along to the customer. The carrier in the originating country has no direct control over the settlements amounts charged by other countries. In the United States, carriers have some leverage because the FCC negotiates the international settlements rates on behalf of all U.S.-based carriers, a practice that prevents whipsawing as the other countries play off one carrier against another.

Beyond simple FCC pressure, however, what else could be done to lower high international settlements rates to a particular country? Until recently, not much. All recognized countries are sovereign states, and current international principles prevent intervention or interference in situations of a purely domestic policy, a principle rarely violated even in cases where humanitarian issues are raised. Another country can protest that settlements rates are too high, but short of invading, what else could be done?

Perhaps Internet bypass with IP telephony is just a convenient club to beat certain countries with perceived restrictive policies and environments and high settlements rates over the head. In this scenario, IP telephony is encouraged as an alternative to the global PSTN for no other reason than to help jawbone the international settlements rates charged by a country into line with what some other country (usually the United States is mentioned here) thinks they should be. (The term *jawbone* refers to threatened actions used for persuasion purposes up to but never including threats of violence.)

IP telephony jawboning would proceed as follows. Country A thinks country B's settlements rates are extortionate and should be lowered to what country A considers to be fair market value. Country B declines, and country A threatens action. Perhaps country A decrees that domestic carriers in country A will pay no more than 10 cents per minutes to country B instead of the customarily assessed 20 cents per minute. This might work, but country B still has ways to retaliate, and country A may be portrayed as the bad guy in this scenario.

Without IP telephony, country A is at the limit of pressure once the decree has been issued. However, now add IP telephony to the mix. Country A can create a climate where ITSPs prosper, and since country B has realized the value of the Internet, links to ISPs in country B exist and continue to multiply. If enough calls are now international Internet bypass calls, then the loss of revenue to country B could be staggering. Thus country B mends its ways, lowers its settlements rates, and pleases country A.

But what does this mean to IP telephony? There is now much less savings to be realized from using international Internet bypass, since the price differential is much less. And maybe the quality is much worse. So why not go back to using the global PSTN? Perhaps country A even tightens the reins on the IP telephony climate now that the “real” purpose of IP telephony has been achieved.

How much should VoIP proponents worry about this hidden IP telephony agenda? It varies based on how confident the VoIP people are that IP telephony will be able to provide the same voice quality as the global PSTN. If VoIP quality remains poor, people will go back to the PSTN unless the cost savings continue. If the quality is there, people may still continue to use IP telephony even without the bypass price break. Obviously, QoS guarantees on an IP network are the key elements here. The time has come to take a closer look at this whole issue.

CHAPTER 12

Quality of Service in Voice over IP

One of the main reasons that the global public switched telephone network (PSTN) evolved as a circuit-switched, channelized, "all the bandwidth all the time" network is simple to understand. Simply put, perceived voice quality depended heavily on maintaining a low and stable delay across the network. The easiest way (and for a long time, the *only* way) to guarantee this low and stable delay (jitter limits) was with circuit switching, channels, and time-division multiplexing. Bandwidth and delay are two of the most important quality of service (QoS) parameters that characterize any network, and some would say the only two that really count.

However, the PSTN is used for more than voice today. Faxing has been called "error-tolerant voice" and is not as dependent on stable delays as is voice. Data networking is more sensitive to restricted bandwidth than to low delays and jitter limits, one of the main reasons that full-blooded data networks were built separately from voice networks in the past.

Today, of course, networks of all kinds are being used for voice and data, and some are also used for faxing and video. Private networks can be built to specifications to provide the proper bandwidth and delay characteristics needed, but public networks are attractive for their easy connectivity aspects. As the Internet and Web become more and more indispensable for life in general, the pressure to use one universal Internet Protocol (IP) network for everything has become a tidal wave.

Unfortunately, popularity and desire will not make QoS concerns on an IP network go away. IP itself is a "best effort" or "unreliable" network service. Thus IP makes a determined effort to deliver packets, but if the packets disappear between or in routers for any reason, it is not IP's job to recover from this situation by keeping copies of packets in the routers or employing any one of a dozen other error-recovery procedures. And the term *unreliable* applied to IP is a puzzler also. After all, if a voice switch fails, all calls in progress are terminated immediately. If an IP router on the Internet fails, however, packets can be dynamically rerouted around the failed router quickly, although some packets may be lost initially.

Why then is the PSTN considered "reliable" and the Internet "unreliable"? Mainly because use of the term *reliable* in this context applies not to the network's availability characteristics but to whether the network can be *relied on* to deliver the QoS needed by the user applications. In this sense, the PSTN is much better than an IP network when it comes to being relied on for any QoS guarantee at all, not only voice QoS. Thus IP (and the Internet and Web) provides an *unreliable* network service between and among routers.

However, if an IP network has no QoS assurances for user applications at all, how can the Internet be used reliably for voice? Can users simply add devices to the network to give them the QoS they need? What role, if any, should the network play? This chapter addresses all these issues and more.

Just What Is Quality of Service?

Everybody talks about QoS and networks, but QoS is one of those concepts that people throw around as if everyone understands exactly what is meant when the QoS topic comes up. When pressed, definitions of QoS usually fall into the very simple bandwidth-only category. However, if bandwidth is all that there is to QoS, why not just say *bandwidth* instead of QoS and be done with it? Some definitions add delay parameters, and a few even go beyond this.

This section defines QoS parameters from the most inclusive viewpoint possible. This means that other definitions may exclude some of the QoS parameters listed here. However, this approach ensures that there are no QoS parameters that could be *added* to a QoS definition. For educational purposes, therefore, the inclusive approach makes the most sense.

QoS itself needs to be given a definition. For the purposes of this chapter, QoS is taken to mean *the ability of networks to guarantee and maintain certain performance levels for each application according to the specified needs of each user*. Some of the terms used in this definition are the keys to understanding what QoS is from a network service provider perspective. The network service provider guarantees the QoS to the user and takes steps to maintain the QoS level as network conditions change due to congestion, equipment and link failures, and so on. QoS is given to each application that users can run, and the proper QoS for a given application can only be specified by the user, since only the user knows what QoS is needed by the application. Not that users automatically know what the network needs to provide for the application, but like the requirements for PC software applications ("64 MB of RAM, 100 MB of hard drive space..."), the user has access to the information. Certainly, the network cannot presume to know the proper QoS needed by a user application.

Conceptually, QoS is a resource of the network, or rather a grouping of network resources such as bandwidth and delay. This concept of QoS as a network resource is shown in Figure 12-1. QoS is "in" the network, and the proper QoS is provided by the network to each individual application as needed.

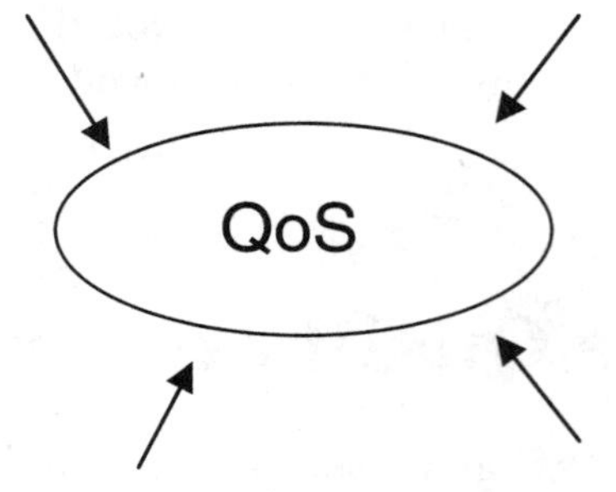

QoS in network; all applications do is ask for it (PSTN approach)

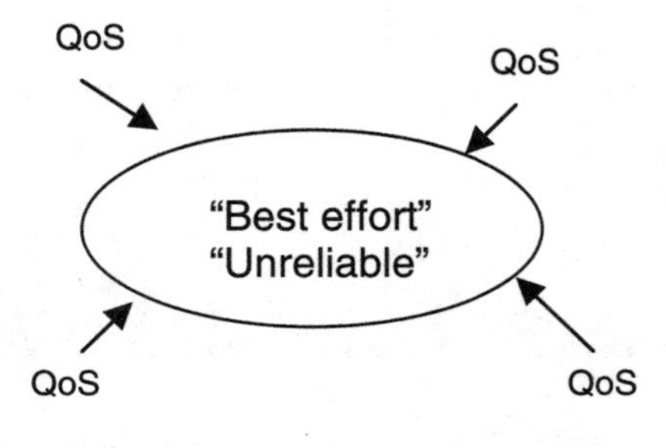

No QoS in network; all applications must do their own (IP/Internet approach)

Figure 12-1 Build QoS in or add QoS on?

The specification of the actual values of the QoS parameters is given to the network service provider in one of two ways. In a permanent virtual circuit (PVC) environment, the values of the QoS parameters simply can be written on a piece of paper and handed to the network service provider representative. Faxes, e-mails, and even telephone calls can be used to provide the same information, as long as the customer and service provider agree on the method or methods used. The QoS is available on the PVC when the PVC is ready, at what is called *service provision time*. In switched virtual circuit (SVC) environments, the QoS parameter values are given to the service provider in the call setup signaling message, which is a key part of the signaling protocol used to support switched services on the network. Both methods can be used in any combination on a network. The PVC method allows QoS to be provisioned more leisurely, whereas the SVC method requires that the QoS on a given connection be established on the fly.

If a network is totally optimized for one type of service or another, the users have little latitude in specifying QoS parameters. For example, on the PSTN, optimized for voice, there is no need to specify the bandwidth of delay needed by the telephone call. All voice calls get the QoS mandated by the relevant national and international standards for voice telephony. If modems are used to place data on the voice connection, the parameters provided by the PSTN still cannot be violated. The PSTN is well suited for voice because of this but not well suited for many applications in use today, especially with respect to bandwidth. However, locking all QoS parameters on the PSTN to those best suited for voice makes call setup and routing quite simple, effective, and quick.

Granularity

Potentially, there are as many different QoS levels as there are applications. Applications vary greatly in even their simple bandwidth needs. Digital voice can require anywhere from 8 to 64 kb/s. Video looks okay at 1.5 Mb/s but much better with 6 Mb/s at its disposal. Web access and file transfers can consume as much bandwidth as available, but only intermittently, and so forth. However, bandwidth on the PSTN and data networks derived from leased lines based on PSTN circuits tend to be restricted to increments of 64 kb/s. An application that needs 100 kb/s either struggles to function with 64 kb/s or wastes 28 kb/s at 128 kb/s. This is the drawback of "all the bandwidth all the time" circuit-switched networks.

A packet-switched network can parcel out bandwidth more gracefully for bursty data applications, but this is not the whole story. A network must be able to give each application the QoS it needs, whether the bandwidth needed is constant or not. The ability on the part of a network to grant applications QoS guarantees such as bandwidth guarantees is referred to as the *granularity* of the network's QoS.

Granularity is an important aspect of QoS. Granularity determines how finely users can specify QoS parameters for specific applications. If the QoS a network affords is not granular enough, this can limit what users can accomplish on the network. As a simple example, consider a network service provider that has established various *classes of service* for user applications. Sometimes classes of service are used as a synonym for QoS, but this discussion uses the term *class of service* to mean a collection of QoS parameter values capable of supporting a certain class, or category, of applications.

The service provider may offer a voice class of service on a packet network that guarantees 64 kb/s of bandwidth between end points when needed and a 100-ms delay with less than 10 ms of jitter. This is okay as long as all voice users need 64 kb/s. But what if new packet voice applications on the network require only 8 kb/s? Or only 4 kb/s? Since the user is guaranteed 64 kb/s, this bandwidth amount must be subtracted from the total amount of bandwidth on the network as a whole. This is so because the network can never tell when the 64 kb/s might be needed. Thus the user is paying for unused bandwidth, and the service provider is reserving bandwidth that can be used to support other users.

Greater granularity would allow voice users, even in the same class of service, to specify their bandwidth needs more precisely. This precision is only achieved at the price of complexity in the network, which is the main reason for limiting QoS parameter values and establishing classes of service in the first place.

Performance Guarantees

What if the network fails in its attempt to guarantee and maintain the proper QoS for a given user application? This is up to the agreement between the user and the service provider in the case of services governed by contract or the service provider and the regulator in the case of tariffed services. For instance, penalties are imposed by the individual states in the United States on local exchange carriers (LECs) in the event that circuit-switched voice service falls below the QoS established for that service. This only makes sense because the LEC usually has a monopoly on local service, and the regulator acts on behalf of the users as a whole. Tariffs typically mandate a basic voice quality level, strict limits on repair time (availability), as well as other things. If the LEC fails to provide this performance guarantee to its customers, the regulator can impose fines, mandate rebates, or both as the regulator sees fit.

Performance guarantees are also key parts of contracts for network services between customers (on behalf of the users) and network service providers. Usually, customers pay for services on a monthly basis. The performance guarantee may set up a system of penalties in the form of discounts for a month of service if the service provider fails to meet the performance guarantees in a particular month. In extreme cases, during a particularly bad month filled with network problems, one or more customers may receive free service (such as it is).

Service-Level Agreements

A network performance guarantee in a contract service environment is usually embodied in a formal service-level agreement (SLA) established between customer and service provider. The SLA may be part of the service contract itself or an entirely separate document. The SLA establishes customer expectations and sets service provider penalties, if any. SLAs also provide a convenient way for customers to compare service offerings from different service providers.

SLAs are a common part of frame relay or ATM public network services. Few customers would consider committing to a major frame relay or ATM service contract without an SLA firmly in place. In fact, the Frame Relay Forum has a standard implementation agreement in place, technically binding on all Frame Relay Forum members, that sets up a standard SLA agreement framework between customers and service providers.

What has all this discussion of QoS granularity, performance guarantees, and SLAs have to do with voice over IP (VoIP) or Internet telephony? The point is that while QoS tuning and guarantees have been an active area of exploration among providers of many public network services, the Internet in general has remained relatively unscathed when it comes to QoS concerns. Part of the reason is that everybody knows that the IP-oriented Internet is a "best effort" network that is "unreliable" when it comes to QoS guarantees. However, even if all ISPs suddenly became totally committed to QoS overnight, there remains no easy or even accepted way to add QoS to an IP network at the IP level.

The closest an IP service provider can come to a QoS guarantee or SLA between customer and ISP is with what is known as a *managed IP* network service. The term *managed* has nothing whatsoever to do with management of the network by the service provider on behalf of the customer. So what exactly is being managed in a managed IP network? The QoS that the IP network provides. This typically is accomplished by isolating the routers, links, and so on used to provide IP service to a particular customer and using these resources on a more or less dedicated basis for that customer alone. In some cases, routers and links are still shared, but only among the pool of customers that have contracted for managed IP services.

Most large Internet service providers (ISPs) offer both general public Internet connectivity and managed IP services. The managed IP portion of the ISP's network is usually used to connect only the sites controlled by the customer. No one can easily guarantee bandwidth or any other QoS parameter on the public Internet, which basically consists of interconnected ISP "clouds" of widely varying bandwidth and other resources. Only by limiting site connectivity to sites serviced by the same ISP can the ISP realistically offer a managed IP service. Links to the global, public Internet can be provided as part of the managed IP service, but all QoS guarantees disappear on this portion of the network, of course. This limited connectivity featured by managed IP network services still can be used to advantage by customers, however. Virtual private networks (VPNs), for example, actually benefit from the QoS guarantees and limited site connectivity that are the hallmarks of managed IP network services.

The problem is that as more and more diverse applications such as voice and video make their way on the global Internet and Web, QoS guarantees become not merely a luxury but an absolute necessity.

However, the Internet as it stands today offers global connectivity and little else in the way of QoS, with the exception of managed IP network services.

The QoS Parameters

So far in this chapter the concept of QoS parameters has been discussed in mostly general terms. Examples have been given such as bandwidth and delay, but no attempt has yet been made to precisely define QoS parameters. Considering the importance of QoS to discussions of VoIP and Internet telephony, however, this section will offer an inclusive definition of six different QoS parameters.

Few examinations of QoS will detail with or even mention all six, but in an effort to be as inclusive as possible, this section defines QoS as the specification of values on an application by application basis for each of six parameters. The six parameters, along with example values of each, are listed in Table 12-1. Because this list includes parameters not always found in many QoS discussions, especially security, a few words about each is in order.

Bandwidth

To some, all discussions of QoS begin and end with bandwidth. This position will be examined in some detail later in this chapter. Bandwidth is definitely the preeminent QoS parameter, but it is hardly the only QoS parameter. Bandwidth in many parts of a network is still a scarce com-

TABLE 12-1

The Six QoS Parameters

QoS Parameter	Example Values
Bandwidth (minimum)	64 kb/s, 1.5 Mb/s, 45 Mb/s
Delay (maximum)	50-ms round-trip delay, 150-ms round-trip delay
Jitter (delay variation)	10% of maximum delay, 5 ms variation
Information loss (error effects)	1 in 1000 packets undelivered
Availability (reliability)	99.99% network availability
Security	Encryption and authentication required on all traffic streams

modity, although schemes that promise "virtually unlimited bandwidth" on fiber are very popular. Unfortunately, disappointments with experiments in "virtually free electricity" and "virtually self-configuring hardware" have left many observers skeptical of ever living in such a bandwidth nirvana anytime soon.

Bandwidth is simply a measure of the number of bits per second available on a network for any application. Bursty applications supported on packet networks literally can have *all* the network's bandwidth to themselves if no other application is bursting at the same time. Of course, chances are, on any network with an appreciable number of users, that other applications are bursting at the same time. When this happens, bursts must be buffered and queued for delivery, introducing delay in the most bandwidth-rich networks.

When used as a QoS parameter, bandwidth is the *minimum* that an application needs to function properly. For example, 64-kb/s PCM voice needs 64 kb/s worth of bandwidth, and that is that. It makes no difference that the network backbone has 45-Mb/s links between major network nodes. Bandwidth needs are determined by the minimum bandwidth available over the network. If the network access is through a V.34 modem that supports only 33.6 kb/s, then the 45-Mb/s backbone does the 64-kb/s voice application no good at all. The minimum bandwidth QoS must be available at all points between users.

Data applications benefit the most from higher bandwidth. These are known as *bandwidth-bound applications*, since the effectiveness of a data application is directly related to the minimum amount of bandwidth available on the network. On the other hand, voice applications such as 64-kb/s PCM voice are known as *delay-bound applications*. This 64-kb/s PCM voice will not run any better with 128 kb/s available. This type of voice is totally dependent on the delay QoS parameter of the network to perform properly.

Delay

Delay is closely related to bandwidth when it comes to QoS parameters. When it comes to bandwidth-bound data applications, the more bandwidth available, the lower is the delay through the network. When it comes to delay-bound applications such as 64-kb/s PCM voice, the delay QoS parameter specifies a *maximum* delay that bits encounter as the bits make their way through the network. Bits can arrive with a lower delay, of course, but the bits must arrive in at least this amount of time.

The relationship between bandwidth and delay in a network is shown in Figure 12-2. In part *a*, $t_2 - t_1$ = delay in seconds. In part *b*, X bits/$(t_3 - t_2)$ = bandwidth in bits/s. (More bandwidth means that more bits arrive per unit time, lowering overall delay. The units of each parameter, bits per second for bandwidth and seconds for delay, are obvious once the relationship between bandwidth and delay is established in this fashion.

Network bandwidth and delay are related and can be calculated in many places along the path through the network, even from end to end. Information transfer takes place as a series of transmission *frames* (IP packets can be used for this purpose also), which are some number of bits long. The time elapsed from the moment the first bit of a given frame enters the network until the moment the first bit leaves the network is the *delay*. As a frame makes its way through switches and routers, the delay may vary, having a maximum and minimum, an average value, a standard deviation, and so on.

Bandwidth is defined as the number of bits in the frame divided by the time elapsed from the moment the first bit of a given frame *leaves* the network until the moment the last bit leaves the network. Actually, this is only one possible measurement point. As frames make their way from access link to backbone network and so forth, the bandwidth the frame has to work with may vary considerably. This is the key to understanding how increasing network bandwidth decreases delay: The bits in the packet or frame arrive that much faster. Conversely, a network could

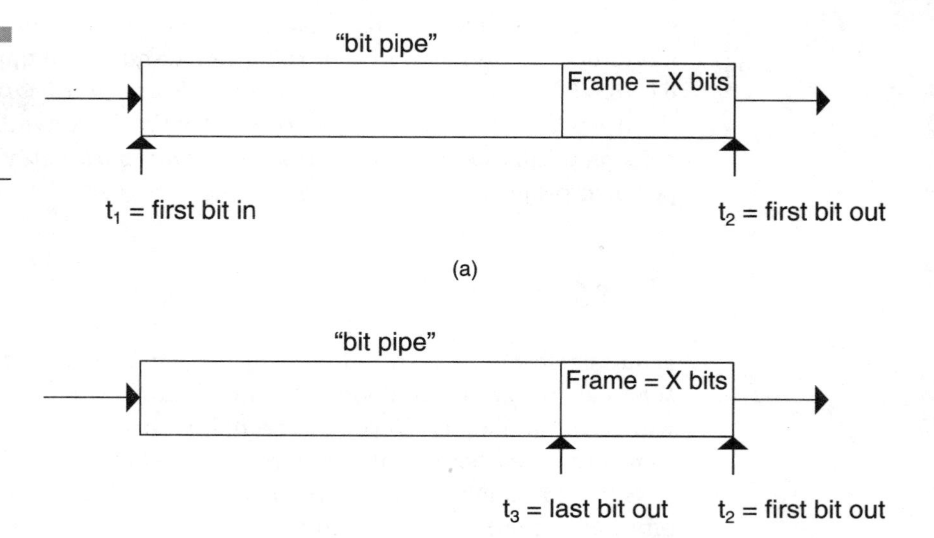

Figure 12-2 *(a)* Delay and *(b)* bandwidth in a network.

have exactly zero delay (bits are instantly "transported" as on "Star Trek" across the network) and still not have enough bandwidth (only 4 bits per second can enter the "bit transporter") to be useful to most applications.

Packet networks give applications varying amounts of bandwidth depending on application activity and bursts. This varying bandwidth means that the delay can vary as well on the network. Congested network nodes can contribute to variable delays as well. However, the delay QoS parameter just specifies a maximum delay and does not set any lower limit on the network delay. If a stable delay is needed, another QoS parameter must take care of this requirement.

Jitter (Delay Variation)

The jitter QoS parameter sets limits on the amount of delay variation an application encounters on the network. Jitter is more properly referred to as *delay variation*, since the term *jitter* also is used in networking to mean low-level timing differences in line-coding techniques. However, the use of the term *jitter* as a synonym for delay variation is common, and the context will almost always distinguish which meaning is indicated. Jitter sets no limit on the absolute delay itself, which may be relatively low or high depending on the value of the delay parameter. This is one reason that jitter is best represented as a related, but separate, delay QoS parameter.

Jitter can be specified as a relative or absolute network QoS parameter value. For example, if the network delay for an application has been established at 100 ms, the jitter may be set to plus or minus 10 percent of this value. Thus any network delay between 90 and 110 ms still meets the delay jitter requirement (in this case, the delay is not a maximum delay, obviously). If the delay is 200 ms, then the 10 percent jitter value allows any delay between 180 and 220 ms and still meets the relative jitter parameter requirement. On the other hand, an absolute jitter limit of plus or minus 5 ms would limit the jitter in the examples to between 95 and 105 ms and 195 and 205 milliseconds, respectively.

The applications most sensitive to jitter limitations are real-time applications such as voice and video. It matters little if the beginning of a Web page makes its way through the network slower than the end of a Web page. The same is true of file transfers. The Internet, given its roots as a data network, has little to recommend it as a network with any limits on jitter at all. Delay variations continue to be the most vexing problem encountered by Internet-based voice and video applications.

Information Loss (Error Effects)

Information loss is a QoS parameter that is not mentioned as often as bandwidth and delay, especially when the network under consideration is the Internet. This is only an acknowledgment of the "best effort" nature of the Internet. If IP packets do not arrive at the destination, the Internet cannot be blamed for losing them. This does not mean that the application will necessarily fail but that if the missing information is still needed at the destination, it is up to the application to ask the sender for a copy of the missing information. The network itself is of no help in this regard, since a copy of the missing information is not available in any of the network nodes in the vast majority of cases.

Why would any network, not only the Internet, lose information in the first place? There are actually many reasons, but most of them can be traced to the effects of errors on the network. For example, if a link fails, then all the bits currently in transit on that link will not, and cannot, arrive at the destination. If a network node such as a router fails, then all the bits currently in buffers and being processed by that network node will just disappear without a trace. Since these types of failures on networks occur all the time, some information loss due to errors on a network is unavoidable.

The impact of lost information varies widely from application to application. Error control over a network is a two-step process that first begins with error detection. The second step is error recovery, which may be as simple a strategy as the sender retransmitting missing information units. Some applications, particularly real-time applications such as those which include the use of voice (or audio) and video, cannot effectively recover from errors by resending missing information units. Data applications are more suitable for retransmitting error-recovery strategies, but even here there are exceptions (military systems, for example, those used to coordinate air strikes and weapons targeting, cannot effectively use a resend error-recovery strategy).

For these reasons, the information loss QoS parameter should not only specify an upper limit on the effects of errors but also should allow users to specify whether retransmission is a viable error-recovery option or not. However, since most networks (especially IP networks) just provide passive transport and error detection, error recovery is typically left up to the application (or user) anyway.

Availability (Reliability)

Networks exist to serve their users. However, networks have a need for preventive maintenance if potential breakdown situations are to be detected and forestalled. A sound strategy of periodically removing a network device from service for routine maintenance and diagnostics for short periods of time can prevent more crippling service outages if such maintenance is delayed or bypassed altogether. Of course, even the most rigorous program of preventive maintenance cannot avoid unpredictable and catastrophic failures of links and devices from time to time.

This reality of the need for preventive maintenance and the fact that equipment does fail despite extensive efforts to the contrary poses an interesting problem for networks. Not all the network will be available for use all of the time. Preventive maintenance can cause scheduled downtime (even if a backup is available, it sometimes takes time to swap devices and links), and equipment failures can cause unscheduled downtime on a network.

Not too long ago, the PSTN had a much harder time scheduling and performing maintenance than data networks. The PSTN had to be capable of carrying calls day or night on each and every day of the year. Some time periods had few calls in progress, such as the hour from 3 to 4 A.M., but there were calls in progress almost all the time. Naturally, there are rules for preventive maintenance on the PSTN. Some things could be done during known traffic lulls, some things could never be done during busy hours or days, and so on.

Data networks had it easier. Most data networks were for business and ran during business hours, usually 8 A.M. to 5 P.M., Monday through Friday. This helped the data network run smoothly. Backups could be done "out of hours," and a full suite of tests intended to uncover problems could be run over the weekend. And if a major component of the network failed at 3 P.M. on Friday, most users actually enjoyed the early weekend break.

The Internet and Web have changed all this. Any global network has to deal with the fact that someone is always trying to work somewhere. And the Internet is perhaps even more useful at home at 10 P.M. than in the office at 2 P.M. (certainly there are fewer inhibitions or prohibitions to deal with at home).

However, if users realize that they cannot have the network at their command all the time (some users do not understand this at all), how much is enough? And when failures do occur, how quickly must service be restored? Both are key aspects of the *availability*, or *network reliability*, QoS parameter.

There are 60 × 60 × 24 × 365, or 31,536,000 seconds in a year. Suppose a network is available 99 percent of the time. This allows a service provider 315,360 seconds, or 87.6 hours, or about 3⅔ *days* of downtime per year. This is a lot of time for a network to be unavailable, about 1 day per quarter. A value of 99.99 percent is much better, and a network with this availability QoS is down for only about 50 *minutes* per year. Of course, the network service provider needs a lot of redundancy and error-recovery mechanisms in the network to achieve this. Table 12-2 shows percentage availability expressed as yearly downtime.

Typical network availability QoS parameters today read something like "99.995 percent, or about 26 minutes per year of downtime. Links restored in less than 4 hours." The restoral factor limits the risk of multiple failures affecting service. Penalties are usually spelled out for missing these numbers in a given month. There is also a difference between network availability and reliability from the individual *user* perspective and the global *network* perspective. Today, entire networks do not fail all at once and render all users isolated (there are occasional but notable and highly publicized cases of total network outages, however). Usually, the availability QoS parameter refers to each individual site or link, but not always. An irate user complaining of a link available only 99.7 percent of the total time in a month could be gently reminded that the 99.99 percent availability advertised and promised applies to the network as a whole. Since this may be the only failure on the entire network during the month, the user has no recourse and no reason to expect a rebate.

TABLE 12-2

Network Availability and Downtime

Network Availability	Total Yearly Downtime
99%	3.65 days
99.5%	1.825 days
99.9%	8.76 hours
99.95%	4.38 hours
99.99%	52.56 minutes
99.995%	26.28 minutes
99.999%	5.25 minutes

Security

Security is a relative newcomer to QoS lists, but an important one. In fact, some would prefer to list security right after bandwidth when it comes to essential network QoS guarantees. Widely publicized threats of hackers and the spread of viruses on the global Internet have moved security to the forefront recently.

Most security concerns revolve around issues such as privacy and confidentiality and client and server validation. A full discussion of networks and security is far beyond the scope of this chapter, but it is enough to point out that security concerns are most often addressed with some form of cryptographic method, such as encryption and decryption. Cryptographic methods are also used on a network for authentication, but these methods do not usually involve any decryption at all.

Briefly, privacy and confidentiality concerns typically are addressed with either private or public encryption techniques. Client validation ("you are who you say you are") is often provided by a simple password, more elaborate digital signature, even more elaborate biometric systems such as retinal scans, or combinations of several methods. Server validation ("you are at the official Rolex Web site") is often provided by a digital certificate issued by a certificate authority and administered by a registration authority.

Whole architectures have sprung up to add privacy or confidentially and validation or authentication to the Internet. The official secure IP protocol, called IPSec, is becoming a key architecture for supporting e-commerce on the Internet and preventing fraud in VoIP environments. Oddly, the global public Internet, frequently cited as the most unsecure network around, has made security part of the IP from the start. A bit in the IP packet header type of service (ToS) field was specifically set up for applications to be able to stress security on the packet exchanges. Lack of consistent ToS implementations among router vendors has been a problem, however.

Users and applications can add their own security to a network, and in fact, this was the way it was done for years. If there was any network-wide security at all, it was usually in the form of a simple network login password. This is no longer an adequate state of affairs as all daily activities slowly make their way onto the Web and Internet. Today's networks need security built in, not added on in a haphazard fashion by each application. Otherwise, the possible combinations of client-server interactions involving security become a nightmare.

A typical security QoS parameter may be "encryption and authentication required on all traffic streams." Alternatively, data transfers may just need encryption, and Internet telephony connections may just need authentication to prevent fraud. The importance of security as a network QoS parameter cannot be underestimated today.

The Key Role of Bandwidth

Some lists of network QoS parameters are very short. Some just contain bandwidth, delay, and jitter as legitimate variables that users and applications can specify on the network. The rest are just what the network gives everyone, so there is no need for anyone to specify values for security and so on. The effects of errors and network availability are identified and measured, not specified and provided.

Some QoS lists have only one entry—bandwidth. Now even delay and jitter are just what the network can give, not what users and applications need or desire. To proponents of this bandwidth-only approach to network QoS, however, this is not a limitation, but actually an improvement in network QoS. Given enough bandwidth, this philosophy goes, all the other QoS concerns fade into the background.

This is a powerful argument, and one that has a lot of logic behind it. The effects of increased bandwidth in terms of lower delay has been discussed already in a previous section. The same line of reasoning can be applied to each of the QoS parameters in turn, with all of them turning out to be insignificant given enough bandwidth on the network. Users can just choose what they want to do for delay, jitter, and so on, and the enormous network bandwidth available will take care of everything. There is no need to specify any QoS parameter except bandwidth to the network.

For example, maximum bandwidth allows packets to arrive almost all at once, minimizing delay. There are never any queues, so jitter disappears as well. Errors are not a problem because the user can send everything multiple times at once, and use of multiple routes minimizes failure concerns. Security is a breeze because the user can diversely route individual bits if they like on multiple channels and paths. Any objections to this logic can be dealt with using more sophisticated lines of reasoning.

Critics of this line of thinking have pointed out that most proponents of the "only bandwidth" school happen to have a vested interest in building, providing, or using networks that basically consist only of leased private lines and other types of links. After all, if bandwidth is all that is important, then when all is said and done, whoever has the most bandwidth wins. Fiber and wavelength division multiplexing equipment vendors and service providers with huge fiber backbones tend to be the most vocal proponents of the "only bandwidth" QoS philosophy. The only thing that matters in a network, critics point out, is what these companies happen to be selling. A very fortunate coincidence.

There is no need to take sides in the debate. Both approaches have a lot of merit. Until bandwidth is really free, however, some other additions to the QoS list beyond bandwidth will have to be considered.

Adding QoS to a Network

If a network is lacking in adequate QoS in any area, then users must add their own methods of providing the QoS needed to what the network has to start with. Mention has already been made of adding security on just such an application-by-application basis. When it comes to VoIP on the public Internet, jitter buffers are commonly employed at the receiver. This process can be expensive, time-consuming, and haphazard. It is usually better if the network can deliver at least some QoS guarantees in this realm.

Of all the QoS parameters, the most difficult for users to add on for themselves is delay. If the fastest that bits can travel through a network is 5 s, then there is little a user can possibly do to lower this amount. And voice just happens to be the application that is most sensitive to network delay. Even one-way video, once it "fills the pipe," so to speak, can be used with relatively high network delays.

However, if users are reluctant to spend money adding their own QoS to each application, what should service providers do to improve the network QoS? Users cannot lower network delays, but should service providers be expected to spend money to lower delay on the entire network when only the applications that require a low delay such as voice will benefit? Why not just take that money and build another network? This is how separate voice and data networks came to coexist for so long in the first place.

This is really just another rationale for the "only bandwidth" QoS approach. The simple fact is that it is more cost-effective to address network QoS shortcomings by adding bandwidth than doing so any other way. Add enough bandwidth, and at least delay and jitter will improve. If the bandwidth is added in just the right way (using self-healing fiber rings exclusively, for example), even errors, availability, and security are improved. And the effects of adding bandwidth can be reliably predicted. Better to spend money on more bandwidth and achieve known results than to spend money on an innovative way to reroute around failures and find that users are no better off than they were before.

Build It In or Add It On?

Thus there are basically two choices when it comes to getting QoS for users and applications. The two alternatives are to build the QoS into the network or add the QoS onto the network. This is no more than the essence of the whole debate that revolves around having QoS within a network and adding QoS application by application outside the network.

These two approaches to QoS were discussed earlier. The first approach, where the users tell the network what they need, is the approach taken by most PSTN service providers. The second approach, where users add their own QoS and rely on the network for little more than connectivity, is the Internet philosophy. After all, the Internet thinking goes, How on earth would a service provider know exactly what every application I could run needs in terms of QoS? The granularity alone would be a nightmare to implement.

The “QoS in the cloud” reply is that with very few exceptions, all applications fall into one of a small number of major categories. Thus only three or four classes of service (groups of QoS parameter values) are needed, not hundreds.

The debate certainly cannot be resolved here. All that can be done is to point out that there are only three ways for a service provider to deal with congestion and resource shortages on a network above and beyond the simple expedient of adding bandwidth. And only one of these methods involves spending potentially huge sums on methods to add QoS specifics to a general network like the Internet. These approaches are shown in Figure 12-3.

A service provider can

1. *Ration access and/or services to users.* This approach requires the service provider to police the network and find out who is hogging the network resources. There is no easy way to do this at present on the Internet, except to try to detect idle dial-up modem users and kick them off, which is very unpopular with the user community, naturally.

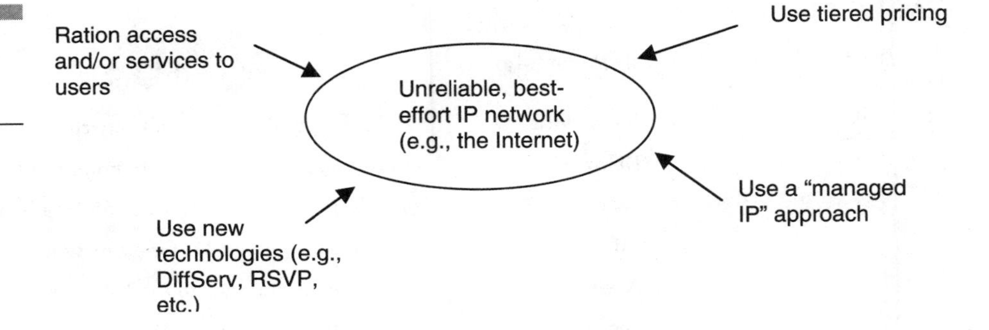

Figure 12-3 Adding QoS to a network.

2. *Use tiered pricing and/or managed IP.* This approach makes users pay more for using more network resources. Static routes can be set up specifically to connect sites using VoIP to minimize delay for a premium above and beyond the ISP's connectivity charge. Some ISPs guarantee only three hops between any two sites as long as they are the ISP at both end points. In its extreme form, tiered pricing includes fully managed IP routers and links dedicated to a single or limited number of major customers.
3. *Use new technologies.* Users are not fond of rationed use, and customers hesitate to commit large sums for managed IP networks (why not build their own?). The ultimate QoS fix is to use one or more new technologies to make an IP network as good as the PSTN when it comes to supporting voice. The question is which new technology? There are many, and they will be surveyed later in this chapter.

If QoS additions beyond simply adding bandwidth or network capacity are to be made by a network service provider, the new technology approach is the best long-term solution. For now, the managed IP approach will have to do.

What can users do to an IP network to supply their own QoS if adequate QoS for applications such as Internet telephony are not available from an ISP? Table 12-3 lists some of the things that can be done to add the QoS parameters available from an ISP.

Most of the user actions are self-explanatory, but keep in mind that these actions are applied in this chapter mainly to VoIP concerns. Seldom does 4- or 8-kb/s digital voice cause a pure bandwidth problem for users on the Internet, except under very congested conditions. Inverse multiplexing may be needed to support many VoIP traffic

TABLE 12-3

Adding Application QoS to an IP Network

To Improve This QoS Parameter . . .	Users Can . . .
Bandwidth	Inverse multiplex to add capacity for many voice calls
Delay	Not do much except try to minimize router hops (but how?)
Jitter	Add jitter buffers (almost universally done now anyway)
Information loss	Add some forward error correction to voice packet (rare)
Availability	Use multiple links to the ISP or even use multiple ISPs
Security	Add their own authentication and encryption methods (common)

streams rather than any individual VoIP connection. Jitter buffers are almost universally used with any VoIP scheme whether the Internet is involved or not because only the circuit-based PSTN has stringent enough limits of network jitter built in to satisfy voice. Error effects can be minimized somewhat by using forward error-correcting (FEC) codes, but this is rarely done. FECs require compatibility and interoperability and thus limit user choice, and FECs are helpless when entire packets just disappear on the Internet. Availability is enhanced by employing multiple links to an ISP or even by using multiple ISPs. Finally, security traditionally has been a user QoS concern, and VoIP and Internet telephony can use any one of a number of schemes for privacy and authentication, although interoperability is still an issue.

VoIP Router Concerns

When it comes to VoIP, there is more that needs to be addressed in terms of QoS than the six QoS parameters established earlier in this chapter. There are issues that emerge simply because the Internet and all IP networks are essentially collections of connectionless routers that route each packet independently based on local network information only. Thus a router can send a packet to an adjacent router that can only forward the packet onto severely congested links, much to the dismay of users trying to deal with the resulting delay, jitter, and packet loss. This is just the way routers work in most cases. The global information router's share is just basic connectivity information unless special tables are built and consulted, which slow the whole routing process down anyway.

Above and beyond packet delay, jitter, and packet loss, there are purely router concerns in a network using VoIP. In fact, packet loss on an IP router network is not just from the effects of errors on the link or congestion in the router. IP packet headers contain a *time to live* (TTL) field that limits the number of hops (routers) a packet can pass through before the packet reaches its destination. Each router decrements the counter (despite its name, TTL is strictly implemented as a hop count) before the router processes the packet. If the TTL field is ever 0 after the decrement, the packet is discarded and the sender notified. This is not much help for real-time applications such as VoIP that cannot resend the packet.

Most VoIP techniques generate from 20 to 40 packets per second for each call. Each packet contains from 10 to 50 ms of voice, with the actual values depending on packet size and coding method. It is not unusual for these packets to traverse 20 to 30 routers on their way to the destination. None of the routers need to be aware that any packets contain voice. Most routers, especially older routers, just route packets regardless of content or application as fast as they can in the order that the packets arrive. Thus voice packets can get tossed just like data packets. Voice is still useful with between 8 and 10 percent packet loss, depending on user expectations, and 3 to 5 percent packet loss can be compensated for by sophisticated coders. For fax applications, 4 percent packet loss is about the limit before complaints begin. And almost all faxes sent over IP with packet loses higher than 15 percent are useless.

Router-specific concerns when using VoIP fall into one of four categories:

1. *Routing asymmetry*. All ISPs will attempt to minimize the time, and thus the routing effort, that a packet spends within their cloud. Thus, if the end points of a VoIP call are serviced by different ISPs, the outbound and inbound delays could be very different, or asymmetrical. This is so because each ISP tries to pass the packet to the destination ISP as quickly as possible, but each ISP may have different traffic loads, router hops between gateways, and so on. The trouble is that most IP tools report one-way packet loss and delay as simply one-half of the round-trip delay and packet loss. There can even be different transit ISPs used in each direction. Studies have shown that 50 percent of the time the paths went through different cities outbound and inbound and 30 percent of the time even used different transit ISPs.

2. *Routing loops*. Routers could forward a packet from router to router and end up right back where they started in some cases. This happens when network link and router failures happen before router tables are all properly updated with this new information and therefore consistent (also called network *convergence*). The packet could still get where it is going, because once the packet arrives back at the router at the start of the loop, the tables are now correct. If the loop is not eliminated rapidly, the TTL counters expire, and the packet is discarded. Most Internet router loops last less than 3 hours, bad enough for VoIP, but some have lingered for about 12 hours. Loops tend to cluster geographically and center around the major Internet exchange points in the Washington, D.C., area.

3. *Route flapping*. Rapidly changing routes on the Internet are referred to as *flapping routes*, or sometimes *route flutter*. In extreme forms, routes can flap between sending packets from the United States to Australia through Japan or through Europe minute by minute. The attractive load balancing achieved in this way for data applications causes extreme jitter for voice and was relatively common in some IPv4 routing methods. Today, newer routing protocols used between ISPs have all but eliminated route flap.

4. *Route instability*. Routers can dynamically reroute around link and other router failures. Thus an ISP can have a truly horrendous track record when it comes to individual router and link availability and yet data users do not mind at all, since file transfers still go through. However, the stability of the path through an ISP's router network is critical for voice. Are the routes the same day in and day out? Month by month? The probability of a route being the same over time is called route *prevalence* on the Internet. How long it is before the router changes the path used is called the route *persistence*. Think of prevalence as a short-term measure of path stability on the order of packet times and persistence as a long-term measure of path stability. Naturally, a prevalent and persistent path through a router network is most desirable for VoIP. On the Internet, most routes are very prevalent (thousands of packets follow the same path), but route persistence can vary from seconds to days depending on the ISP.

There is not much that users can do to prevent these router actions from degrading VoIP calls short of changing ISPs. However, first the problem has to be identified and isolated to a particular ISP. What can VoIP users do to measure and identify these router issues?

Fortunately, there are a number of tools available. These are the same tools used for general IP network management, but now they are applied specifically to VoIP issues. Anyone concerned with VoIP QoS on a router network can and should use these tools to measure router-specific QoS performance in the four areas just listed.

Voice gateway IP addresses should be pinged routinely to measure round-trip delay and packet loss. Pinging in both directions, if possible, can detect routing asymmetries. Traceroute should be used periodically in both directions between VoIP end points to detect route loops and flapping. Repeated traceroutes will measure route prevalence and persistence as well.

Adding QoS to IP

Only so much can be done with a router network to provide VoIP with the QoS it needs. Sometimes, all that can be done is to use new technologies to add QoS to the hosts or routers or both. When it comes to QoS, the Internet starts with nothing at all. All agree that the Internet provides global connectivity and nothing when it comes to QoS. QoS on the Internet is such a joke that an RFC issued on April 1 (April Fool's Day) was entitled "IP over Avian Carriers with Quality of Service" (RFC 2549). The joke was as much that QoS could be added to IP as easily as carrier pigeons could be used to deliver IP packets.

Most Internet observers would agree that one of the following methods will become the accepted way to add QoS to an IP network. No clear-cut leader has emerged, however, and the players change from time to time as well. This list is not intended to be exhaustive, but it is representative of the variety of approaches that have been tried or proposed in the past few years. This list is in alphabetical order and is not intended to champion any one method over any other.

Committed Access Rate (CAR) This method is a function of Cisco "switch routers." The vendor-specific approach taken here is not as much of a limitation as it might seem at first, since the vast majority of routers on the Internet (and off) are Cisco routers anyway. However, not all Cisco routers can run CAR. CAR limits the amount of bandwidth consumed on a link by any given application. Thus, on a 1.5-Mb/s link, CAR can limit

Web access to 50 percent of this amount, leaving 50 percent for other applications such as voice. CAR does not so much add QoS as limit the competition for bandwidth. Given the primary position that bandwidth holds in the QoS list, this is probably okay. CAR can be added to an access router and greatly improve network performance even if other routers do not know about or implement CAR at all.

Class-Based Queuing (CBQ) This method is a public-domain scheme proposed by the Network Research Group at Lawrence Berkeley National Laboratory. Thus anyone is free to implement this traffic-management technique. CBQ dwells at Layer 3 of the access router connecting local-area network (LAN) and wide-area network (WAN). CBQ divides all user traffic into categories and assigns bandwidth to each class. The classes can be individual streams of packets or represent an entire category of applications, departments, users, or servers. The classes themselves can be established by configuring CBQ by combinations of IP address, protocols such as TCP or UDP, and ports that represent the applications such as file transfer, Web access, and so on. CBQ can relieve the bottleneck between LAN and WAN, is very flexible, and does not require massive changes to the entire Internet infrastructure.

Class of Service (CoS) In this chapter, a *class of service* is a grouping of one or more of the values of the QoS parameters to represent a whole category of applications. However, CoS is also a newer LAN concept defined in the IEEE 802.1p specification. This specification is used in the creation of virtual LANs (VLANs) that can span sites connected by a WAN yet still function as a single, unified LAN. CoS uses 3 bits in a LAN frame header, such as the Ethernet frame header, to assign seven priority levels to LAN frames. CoS levels can be mapped to IP type of service (ToS) levels or supported in routers with a number of other mechanisms.

Differentiated Services (DiffServ) This technique, discussed more fully in Chapter 7, is closely tied to VoIP and Internet telephony and has been publicized widely. Much of this has to do with the fact that DiffServ has the backing of IETF. DiffServ redefines 6 of the 8 bits in the IP header type of service (ToS) field to allow the ToS bits to be used for service differentiation. These 6 bits can be combined to form 64 classes of service, which represent various categories of applications and would be standardized among all ISPs and routers. The standard nature of DiffServ is quite attractive, but of course, all routers must be able to understand and honor the DiffServ QoS categories. DiffServ has no absolute QoS performance guarantees. The best DiffServ can do for VoIP, for example, is guarantee that voice packets are queued first to an output port.

IP Precedence One of the problems with QoS on the Internet is that there are usually many ISPs involved with getting an IP packet from place to place. One ISP usually has no idea what traffic is important when it arrives from another ISP's network. IP precedence allows three of the ToS bits in the IP header to be set with the values 0 through 7. This ranking determines the priority of the packet flow as it leaves one ISP network for another, with 7 being the highest priority. Thus the next ISP can handle the packet with the priority indicated. This method conflicts with DiffServ plans for the ToS field and requires all ISPs to understand this use of the bits.

Multiprotocol Label Switching (MPLS) This method is also an IETF standard, but it can work easily with the DiffServ approach. DiffServ establishes a mechanism for identifying IP CoS but more or less leaves the implementation technique up to the service provider. MPLS provides one such possible mechanism by requiring the routers to be *Layer 3 switches*. There are many ways to make a router into a Layer 3 switch, and one way is to attach the router to an ATM network and effectively make the router into a ATM switch. Based on a Cisco method called *tag switching*, MPLS requires that ISPs build a new MPLS infrastructure to process the labels used in MPLS and so preserve the characteristics of an IP router and an ATM switch all in one device. MPLS will address issues of privacy and scalability as well through the use of virtual circuits (ATM is a virtual circuit network) and packet processors that operate as "wire speeds."

Per-VC Queuing IP routers are frequently connected by virtual circuit (VC) networks such as frame relay and/or ATM. Many frame relay and ATM switch vendors used to implement a single output buffer for all traffic bound for the same output port. Per-VC queuing uses a separate output buffer for each virtual circuit. Each buffer can be given a priority, so voice virtual circuits (for example) can be given precedence over other data-carrying virtual circuits. This method establishes no firm relationship between the IP flows and the virtual circuit numbers themselves, so the priority flows must be identified by some other mechanism.

Policy Routing This is a concept that has been around for a while and is built into many newer routing protocols such as Open Shortest Path

First (OSPF). The network administrators must make a choice of one or more *policies* that are applied when the routers build their routing tables. For instance, security could be one routing policy that can be used to instruct the router to choose the most secure routes first (such as links using encryption) and choose the least secure routes only as a last resort (such as links on microwave or other broadcast media). Each policy requires a separate routing table to be built and maintained by each router. Usually, the ToS field in the IP header is used to decide on the routing table to be used for a particular packet. To be effective, policies must be applied consistently on all routers and using the same principals.

Quality of Service (QoS) Queues Also known as *class of service (CoS) queues*, in this method router and switch vendors establish a small number of queues for each output port and parcel out traffic into queues based on QoS needs. This is a type of "per-VC queuing without the VCs." Without VCs to identify QoS needs, the QoS need must be set for a particular packet stream by another mechanism such as the IP packet type of service (ToS) field. This ToS can be used to map the packet into a QoS class of an underlying network infrastructure. ATM network switches usually have four queues for output traffic, but the granularity of these QoS queues for IP packet delivery is of limited utility because all IP packets tend to fall into the same ATM QoS category.

Random Early Discard (RED) This method relies on rules based on probability to instruct a router to begin dropping packets when established queuing thresholds are crossed. For example, an RED router may begin to randomly toss packets when an output buffer reaches 80 percent of capacity. The goal is to prevent buffer overflow and the chances of wiping out a large number of high-priority packets from a single source. Thus, instead of taking a chance on losing a lot of voice packets, an RED router tries to lose only a few packets from a number of sources, which may have lower priorities, when congestion begins. RED can be combined with other QoS strategies and need not be employed in each and every router to be effective.

Resource Reservation Protocol (RSVP) A few years ago, RSVP was the leading contender for a standard way to add QoS to an IP network. An IP host supporting RSVP could request very specific QoS parameter values (64 kb/s, 100-ms stable delay, etc.) from the network, and RSVP routers could supply the QoS needed. Thus RSVP requires changes not only in the routers but also in all the hosts, unlike most other QoS methods, which only are applied in routers. RSVP actually reserved the resources requested, so a 1.5-Mb/s link could accommodate up to twenty-four 64-kb/s requests and no more, for example. RSVP was unusual in that the receiver (server) was the device requesting the QoS, not the sender (client). There was no mechanism for making the hosts give back the resources to the network in any time frame, which made it difficult to scale RSVP into an environment of thousands of bandwidth-hungry hosts. Much of the momentum and emphasis on RSVP has shifted to DiffServ.

Type of Service (ToS) The IP header (IP version 4, that is) contains an 8-bit field called *type of service* (ToS) that was supposed to be used to indicate packet priorities in several QoS areas. Router vendors routinely ignored the ToS field because IP host software implementations never really set these bits in any consistent fashion. IP was always "best effort" anyway, until some vendors began to use the field for their own purposes. The ToS field is redefined in DiffServ.

Traffic Shaping Many IP routers are linked by frame relay and/or ATM networks. In ATM, IP packets entering an ATM network are *shaped* at the access device to prevent a huge burst of traffic from congesting the backbone network. Shaping involves accepting a burst from an input device, buffering the traffic, and then "smoothing" out the flow so that the burst is distributed over a longer period of time, a time period based on configuration parameters. Traffic bursts above a certain limit are ignored by the input device, again based on configuration. Traffic shaping in a frame relay network is part of the committed information rate (CIR) and excess information rate (EIR) concept.

Weighted Fair Queuing (WFQ) This method also can be combined with other techniques and is frequently mentioned in MPLS discussions. WFQ applies to the bandwidth an application receives on an output link. Each packet stream that WFQ is applied to is buffered separately and receives bandwidth on a variable, weighted basis. For example, 100 data packets and 100 voice packets could arrive on two input ports in the same time frame and be queued to the same output port. Ordinarily, the packets would be queued together and output sequentially without regard to priority. However, WFQ will output the voice packets first and then the data packets. This queuing method is *weighted* in favor of the voice packets and yet *fair* because the 100 data packets must still be output before any subsequent voice packets.

Weighted Random Early Discard (WRED) This is a variation on random early discard (RED). A pure RED router just randomly selects packets to drop when some buffer threshold is reached. Of course, the packets selected may turn out to be the very high-priority packets that the network is seeking to free up buffer space for. Thus WRED tries to identify the low-priority traffic and randomly discard those packets when congestion occurs.

This little survey of 15 ways to add QoS to an IP network could easily be extended. However, this is enough to show how active, and fragmented, the whole field and approach has been. Some methods must be added wholesale to all routers, some apply mainly to access routers, some apply mainly to backbone routers, some require a frame relay or ATM infrastructure, and yet others require changes to hosts and routers alike. All in all, it will be a while before any large-scale IP network, let alone the global Internet, becomes as reliable from a QoS standpoint as the PSTN. But no one expects this fact to slow VoIP or Internet telephony for a moment.

The next chapter will show how the rush to VoIP and Internet telephony is quite clearly completely independent of IP QoS.

CHAPTER 13

IP Telephony Case Studies

The multitude of standards, products, and solutions that have appeared on the Internet Protocol (IP) telephony scene testifies to the fact that there is no such thing as a "one size fits all" solution when it comes to implementing voice over IP (VoIP). While all IP telephony solutions have the Internet Protocol (IP) in common, wide differences can exist between almost every other attribute of hardware and software. The H.323 specification helps to set a suite of protocols to be used for VoIP but does nothing to say how they should be combined into product offerings or implemented in production networks. This inconsistency is most pronounced when examining products from differing vendors but is even apparent within product lines, as gateway manufacturers try to offer products that meet the needs of all types of organizations—from the small enterprise looking for a pair of gateways to the global Internet service provider (ISP) wishing to add voice to its suite of services.

This chapter briefly studies the needs of three fictitious organizations, each of which wishes to implement VoIP in order to address its own particular set of issues:

Gaba & Sweeney Engineering is a construction consulting firm that needs a way to provide economical integrated voice and data services between its offices.

Helena Fabrics is a multinational manufacturer of textiles that needs a phone system for its new corporate campus.

CCK Networks is a national ISP that is interested in leveraging its enormous investment in IP backbone technology by offering inexpensive long-distance voice and fax services.

These three companies, while sharing a desire to save or make money by providing voice and other services via IP, differ greatly in the specific business objectives they seek to address with the new technology. To achieve these goals, they will implement widely varying solutions from a number of different vendors.

The three solutions presented in this chapter serve as general examples of approaches to the problems faced by the companies and are not meant as endorsements of particular products or vendors.

Gaba & Sweeney Engineering

Current Situation Gaba & Sweeney Engineering (G&S) maintains seven offices: a headquarters in New York City, with branches in Washington, D.C., Atlanta, Dallas, Chicago, Los Angeles, and Salt Lake

City. The firm provides engineering consulting services to builders and designers of residential and commercial buildings. Each branch office is somewhat independent in terms of its customers, projects, and staff but uses a research service and specialist consulting engineers provided by and located in the head office. Companywide administrative functions such as billing and payroll are also handled centrally from the New York office. Business is booming, but while all offices are busy with current and planned projects, there are no plans to add new offices in the near future.

This situation results in a voice traffic pattern that is tilted heavily toward interoffice use, which has been noted by the IT director and CFO. Since Chicago is the second-busiest office and serves as a de facto second headquarters, there is a private voice (tie) line supplying four voice channels between New York and Chicago, which the company purchased on the advice of its PBX manufacturer. While G&S has explored the idea of tie lines between the other offices, it has always shied away from the commitment of a tie-line contract, given the high cost and uncertainty of the correct way to proceed with implementation.

Currently, G&S obtains all its long-distance telephone service, including the tie line, from AT&T, with whom it is in the second year of a 5-year contract—a fact that causes no small amount of aggravation for the CFO, who negotiated the per-minute price herself. The resulting price schedule was quite competitive for the time but has begun to look less and less so as discount long-distance providers begin offering service—some of it via VoIP—that directly competes with AT&T's service.

The headquarters and Chicago offices lease Lucent Definity PBX switches to provide phone service to the 70-odd phones at each location. These switches connect to their respective incumbent local exchange carrier (LEC) via a T-1 circuit, providing 24 voice lines, and each has the capability to add two more T-1 connections should they be necessary. Each of the remaining four locations has an office phone key system, purchased without regard to standardization and therefore of varying brands, capability, and age. While none of these key systems could be considered "carrier class," they are relatively reliable, provide the services that the office workers need, and are suitable for the offices they serve.

On the data side of the house, G&S uses, as do many engineering firms, somewhat sophisticated modeling and computer-aided design (CAD) software on its engineering workstations. The size of the files generated by these software packages can be quite large, and the regular necessity of transmitting them from office to office has resulted in G&S deploying its own hybrid wide-area network (WAN) strategy to handle the load of this mission-critical data. The WAN topology consists of six

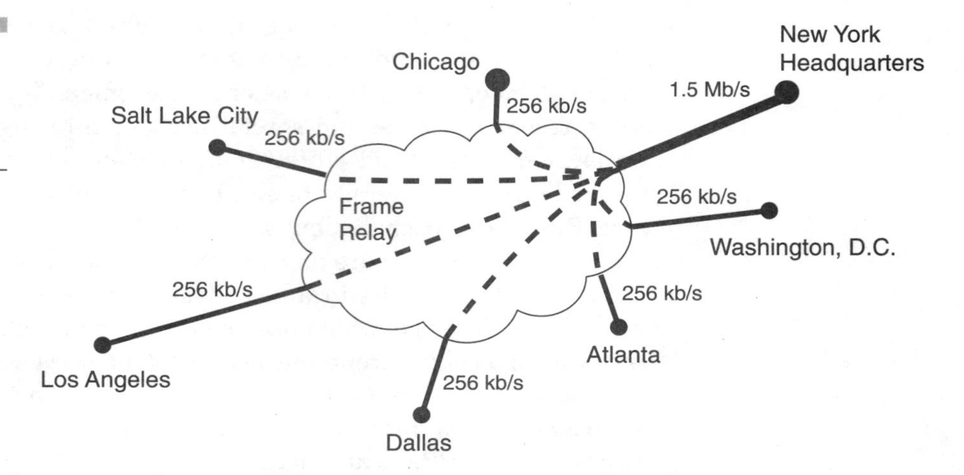

Figure 13-1 Current Gaba & Sweeney data network.

frame-relay connections provided by Sprint, with 1.5 Mb/s at headquarters and 256 kb/s at each of the branches. Circuits describe a star pattern, with New York as the hub. Since the WAN currently supports what is essentially a file transfer service, the frame-relay circuits are provisioned with 0 committed information rate (CIR) (Figure 13-1).

At headquarters, each of the engineering workstations and servers benefits from a newly installed switched 100-Mb/s Ethernet local-area network (LAN), while the administrative and other staff PCs share a number of 10-Mb/s Ethernets. With the exception of the Chicago office, which recently has invested in an Ethernet switch, all the branches use shared 10-Mb/s Ethernet for desktop connectivity. All branch Ethernets are connected to the frame relay network by IP routers—some from Cisco Systems, but with the older ones from a variety of other manufacturers.

While the company has been relatively unsophisticated in its approach to telephony, the IT department at G&S is proud of its data network. The company has an excellent IT staff of six, two of whom are Cisco-certified internetwork engineers and all of whom are interested in the possibility of using new technology to solve business problems. Management has given the IT department a mandate to reduce the cost of telephone service, and—given the current buzz in the industry—the IT director sees IP telephony as a potential component of a solution and one that is amenable to the IT staff's interest and expertise.

Rejected Approaches Initially, G&S identified three possible approaches to the goal, all of which were rejected on the following grounds:

Obtain a new long-distance contract. It would have cost $20,000 to buy out the current long-distance contract, and that expenditure would do nothing to improve the situation other than allowing G&S to solicit bids for the same type of service all over again. While a new contract undoubtedly could be negotiated at a somewhat lower rate, the cost of buying out the current one would erase the savings.

Add more interoffice tie lines. This would decrease the amount of long-distance calling charged per minute, but the large amount of inter-branch office calls (particularly between Chicago and the other branches) would require a large number of circuits or require all calls to go through New York. The interstate nature of these lines makes them particularly expensive, and this option was rejected on cost grounds and because of the lack of flexibility it imposes.

Desktop telephony. Some members of the IT staff proposed a desktop telephony model solution, involving either Ethernet telephones or PC telephony software to allow workers in any office to call other offices over the frame relay line. This approach looked promising when demonstrated by the IT department, which completed a detailed traffic analysis showing that there was more than enough bandwidth available over most network links to support the current amount of interoffice calling. When trials were conducted between three offices on a volunteer basis, however, users were dissatisfied with the "call from your PC" model. The Ethernet phone solution was accepted by users more readily, but at $300 each, the capital cost of the phones and the logistics necessary to install them and a calling center were large enough obstacles to kill the project.

A Solution A number of characteristics of the G&S situation help point toward possible solutions:

A limited number of offices. With only six branch offices, and this number not expected to grow soon, the call control and routing of interoffice calls are relatively simple.

An existing high-quality WAN. The frame relay backbone connecting the six offices provides enough bandwidth for bursty data traffic but at the same time sits idle much of the time. It is also extremely reliable, never having had an outage in the life of the service.

The CCIE certification of two IT staff members. This Cisco Systems certification essentially means that those who hold it have a proven ability to maintain and troubleshoot networks comprised of Cisco products.

Given the preceding qualities, the IT director determined that VoIP, running transparently between back-to-back gateway routers at each site, could serve the mandate. He solicited bids from a number of vendors but ultimately chose a solution from Cisco Systems, based on the features of the products, the reputation of the vendor, his relationship with the value-added reseller (VAR), and—not least of all—the training and certification of the staff who would implement the solution.

The fact that each office is already equipped with an office telephone system that functions well and with which the staff is comfortable means that there is no reason to replace that system. Ideally, in order to make the changeover to the new VoIP long-distance service, users should have to learn as little as possible in the way of new procedures, and replacing the phone systems would add to this administrative burden as well as incur significant unnecessary expense.

Therefore, the G&S IT staff decided to purchase one Cisco 3600 series and five 2600 series VoIP routers, using them to replace the current router infrastructure. The existing routers will be retired and traded in to the dealer or used elsewhere in the network. The new routers will provide the same data service that the old ones did but also will connect—via an analog or T-1 interface module—to the existing office phone systems (Figure 13-2).

While all office staff are used to dialing 9 to get an "outside" line, in

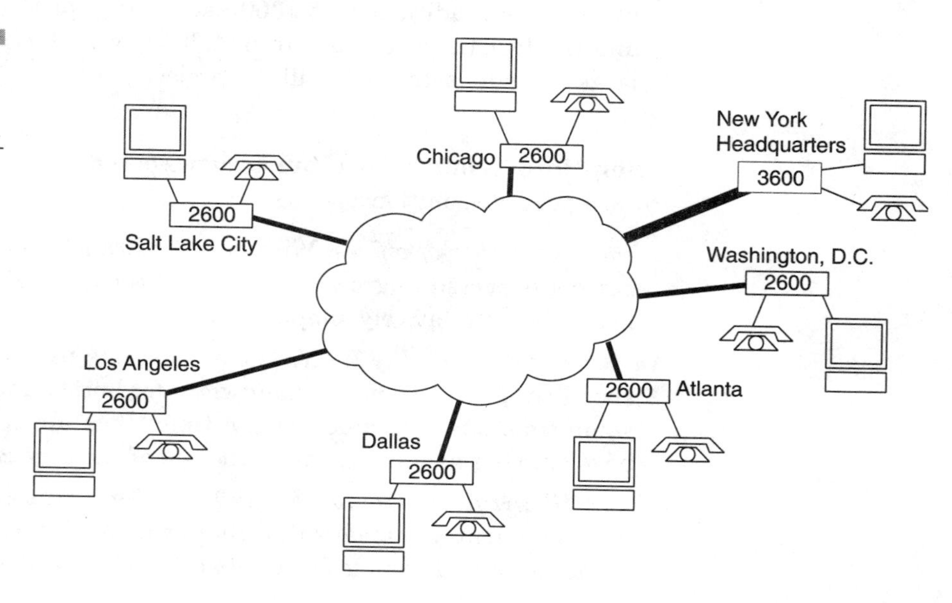

Figure 13-2 Gaba & Sweeney VoIP WAN solution.

order to place an interoffice call, users pick up the handset and dial 8. (New York and Chicago staff are already used to this when they call between these two offices, because this is how the tie line is accessed.) This instructs the telephone system to route the call to the VoIP router interface, which returns dial tone to the calling phone and waits for DTMF tones from the user indicating where to direct the call. When a router receives an incoming call over the IP network, its job is simply to hand the call to the local PBX or key system, which then handles the call as though it originated in the local office, ringing the indicated extension or sending it to the local voice mailbox.

Since the amount of voice traffic over the network is not expected to cause significant congestion, G&S decided to leave the permanent virtual circuit (PVC) map of its frame relay service as is, at least for the time being. This results in all interbranch voice traffic being routed through the New York office, one of the reasons that the faster 3600 series hardware was deployed there. If congestion becomes a consideration, the move could be made to a fully meshed topology in the future. Meanwhile, the smaller size of the routing table makes configuration and maintenance easier

The simplest Cisco voice routing architecture, and that which G&S has chosen to implement, is to configure *dial peers* that associate a dialed extension and number with a particular IP address—that of the gateway router located at the office being called. This mapping is accomplished statically at the router command line. An enterprisewide table may be built that all routers share. While the table must be maintained manually at each router, the static nature of the network (and the fact that there are only six routers) means that it should not change often.

The Cisco architecture allows the use of the IP precedence field of VoIP packets, by assigning higher values to VoIP traffic. This allows the networkwide prioritization of voice over data and should resolve issues of contention between the two types of traffic.

Helena Fabrics

Helena Fabrics (HF) is a textile manufacturing company, with design, sales, and administrative offices in Kansas City and production facilities in Minnesota, Mexico, and Honduras. The headquarters, which now numbers roughly 300 people, is in the planning stages of relocating to a three-building campus after having outgrown its previous office building.

HF has used, for the past 6 years, Centrex service from US West as its office telephone system. This is not because it was the most economical solution, but rather because the rapid growth of the business over that time left management too busy to effectively manage the telephone. The ease of simply saying "I need three more phones" to the local phone company proved a compelling reason to outsource not just the design but also the facilities for phone service.

The down side of this solution, of course, is the cost. Centrex service can grow to almost any size but tends to be most economical for offices of 100 or so telephones. The service charges themselves are bad enough, receiving no economy of scale as one would with a centralized PBX system. Worse, however, are the costs and inconvenience of adds, moves, and changes, which require administrative action on the part of the office staff, a call to the phone company, and the inevitable delay in getting service.

HF's chief technology officer (CTO) therefore decided that the move to the new campus environment would be a good time to switch to a PBX, effectively moving management of the office phone system away from US West and to his staff. The question remained, however, as to what type of PBX system to implement.

The parameters for the deployment were enviable, given that the buildings were constructed expressly for Helena Fabrics, and the CTO was in a position to offer early input into the data and voice wiring plans. From the start, he was determined to support the highest data rates economically feasible in all parts of the network and therefore had category 5e cables installed in all horizontal locations, multimode fiber (MMF) in all vertical locations, and redundant MMF runs between the three buildings. As shown in Figure 13-3, the backbone of the network is entirely switched, with a combination of 100- and 10-Mb/s links. Interbuilding runs are accomplished via Fast EtherChannel, which allows the inverse multiplexing of 100-Mb/s links into 200- or 400-Mb/s links. Each building is connected to the others via these links terminating on Cisco 5000 series switches. When his VAR salesperson suggested that he evaluate VoIP as a campus telephony solution, the CTO was surprised, but his confidence in the quality of the Ethernet infrastructure made him agree to listen to a sales pitch.

The VAR demonstrated a solution using Cisco System's Ethernet phones, Cisco CallManager software, and a digital Cisco access gateway. This system supports business-class PBX services such as hold, transfer, caller ID, and call forwarding. These features come in an H.323-compliant package, which—as the salesperson pointed out—would allow HF to interface with Microsoft NetMeeting clients over the Internet for inex-

Figure 13-3 Helena Fabrics campus LAN.

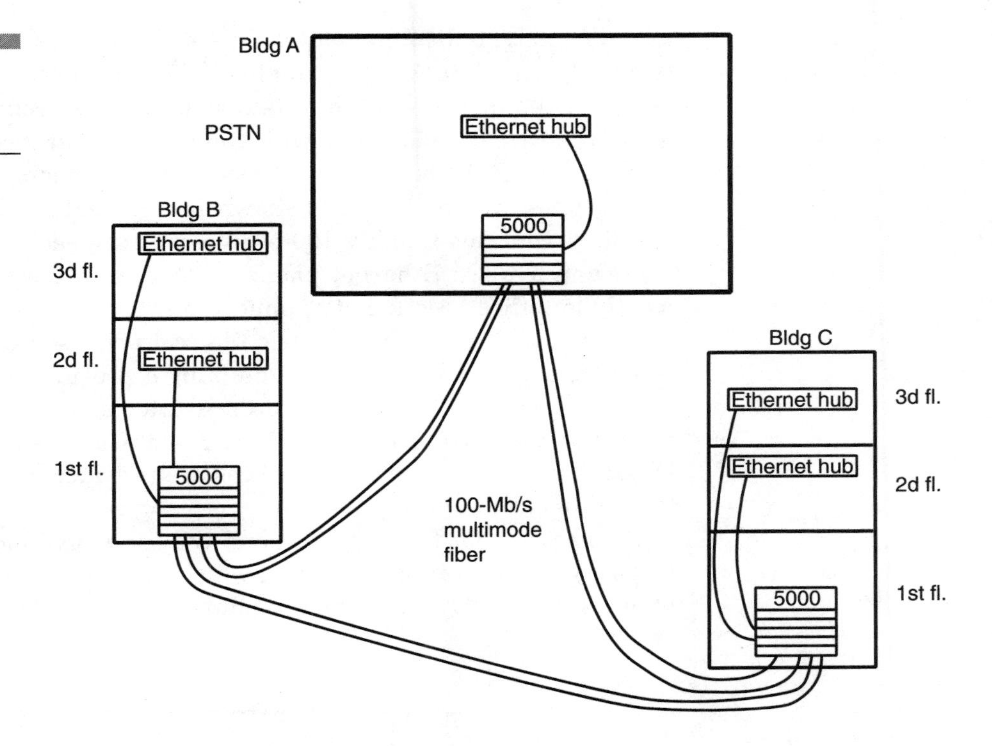

pensive long-distance calling. While this functionality is not perhaps ready for enterprise use, it remains an attractive long-term option for HF, which spends thousands of dollars monthly on long-distance calls. Since reliable Internet access is not available at some of the factory locations, though, interoffice VoIP will have to wait a while.

At the center of the Cisco system is the CallManager software, which runs on a Windows NT server. CallManager acts as the central management console for all VoIP devices in the system and is used for configuring phones, routing plans for the gateways, and all other features. It also provides call-management services for all Cisco devices on the network. CallManager is implemented as a collection of active server pages (ASP) running on the Internet information server (IIS) integrated into NT, which allows management to be completed via a Web browser from virtually any PC. Configuration data are stored in a database on the NT system.

The performance of the demonstration system, with its features, robustness, and excellent voice quality, impressed the CTO. After careful consideration, he decided to purchase a LAN telephony system based on the demonstrated components. This decision was a bold one, for the telephony and data worlds had always been separate at HF. In the end, however, the qualities of the Cisco solution were convincing enough to make the case—particularly with the added assurance of the VAR that as an alpha customer, he would receive top-notch support services.

The final system design is shown in Figure 13-4. Each of the three buildings contains roughly 100 people, who are each supplied with both a PC and a Cisco Ethernet phone on their desktop. The PCs are used mostly for office productivity applications at the workgroup level, which are supported by building-centric file and print servers running Novell NetWare 5. An AS400 located in building A provides access to the corporate database of customers and orders. Internet connectivity is provided via a Cisco 4500 router and a Linux e-mail host—also in building A. DHCP address configuration services are provided on three servers, one located in each building.

Connections for fax machines and the occasional modem are handled by Cisco analog station gateways, which provide 2, 4, or 8 plain old telephone service (POTS) interface modules per unit. These gateways allow

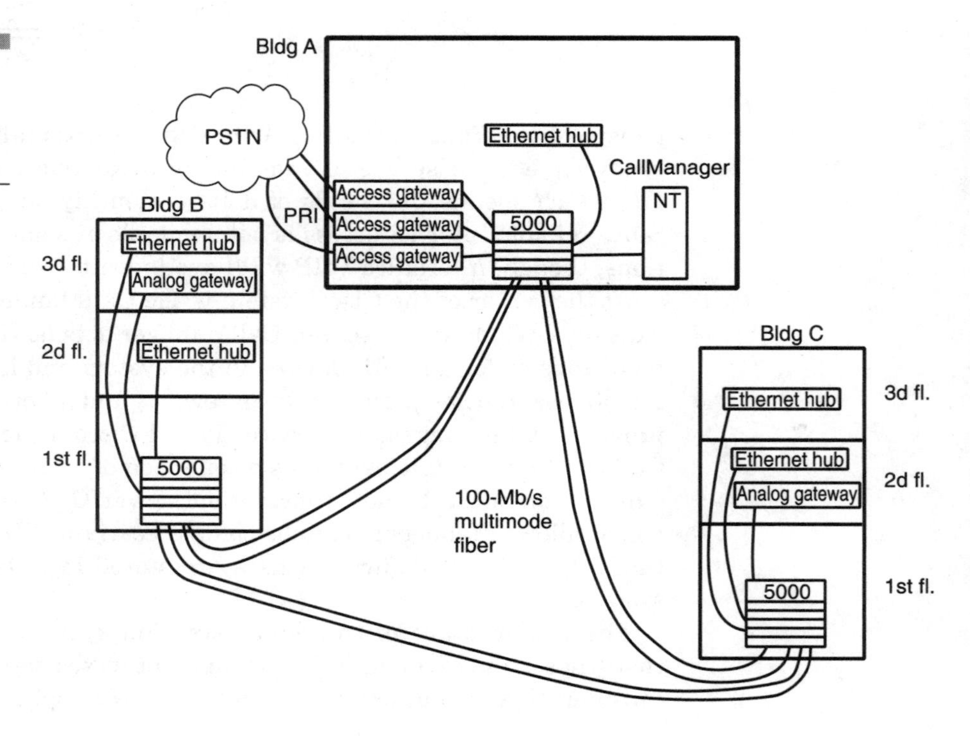

Figure 13-4 Helena Fabrics campus telephony solution.

traditional analog devices such as fax equipment and standard handsets to appear as H.323 devices to the call manager.

The primary concern in the design of the new telephone system was reliability, which made the placement of the CallManager server critical. In the end, it was decided to locate the server centrally in the building A data center and to host it on an enterprise-quality server with redundant power supplies, hot-swappable hard drives, and other features designed to make the server suitable for mission-critical applications.

Connectivity to the local phone company is handled by three Cisco digital access gateways, which connect to the outside via three ISDN PRI lines, providing 69 outgoing and incoming lines. CallManager provides all connectivity and call-routing services to the devices on the network. When a call is placed between two devices (Ethernet phone to Ethernet phone or Ethernet phone to access gateway for calls to the outside), the call setup is negotiated through the CallManager. Ethernet phones pass dialed numbers and commands to the CallManager server, which locates the destination and then passes the incoming call to it. Once the call has been negotiated between end points, all subsequent voice traffic passes directly between the two systems. CallManager also provides voice mail services, hosted on the same server—allowing the internal handling of busy conditions.

CCK Networks

CCK Networks is a business-only ISP with a network connecting the United States, Europe, and South America. Its primary businesses are Internet connection and managed IP network services between their 17 points of presence (POPs). While sales are strong, a recent marketing study has indicated that existing customers would like to have the option of low-cost international voice services between the same locations that their managed IP network serves. The CCK IP networks are provided over a simple series of dedicated ATM PVCs, whose interfaces operate at OC-3c (155 Mb/s) speed. It is the opinion of the technical staff that the network is capable, at its current speed, of supporting near-toll-quality voice for all the overseas calling of the customers who are interested in purchasing such a service.

CCK therefore has decided to trial a dial-access toll-bypass service offering good-quality voice over its existing WAN infrastructure. The service will be made available in six cities initially, with a potential expansion to the others if the trial proves successful. The customers who have indicated the most interest in VoIP have locations near the New York, Hamburg, London, Caracas, São Paulo, and Rio de Janeiro POPs. Therefore, these six POPs will be the first to have gateways deployed (Figure 13-5).

Competitive bids were solicited from gateway system vendors, and CCK selected VocalTec Communications as the winner of the contract, based on its length of time in the VoIP industry, focus on provider networks, and competitive price. VocalTec presented a solution based on their Ensemble architecture, consisting of a VocalTec gatekeeper, six VocalTec telephony gateways, and a VocalTec network manager to provide operations, administration, management, and provisioning (OAM&P) for the network.

VocalTec gatekeeper is a Windows NT-based, H.323-compliant gatekeeper that provides authenticated call establishment, call routing, and security. It also provides an interface for the collection of call detail record (CDR) information, allowing the centralization of accounting and billing information. The gatekeeper is hosted in London, colocated with one of the gateways. The gateways, located in their respective cities' POPs, provide dial-in capability over the local PSTNs, via multiple T-1 or E-1 interfaces, and send packetized voice and fax to their POP routers via three switched 100-Mb/s Ethernet interfaces.

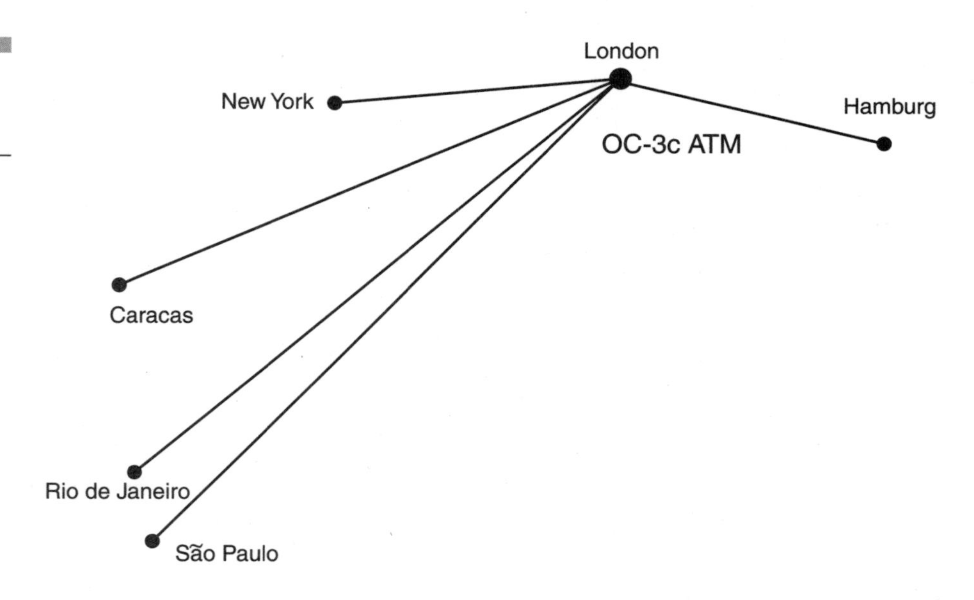

Figure 13-5 The CCK network topology (partial).

The VocalTec network manager provides central device configuration and control for the gateways and gatekeeper. It also provides status monitoring for all devices and—more important—monitors the provisioning, health, and capacity of all lines. Figure 13-6 shows a schematic diagram for the London CCK POP.

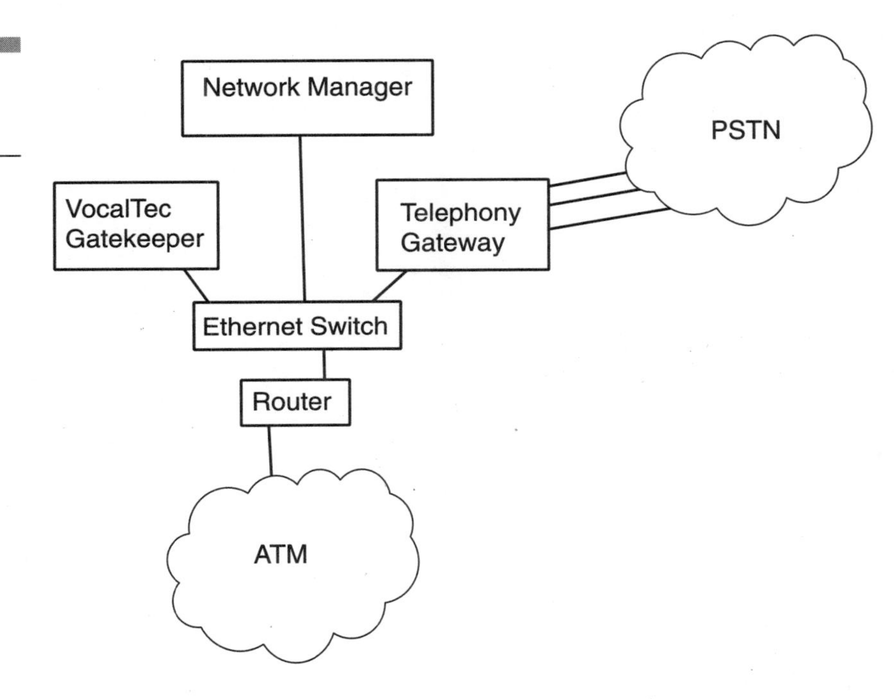

Figure 13-6
The CCK London POP.

CHAPTER 14

The Future of Voice over IP and Internet Telephony

Predicting the future is a thankless task. If one sticks to safe predictions based on extrapolating current trends, the result is accurate but fairly useless. It does not take a genius to gaze into a crystal ball and announce with proper gravity, "You know, this Internet thing could get bigger." On the other hand, risky predictions based on nothing more than a gut feeling can be wildly off target but still useful. Many predicted that PC computing would be common, and multimedia would be added, but no one saw the Web coming. Yet the path to the Web was made easier by following the path that these visionaries blazed through the green-phosphor world of DOS.

In the long run, in fact, it is the conservatives that tend to be ridiculed, not the visionaries. Many snicker at former Digital Equipment Corporation (DEC) chief Ken Olsen's statement regarding early PCs to the effect that "no one would want one in their home." But Olsen was contemplating the line-by-line scrolling monochrome world of the early PCs, which did not do much of anything without at least some rudimentary programming skills. In the context, Olsen was exactly right. No one would sneer at anyone examining the first Wright airplane and concluding, "That won't carry 300 people from New York to Atlanta six times a day." But predictions, by their very definition, can only take into account current situations. What is obvious today was hidden deep behind a cloud of unknowing only yesterday.

This said, this chapter will attempt to predict the future voice over IP (VoIP) and Internet telephony. This chapter will do so while trying not to be so conservative as to be obvious nor so liberal as to be completely off base. Only time will tell for sure.

Oddly, the future of VoIP may not be the same as the future of Internet telephony. VoIP requires the Internet Protocol (IP), and Internet telephony requires the Internet. Right now, the IP and the Internet are inextricably intertwined, but this does not necessarily mean that they will stay this way forever. As recently as 1990, in fact, the backbone of the Internet was the NSFNet run by the National Science Foundation, and there were few "ordinary citizens" on the Internet. The Web did not exist. And IP was about one-third the size it is today in terms of specifications (RFCs).

The position of IP as the network protocol of choice at the client and server ends of any network interaction seems assured for the foreseeable future. Even today, however, IP is pushed onto frame relay and ATM backbones. Does this mean that one of these core network technologies may someday displace IP as the user? And the Internet is an excellent vehicle for global connectivity. However, access to the Internet depends on the connectivity provided by the public switched telephone network

(PSTN) in many cases. Does this mean that the PSTN has the upper hand in the long run? The answer to both questions is no, and this is why.

The biggest factor that both IP and the Internet have in their favor is their adaptability and agility in the face of change. IP began as a variation on the X.25 packet theme but now has been adapted to run with real-time video, multimedia flows, gigabit networks, wire-speed routers, and so on. Try any of this with X.25. Part of this extensibility of IP is purely a "brute force" factor. For every one researcher working on X.25, there are hundreds, maybe thousands, working on IP. The same is true of IP and frame relay or Asynchronous Transfer Mode (ATM). It is not by chance that IP came out of the universities.

The Internet has changed as well, as much if not more than IP. The Internet was a valuable resource as the ARPANet, as the NSFNet backbone (when the term *Internet* was first introduced), and now as the Internet (although a better term today probably would be *ISP-grid-net*). Even if the Internet changed its structure yet again, everyone would still call it the Internet anyway.

This "more of the same" future of VoIP and Internet telephony is a safe but boring prediction. Surely there must be something that can be discerned today regarding the future that is not obvious to all. And even if the predictions are wrong in detail, perhaps the outline of the future described will be enlightening enough.

The User Future

What does the future hold for users in the VoIP and Internet telephony worlds? Will people simply toss out their telephones and sit down at the household PC when they want to converse with others? Possible, but unlikely. Specialized devices exist for two reasons: First, it is technically impossible to combine functions in a cost-effective package that will sell, and second, people like specialized devices. The second factor always dominates marketing efforts. For example, washer-dryer combinations have existed for years, yet beyond space-restricted environments, separate washer and dryer products totally dominate the marketplace. The overworked term *convergence*, which is applied to telecommunications for everything from video and data services (cable modems) to user devices (the Web-enabled TV), is not always a brilliant idea. Cable modems use the cable TV network for both data and video services, but not many people mistake their PC for their TV set.

Consumers can be extraordinarily fickle from a service and device standpoint when it comes to picking and choosing what will succeed and what will fail. The service and/or device has to work not only as advertised but also *as expected*. How can VoIP and Internet telephony be packaged to live up to user expectations?

Typical Package

There is a very good reason that future VoIP and Internet telephony products cannot continue to be based on the PC platform, at least as the PC platform currently exists. A 5-year-old can make a telephone call. Few 5-year-olds, despite highly publicized efforts to bring the PC to preschool, can boot up a PC. Most preschool PCs have specialized keyboards and interfaces, but telephones are instantly recognizable and useful by almost everyone.

To be successful, therefore, VoIP and Internet telephony products must look and feel like a telephone. In some cases the "telephone" may be the same type of "virtual telephone" that looks just like a wall phone and sits comfortably in the corner of the Windows desktop ready to take and make voice calls, but this will be the exception rather than the rule. This VoIP and Internet telephony product architecture is shown in Figure 14-1.

At the end of the access line is the network interface, same as before. But now the same house wiring is shared by the telephones and PCs on a more or less equal but not exclusive basis as with circuit-switched voice. The telephone is now just a specialized packet voice device. The PC is still a general-purpose packet device on the network. This architecture is not integrated services digital network (ISDN) but packet VoIP. And there is no "real" voice switch at the service provider end of the access

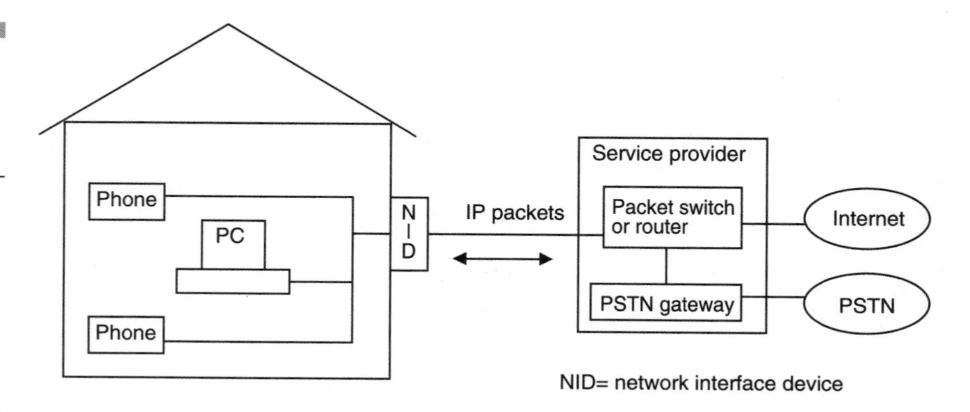

Figure 14-1 VoIP and Internet telephony product architecture.

line. Voice packets to or from the circuit-switched PSTN must pass through a gateway at some point.

This separate device package for VoIP and Internet telephony products has a number of advantages. The ability to handle telephone calls is not dependent on the presence of a running PC. The packet processing power needed to digitize and encrypt the voice does not drain the processing power of the PC. The audio delivered by the telephone need not be as robust as the theater-quality audio delivered by the PC. Compatibility is less of a concern, and vendor independence is maximized. And so on.

A possible variation on this theme is to put the VoIP hardware into the PC, allowing the voice packet processing to proceed more or less independently of the PC processing. In this scenario, the telephone handset could still be a separate device, but one that plugs directly into the PC board or card, not into a normal wall jack. Ironically, awkwardness of use, in the sense of needing the PC up and running, could be an *incentive* for these packages. These packages are easier to secure against unauthorized use through the use of passwords, although the separate device also could be secured in this fashion.

The *least* likely VoIP and Internet telephony product package is currently the most common. This is the pure software on a PC approach. Besides being awkward to use, it makes no sense at all to steal processing cycles from a PC when a voice call is in progress. This argument applies no matter how powerful PCs become (even beyond the gigahertz range) because even small drains can quickly add up to a considerable load on a processor (gigabyte hard drives once seemed inexhaustible as well).

1-kb/s Packet Voice

What's the big deal about packetized voice at 1 kb/s? Voice has gone from 64 to 8 kb/s without too much of a problem. Why should the move from 8 to 1 kb/s be any different? What should be added here is "1-kb/s packet voice *that sounds like it should*."

Early VoIP and Internet telephony products from the mid-1990s sounded, well, pretty horrible. The voice was distorted and clipped, noisy and barely understandable. Improvements in echo cancellation, noise suppression, compression, and so forth will continue until the maximum bandwidth consumed by a toll-quality voice call will be 1 kb/s at most. And during periods of silence, given the packet nature of VoIP and standard silence suppression and voice activation, the bandwidth consumed will be nil. Thus, on average, the bandwidth consumed by voice will be about 400 kb/s, in one direction at a time, and will never exceed 1 kb/s in either direction at any point.

This emergence of acceptable 1-kb/s packet voice will change everything, especially on the business side of the VoIP house. To get absolute toll-quality today from VoIP, about 150 kb/s is needed. This represents the 64 kb/s needed *in each direction* (there is no silence suppression or voice activation) by G.711 voice encoding, as well as the 22 kb/s or so overhead for H.323 and IP itself. Thus, adding toll-quality VoIP to a network is not painless and easy to do in a world where analog modems operate at mostly 56 kb/s and even digital access lines really run above 64 or 128 kb/s.

However, 1-kb/s packet voice will not only be painless to support on almost any network, it also will only make sense to do so as long as the expected quality is there. Right now, many network administrators are reluctant to add packet voice to network portals already overburdened with Web pages and electronic transactions. But 1-kb/s packet voice is as much a psychological goal as a technical goal. Many administrators will be persuaded by the "it's only 1 kb/s" line of reasoning from management and users alike or, at the very least, be more open to listen to packet voice arguments than they were before. And since most commercial businesses are trunked about 5-to-1 (one outside trunk for every five inside telephones), up to 40 people in an office could be supported with only 8 kb/s of VoIP bandwidth. With 64-kb/s voice, this bandwidth/trunk need is 512 kb/s. Even 8-kb/s voice swamps the "normal" 64-kb/s access line. This progression is shown in Figure 14-2.

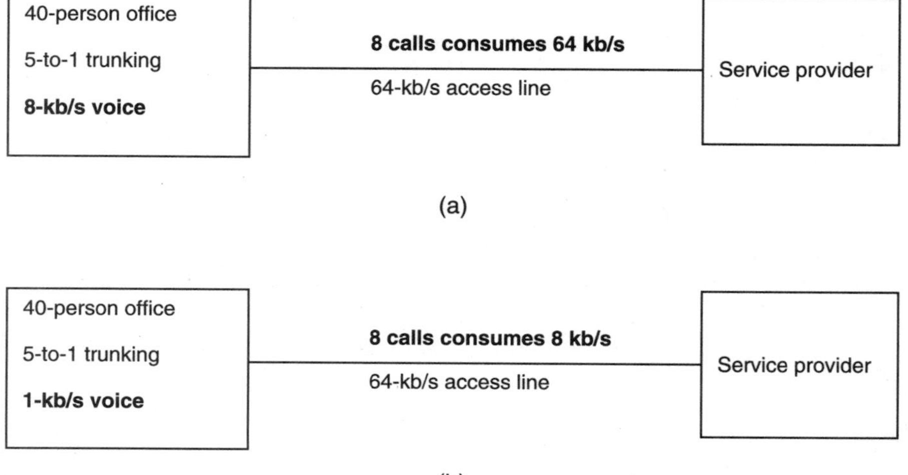

Figure 14-2 The impact of 1-kb/s voice. *(a)* Today: No room for much data. *(b)* Tomorrow: Plenty of room for data

Thus voice in this hypothetical 40-worker office goes from consuming 512 to 64 to 8 kb/s. And as the price of bandwidth falls as broadband access methods emerge into the mainstream, such as ADSL, G.lite/UADSL, and HDSL2, voice becomes even less of a network burden.

A potential problem is fitting 1-kb/s voice under the current standards umbrella. H.323 would have to be revised to allow the new packet voice at 1 kb/s, but this will not solve everything. H.323 already allows a lot of different coding techniques. Thus, even if both sides of a VoIP call are using standard H.323 products and can establish an Internet telephone call between them, the call still cannot be completed successfully unless both sides are using compatible coding methods.

The TelePC and the Ethernet Phone

Just because advances in miniaturization and electronics now mean that PCs can be used for IP telephony and telephones can be attached to local-area networks (LANs) and have some data capabilities does not mean that these ways of using IP telephony will become common. However, this does not mean that such products will not appear and be useful either. Both the telePC and Ethernet phone will have a place in the VoIP future.

The telePC will not be the size of a current desktop PC or even a small laptop. Most likely, the telePC will be the combined size and shape of the current "palm top" crop of products and a small cell phone. How many people carry and use both anyway? Specialized devices make sense when they perform different functions, but not when their forms and functions overlap to a considerable degree.

Think of the telePC as a cell phone with a palm-top built in. Want to make a telephone call from the golf course? Fetch up your Web page from the telePC, and access the telephone directory section (do not forget the authorization code). The numbers are organized by name, place, and company (how many times are numbers and only simple information like "the sales rep from the National Gear Company" available?). If the number is not in the telePC, it may be on the office or home PC. The telePC can get these Web page directories as well. Or access the national Yellow Pages to replace that putter you just tossed into the pond. No matter, just point, and the call is placed. Since this is a PC, if the number is busy, the PC can keep trying. You could even record a message that says, "I'm putting right now. Hold on a second."

Then consider the Ethernet telephone. The rationale behind this package is simple. Why should companies have to install and maintain sepa-

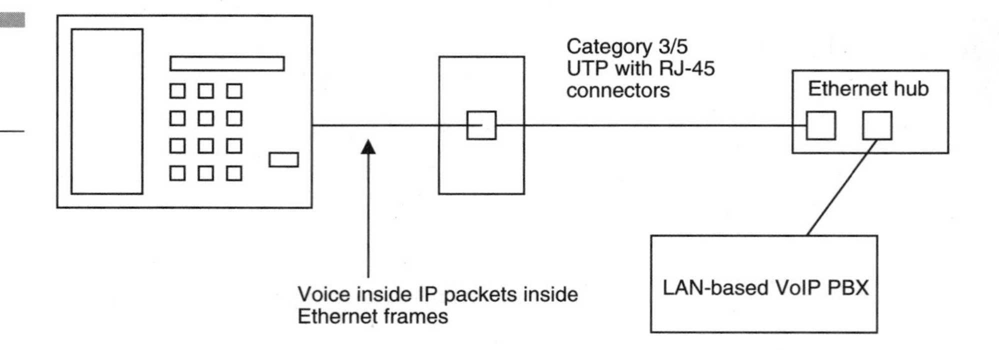

Figure 14-3 The Ethernet phone.

rate wiring systems to every work location for both voice and data? Even if both voice wiring and LANs use unshielded twisted pair (UTP), the systems are usually separate because of the difficulty of mixing voice circuits with data packets efficiently. However, once voice becomes IP packets just like everything else, the voice wiring system becomes totally superfluous. Use the voice wire to pull new data cable or fiber in.

The Ethernet phone looks just like a normal office handset, with speed-dial memory, display area, redial and hold buttons, and so on. Instead of a normal RJ-11 modular voice connector on the back to plug into the wallplate to reach the PBX down the hall, however, the Ethernet phone has an RJ-45 UTP Ethernet (actually 10base-T) connector on the back and plugs into the LAN wiring just like a PC does. Figure 14-3 shows the form and architecture of the Ethernet phone.

Naturally, the PBX itself is now a LAN-based device as well. Nothing surprising here. Ethernet telephones are available now. Thus the prediction involved here is not the mere existence of the Ethernet phone but the popularity of the device in a business situation. Home use, lacking a PBX, is unlikely to be common. The residential market is what the telePC is for.

Touch, Talk, See

There is no paradox in saying that the most common Internet telephone device will look just like a telephone. To be accepted, the device more or less has to resemble what has gone before. However, when it comes to creating new devices with new function never before experienced, the field is wide open. It is fairly easy to see where this smaller/better trend is going. New devices not only will be used to display data and talk to others, the devices themselves also will be talked to and talk back. The devices will be made to touch and feel, not pound on a keyboard. And users will see sights on these devices never before dreamed of.

None of this is new in and of itself. All these user interaction features exist, just not in the same device. Palm-tops are written on. GPS mapping software talks to drivers ("Slow down! Your turn is coming up on the left"). Hand-held digital versatile disks (DVDs) show the latest movies. Why not just combine it all into one small, wireless, hand-held device and be done with it? Not only talk *through* the device but *to* the device.

This may still be a ways off, both with regard to technology and especially with regard to price. However, imagine sitting in an airport and taking out a small 8-in diagonal display device no thicker than a paperback book. A simple headset can be used for privacy. Tell it to check e-mail, and use it to call home. Watch the airport news right on the device as well, or a movie. Everything goes in and out using IP packets.

The Service Provider Future

Free implies that the good has no value at all to buyer or seller. Since this is obviously not true of telephone calls, which can have enormous value in emergencies, there will be a cost, and money to be made, from VoIP and Internet telephony services. The costs might not be direct to the consumer. Emergency 911 calls have the most value and yet are technically free to the user. However, considerable sums are spent to support the infrastructure that enables the 911 system to function.

It is possible, but very unlikely, that IP telephony services will be supported by other sources of funds. Internet service providers (ISPs) can add some PSTN gateways and capacity to handle increased voice packets from their general pool of funds set aside for infrastructure upgrades. The ISPs can attempt to fund the expansion in capacity and equipment from their general revenues accruing from their regular data line of business. However, for many ISPs, especially smaller or regional ISPs, the funds needed are not available through this channel. Loans are another possible source of the funding needed to add adequate resources for massive voice packet loads, but many ISPs are already leveraged to the maximum just to set up and expand their data line of business.

This leaves no other alternative but for the majority of ISPs to fund voice expansion with revenues from the voice line of business. Thus IP telephony cannot and will not be free to users. Bundling voice costs in with traditional data costs of running the ISP business just masks the charges. This bundling does not make the charges go away. Ultimately, IP telephone billing will evolve to furnish the same level of detail users have become accustomed to on the PSTN. (In some countries, itemized and detailed telephone bills remain rare, so the evolution of IP telephony billing will necessarily reflect local variations.)

Many IP telephone service providers are not only ISPs but also traditional telephony companies, whether local exchange carriers (LECs) or international exchange carriers (IXCs). Traditional LEC telephone companies have billing equipment and software in place to track and bill for the most arcane calling arrangements. However, since the ISP arm of the LEC in the United States today is technically a totally separate subsidiary of the regulated LEC, this sophisticated billing system is not available to the subsidiary free of charge. Thus, in many cases, the ISP is forced to develop its own billing package anyway.

The biggest issue to be considered when contemplating the future of IP telephony from a service provider point of view is the role that regulation will play in the ultimate structure of IP telephone service providers. This chapter considers regulation as a reality and does not worry about whether it is really a good thing or bad thing or at which point even good-thing regulation becomes too much of a good thing. This debate could hardly be settled once and for all to everyone's satisfaction in the pages of this book anyway.

It is enough to point out that deregulation is the current trend in telecommunications today. This is not to say that abuses do not occur in a competitive environment. However, current thinking is that any competition, no matter how flawed in practice, is better than the tightest regulation of the telecommunications industry, no matter how perfect in theory.

Regulation will continue to determine the course of competition, however. Since regulation will not go away completely, the issue becomes one of how much competition will flourish under a given regulatory environment, no matter how much lip service is given to deregulation and competition. In the United States, for instance, the competitive LEC environment envisioned by the Telecommunications Act of 1996 (TA96) either has been a total failure or is being realized slowly in a rational, careful manner by concerned regulators, depending on whom one talks to. No one would argue, however, that the speed of implementation of TA96 is totally dependent on the role of the regulators, whose diminishing role was ostensibly what TA96 was all about.

Possible Futures

Regulation makes the most sense in capital-intensive, resource-hungry service industries. Modern regulation started with the railroads. Nobody needed or wanted competition in the form of six sets of railroad tracks leading out of town eating up the landscape and going to the same place. Thus regulation granted local monopolies to the railroads but set their profits and approved their prices. Traditionally, the telephone, insurance, trucking, banking, and other industries with the same capital and resource characteristics have been closely regulated as well.

Despite highly publicized deregulation efforts in the airline, insurance, banking, and other industries in the United States, regulation still plays an important role when a service industry promises "universal service" to the public at large. The public needs protecting from unscrupulous operators seeking to maximize profits while minimizing service. Regulators can establish performance standards and operational procedures that must be followed. The trucking industry in the United States is still tightly regulated with regard to the size and weight of rigs, the hours a driver can be behind the wheel, and so on, all in the name of protecting the public sharing the road with trucks. The trucking industry in the United States is still highly competitive but controlled.

In one sense, even the automobile industry provides a service to consumers: personal mobility. Regulation provides rules mostly involved with personal safety when the public is out exercising this mobility. Thus features such as window safety glass that will not shatter into jagged fragments and seat belts become mandatory on cars. Few automobile manufacturers are overly excited about adding to the cost of an automobile by adding safety features, unless the whole industry has to do so all at once. For example, when safety glass was invented, one large manufacturer in the United States was unfazed by arguments that safer windows would save lives in an accident and therefore should be added to cars. "I am not in the business of saving lives," the story about the CEO goes. "I am in the business of making money for my shareholders." Anything that compromised that principle, as good-intentioned as the idea might be, was to be rejected out of hand. (One competitor *did* embrace safety glass, but only after it was pointed out that accident survivors would have to buy a new car, while fatal accidents generated no such repeat business.)

The point is that neither VoIP nor Internet telephony can realistically expect to avoid regulation. The question is how heavy-handed the regulation will need to be to enforce the behavior that the regulators want.

In addition, the technology used as the favored IP telephony infrastructure could vary as well. IP telephony could be based on fiber-optic cable with dense wavelength division multiplexing (DWDM). Such a DWDM fiber world would offer more bandwidth than has ever been available before, with huge consequences for other quality of service (QoS) concerns (given enough bandwidth, other QoS concerns fade). Then again, the IP telephony world could be based on a host of new wireless technologies, from wireless access to wireless backbones. Many developing countries have no interest in deploying expensive fiber infrastructures for telecommunications but would prefer to migrate straight to wireless scenarios, since wireless is seen as the wave of the future in many cases.

This section will explore the service scenarios first and then consider the technology issues regarding IP telephony.

Service Scenarios

IP telephone service scenarios that consider the future generally fall into one of two categories. In the first scenario, the IP telephony future from the service provider perspective looks pretty much like the current telephone service provider environment. That is, some countries will have one big national IP telephone service provider looking much like a traditional regulated monopoly, many countries will have a kind of "duopoly" where two large service providers do it all (in some cases, this may even be a "triopoly"), and a few countries will have many, but smaller, competitive providers of the basic IP telephone service. In the second scenario, the IP telephony future from the service provider perspective looks more like the current ISP environment in most countries around the world. That is, with a few notable exceptions, all countries have a number of IP telephone service providers, and they function more or less independently of each other and without requiring tight regulation. Free-market pressures are relied on to weed out inefficient service providers and keep prices to customers reasonable.

Changes in technology always provide an opportunity for changes in service providers. No horse stable became an automobile gas station. No railroad became an airline. The continuation of large national service providers in the IP telephony world is not a given. In fact, it is more of a question of survival than continuity.

Besides the service providers themselves, who cares which scenario prevails? Actually, the future structure of the IP telephony world will affect users (customers) and equipment vendors in different ways depending on which scenario becomes reality. And the role of regulators will vary based on these possible futures as well. These differences are worth exploring in some detail.

Bigger Is Always Better Monolith service providers such as the Bell System in the past always had a ready answer to the question of why there was no service alternative to speak of: We are the best there is, so no one else could ever compete. Who else but a huge company could rent an expensive piece of equipment like a telephone to one and all for pennies a day? How could a smaller system guarantee that the longest-distance calls would go through? Why, the interconnection agreements alone needed with even the small independents in those days were enormously complex and would slow call completions to a crawl. Bigger is always better than smaller when it comes to universal service, the argument went, and still goes to some extent.

Today, the very size of a modern national telephone company is the biggest argument in favor of its existence. If everyone needs seamless and reliable IP telephony service, what better way to offer it? The wired infrastructure remains as capital-intensive an industry as ever, and even wireless infrastructures will not be inexpensive if every nook and cranny, mountain and valley of a country needs coverage. The counterargument would be that technology today is standardized and sophisticated enough to allow IP telephony services to be deployed on the Internet model. Aggressive companies can always do whatever larger, more conservative companies can do.

In the "bigger is better" model, the big carriers continue to dominate the IP telephony future. Competition never really takes hold, and the existing global alliances hold firm enough to prevent the emergence of any new players.

The "bigger is better" model has several implications for the overall user experience, equipment vendor strategies, and the role of the regulators. When it comes to users and customers, there is only one game in town. If the IP telephony service is not what the user expects it to be, so be it. There is simply no alternative available. Users can rant and rave, threaten and moan, but if they cannot do without, the users will have to live with the service they receive.

When it comes to equipment vendors, the big carriers will be more comfortable dealing with the equipment vendor environment as it exists today. There are only a few large equipment vendors, and these have a large, global base of operations that can furnish and service equipment worldwide. There are only a half dozen or so vendors of central office switches, for example, an industry structure fostered by the monopolies on local service that have endured for so long. Interoperability in such an environment is not so much driven by standards. Vendors can be very creative, as long as some basic interoperability for fundamental services exists among the vendors. It is one thing to create and maintain interfaces on a one-for-one basis for 5 other vendors. It is quite another to create and maintain interfaces on a one-for-one basis for 50 others. With many vendors, standards support is paramount. With few vendors, features are more important, no matter how focused on a single platform, and drive the marketplace.

In such a world of big carriers, the regulators continue to play their traditional role as "service police." Everything must be detailed in full, from pricing to service definitions, to prevent price gouging and ensure some level of customer and user satisfaction. This role is quite time-consuming and not always effective, but it is the only choice in this scenario.

Competition and Deregulation Prevail The opportunity for fundamental change in the service provider arena afforded by the shift to IP telephony could be realized by a movement toward more competition and deregulation than ever before. IP telephony actually will quicken the pace of deregulation and spur competition, based on the current model and example of the ISPs. A good example of this world is the long-distance market in the United States today. Competition and deregulation have driven the price of long-distance calls in the United States down to razor-thin profit margins and an environment where a company can prosper with little more than a carrier code (101-0666, or often seen in the form 10-10-666) and an advertisement on television ("7 cents a minute!").

The key word in this environment is *collaboration*. Big carriers may still exist, but they are extremely tolerant of the presence of smaller companies and freely interoperate and cooperate with each other. Even big carriers realize that they cannot do everything everywhere all the time, so the bigger companies allowed plenty of space for niche players to operate. These smaller companies offer more specialized or customized services and easily form partnerships with other and larger service providers.

When it comes to equipment vendors, the collaborative world offers a wide-open field. There are buyers for small devices designed for small customer bases and large packages intended for large, national carriers. Some equipment vendors will cater to smaller customers, and others will market to larger service providers. Some vendors will attempt to do it all.

The whole field will be vigorous and dynamic. Such an industry is deeply concerned with standards and interoperability. No one vendor could possibly test interoperability with every other vendor of similar products. Features and options can only be so creative, or else interoperability will suffer proportionately. And market share becomes the determining factor for such creativity, with large suppliers and extremely small suppliers being the most creative.

The role of the regulator is different in the competitive world as well. There is less need to regulate in extreme detail, since competitive pressures will adjust prices and prevent abuses. Regulation is still needed to protect customers from deceptive practices, but many of these regulatory activities involve common sense, such as outlawing ironclad, 99-year service lock-ins and the like. Another important role for the regulator is to prevent buy-outs of competitors that potentially could recreate the monolithic service provider environment this scenario seeks to avoid. In some cases, the regulator may need the cooperation of other authorities (such as the Department of Justice in the United States) to enforce its decisions regarding mergers and acquisitions.

Technology Scenarios

The structure of the IP telephony service provider industry is not the only thing that counts when it comes to possible future scenarios for IP telephony. The fundamental technology that characterizes the network could vary as well. The competition here is IP over fiber-optic cable using DWDM to supply more bandwidth than ever dreamed of in the past versus wireless technologies that promise unprecedented user mobility and flexibility of implementation.

The nice thing about IP is that the Internet Protocol suite is a general-purpose network protocol stack that runs on almost any type of network. IP runs with ease on leased private lines, a public frame relay network, a private Ethernet LAN, and many other types of network. From the IP perspective, these networks form the bottom two layers of the OSI-RM, the physical layer and the data link layer. IP at the network layer supplies the packets for these networks to transport.

The issue is that IP is not tuned to run well on any one of these networks. Such tuning must be done either by adjusting IP (and TCP/UDP) parameters to allow for the different QoS characteristics of these transport networks or relying on implementing specific RFCs to help IP cope with a particular network environment. For example, IP running on a frame relay or ATM network requires implementation of RFCs that allow the Address Resolution Protocol (ARP) to run properly. But both sender and receiver must implement these RFCs in a compatible fashion for this to work correctly.

Thus IP optimized for a fiber-optic cable backbone may look very different from IP optimized for a wireless environment. Which version of IP will be more common? This is the heart of the issue. If the world is mostly IP on fiber, then the conversion tasks will be left up to the wireless IP portions of the network. And if the world consists of mostly wireless IP links, then the conversion burden will fall on those who insist on running IP packets over fiber.

Today, there are those who feel that the future of networking depends on running fiber-optic cable everywhere, even to the home. Then there are those who are equally sure the future of networking depends on wireless links everywhere, not only for access, but also for the backbone network. While the reality may be a mixture of fiber and wireless, most expect one form or another to dominate. These two visions of the networking future are shown in Figure 14-4.

What this means for IP is that the day may come when users, equipment vendors, and service providers have to choose which version of IP to support as their core protocol and which version of IP will become the exception. The burden of conversion will be on the proponent of the less popular variation.

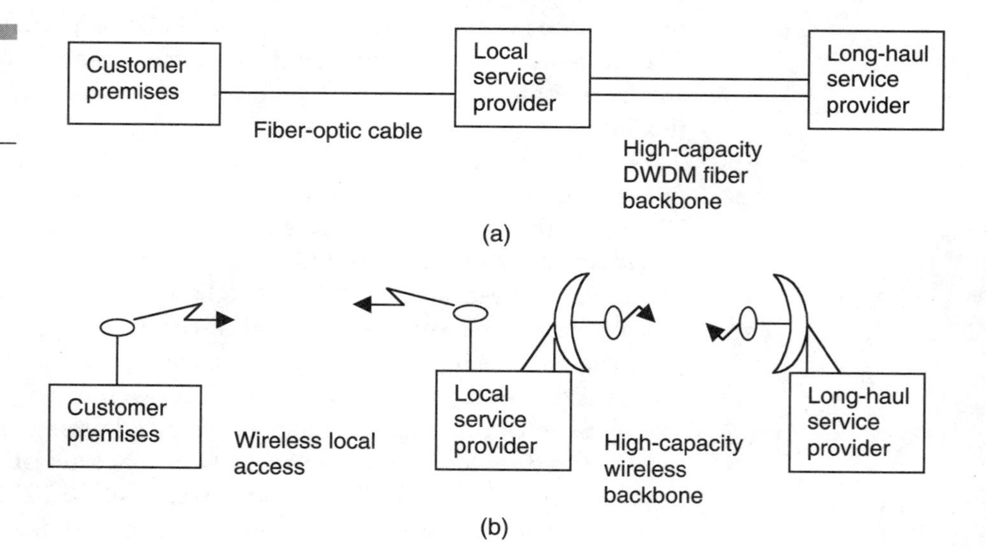

Figure 14-4 *(a)* Fiber and *(b)* wireless futures.

IP on Fiber Conquers All Today, IP packets can be placed on fiber-optic cable, but only by placing the IP packets inside some other structure. Typically, an IP packet is placed inside a stream of ATM cells, and then inside a SONET frame, and finally, the SONET [Synchronous Optical Network, or Synchronous Digital Hierarchy (SDH) outside of North America] frame is sent serially across the fiber link. In some cases, IP packets can be placed directly inside the SONET/SDH frame, a process called *packets on SONET*. However, this retains the SONET frame structure. Today, IP packets can be placed directly on a fiber-optic link in a process sometimes called *packets on fiber, packets on glass, packets on photons*, or even something more exotic. Packets placed directly onto fiber links lose the support in terms of administration, operations, and maintenance (OAM) that SONET/SDH and ATM supply but gain from the lessening of the overhead and processing burden involved in getting packets on and off the fiber.

If IP directly on fiber conquers all, the QoS concern list becomes very short indeed. Bandwidth will effectively compensate for a lack of almost any other network QoS concern such as delay or jitter. Thus packets sent directly on fiber, when used for IP telephony, benefit from almost unlimited bandwidth. These IP versions can add as many features and functions as needed and not worry about overburdening or congesting the network. Delay is not a worry, and jitter is fine through the line-level devices (switches may still add jitter), so only small jitter buffers are needed.

Fiber has very few errors to speak of, so IP optimized for fiber can dispense with resends (maybe). Since fiber is normally deployed in self-healing rings, availability is not a concern. Fiber has good security as well, just being a guided medium where taps are easily detected. Thus IP on photons should only worry about the basics of security.

Wireless IP Rules What if IP on fiber is not the wave of the future? Instead, the future of IP may belong to versions of IP optimized for the wireless environment. Some already exist, although IP on wireless links is nowhere near as popular (or mature) as IP on fiber-optic links. However, this is to be expected for now.

The important point is the IP optimized for wireless looks and acts very different from IP optimized for the fiber world. If wireless IP rules, the QoS concern list stays fairly long. Wireless bandwidth is restricted by regulation. Although researchers are always inventing new ways to increase the number of IP packets that can be sent in a given frequency range, bandwidth will always be more difficult to add to a wireless network than to an existing fiber network for the foreseeable future. Thus wireless packets, when used for IP telephony, have to be careful when using bandwidth. These IP versions cannot add features and functions and do need to worry about overburdening or congesting the network. Some adjustments, such as packet compression, not needed on fiber, may be essential on wireless flavors of IP. Delay on some wireless links, such as geosynchronous satellite links, is now a worry, especially when the added delay for compression and decompression is factored in. Adjustments may be needed here as well. However, like fiber-based IP, jitter is fine through the line-level devices (switches may still add jitter), so only small jitter buffers are needed.

In contrast to fiber, wireless links have very high error rates (sometimes thousands of times higher than their fiber counterparts, depending on atmospheric conditions). Thus IP optimized for a wireless world cannot dispense with resends and may need additional help in the form of forward error correction (FEC) codes, which also add delay and require a lot of processing power. Since wireless links are prone to outages due to atmospheric conditions and sometimes just heavy rains, availability is a concern. Oddly, many wireless links are backed up by fiber rings. Wireless has no security to speak of, since almost anyone with a receiver can intercept a signal and never be detected. Thus wireless IP should have as many security bells and whistles as possible.

The QoS differences and the IP response to them in the fiber and wireless worlds are shown in Table 14-1.

TABLE 14-1

QoS, Fiber-based IP, and Wireless IP

QoS Parameter	Fiber Has . . .	So IP on Photons Should . . .	Wireless Has . . .	So Wireless IP Should . . .
Bandwidth	Virtually unlimited bandwidth	Add as many functions as needed	Restricted bandwidth	Be limited to essentials
Delay	Low delays	Not worry	Some high delays	Make adjustments
Jitter	Good through line-level devices	Use some jitter buffers	Good through line-level devices	Use some jitter buffers
Information loss	Few errors	Disable resends?	More errors	Use FEC?
Availability	Self-healing rings	Not worry	Many problems	Use fiber backup?
Security	Good security	Do the basics	Many concerns	Do a lot

How IP on Photons Would Differ from Wireless IP

In summary, the adjustments to IP needed to implement IP on photons may be so different from those needed on wireless IP that they would amount to different versions of IP. Of course, the result would still be IP. In the same way, Ethernet LANs are still Ethernet, but they can differ remarkably. The term *Ethernet* can be applied to the original, proprietary coaxial cable Ethernet, the common unshielded twisted-pair 10base-T version of Ethernet, the full-duplex Switched Ethernet that mimics some Token Ring characteristics, or the Gigabit Ethernet that is essentially a fiber channel LAN carrying Gigabit Ethernet frames. What is important is the Ethernet label, not the specifics.

IP on fiber and wireless IP, however, will differ in at least the following important ways:

1. IP on photons could be very "chatty"; wireless IP must be careful.
2. Wireless IP should adjust for high GEO satellite delays.
3. Wireless IP should include forward error checking (FEC) (IP on photons should anyway).
4. Wireless IP cannot be fully optimal (outages must be shunted onto fiber systems).

When it comes to the jitter and security QoS parameters, both fiber and wireless flavors of IP are more or less a wash. Jitter buffers will be needed no matter what type of links are used, since most jitter comes from the network nodes and not the links themselves. Although security is technically better on fiber than on a wireless network, most current security architectures emphasize complete security protection for all types of links. The same logic could be applied easily to errors as well, since IP on photons, despite the extremely low error rates on fiber links, often still employs an FEC method to absolutely rule out the presence of even the most infrequent errors, but this is not strictly necessary. (In a pure IP on photons environments, since there is no place to put the FEC information in the IP packet, FEC is often sent on a separate DWDM channel formatted to carry SONET/SDH frames. The sole contents in these frames are the FEC checks on the rest of the channels.)

Finally, wireless systems suffer periodic outages due to atmospheric conditions. Some outages are predictable, such as the solar outage caused when the sun crosses the orbit of a geosynchronous satellite, but others are not, such as heavy rainfall or solar flares. Whenever these outages occur, however, it is common to shunt the wireless traffic onto a fiber link to avoid a service outage. Thus wireless IP could never be completely optimized for wireless links alone.

The networking world could evolve to feature wireless access and fiber backbones. But this just compounds the basic problem of what to optimize IP for. If IP is to be used for everything from voice to video to data, IP is better off being optimized for one world or the other. This presents users, equipment vendors, and service providers with consistent and compatible packages to build the network of the future.

Does the PSTN Fade Away?

Where does IP telephony leave the global, circuit-switched PSTN? Is there still a place for circuit switching in a world dominated by packets? Perhaps not. But this does not mean that the PSTN will ever completely disappear.

The PSTN will most likely continue to provide more and more passive connectivity when compared with active switching services such as voice. For instance, many IP telephony networks will continue to rely on point-to-point leased lines from the telephone companies to send their voice IP packets between routers. The voice switches will have fewer and fewer calls to switch and less and less traffic to carry. Eventually, more links into a central office or local exchange switch will lead to a router than to a voice switch (this will be true no matter who owns or controls the router).

However, it is extremely hard to find any standard network architecture or protocol that is totally extinct, let alone one that has been around as long and is as popular as the PSTN. X.25 networks still flourish in niche markets, Arcnet LANs still soldier on in the face of Ethernet, and so forth. If the PSTN does fade away, the fading will be so imperceptible that it will not be obvious for some time to come.

A greater concern is that those who can afford PCs, Internet access, and so on will migrate to IP telephony, leaving the "technology poor" peoples of the world on the global PSTN. Gateways will prevent their isolation, of course, but no exciting new technologies will be available to them. Nor will anyone seem overly concerned with their well-being or nurturing unless forced to by regulators.

The greatest threat posed by IP telephony is just such a world of "haves" and "have nots." The energy exhibited with regard to ISP access charges, local service competition, access line unbundling, and related topics must be applied to this issue soon to prevent IP telephony from becoming the technology of choice for the elite instead of for the people.

ABBREVIATIONS AND ACRONYMS

ACD automatic call distribution
ACELP algebraic-code-excited linear prediction
ADPCM adaptive differential pulse code modulation
ADSL asymmetric digital subscriber loop
AMI alternate mark inversion
ARP address resolution protocol
AS autonomous system
AT advanced technology
ATM asynchronous transfer mode
BBN Bolt, Baranek, and Newman
BGP border gateway protocol
BIOS basic input/output system
BRI basic rate interface
CABS carrier access billing system
CAR committed access rate
CCS common channel signaling
CD call disconnect
CDR call detail record
CED called station identification
CELP code-excited linear prediction
CERN European Council for Nuclear Research
CGI common gateway interface
CIR committed information rate
CLEC competitive local exchange carrier
CNG calling number
CO central office
CPU central processing unit
CRM customer relationship management
CSU channel service unit
CTI computer telephony integration
DAM diagnostic acceptability measure
DC direct current

DEC Digital Equipment Corporation
DHCP dynamic host configuration protocol
DM delta modulation
DNA digital network architecture
DOS disk operating system
DPCM differential pulse code modulation
DRT diagnostic rhyme test
DS digital signal
DSL digital subscriber loop
DSP digital signal processor
DSU digital service unit
DTE data terminal equipment
DTMF dual-tone multifrequency
DVD digital versatile disk
DWDM dense wavelength division multiplexing
EIR excess information rate
ESF extended superframe format
FAQ frequently asked questions
FCC Federal Communications Commission
FDDI fiber-distributed data interface
FDM frequency division multiplexing
FEC forward error correction, fast EtherChannel
FM frequency modulation
FTP File Transfer Protocol
GB gigabytes
GEIS General Electric Information Services
GEO Geosynchronous Earth Orbit
GIF graphics interchange format
GLP Gateway Location Protocol
GPS Global Positioning System
GSM global system for mobile
GSTN global switched telephone network
HDLC high-level datalink control

HF high frequency

HTML hypertext markup language

HTTP Hypertext Transfer Protocol

IA implementation agreement

IBM International Business Machines

ID identifier

IEEE Institute of Electrical and Electronics Engineers

IETF Internet Engineering Task Force

ILEC incumbant local exchange carrier

IN intelligent network

IP Internet Protocol

IRC internet relay chat

IS information systems

ISDN integrated services digital network

ISN information systems network

ISP internet service provider

IT information technology

ITSP internet telephony service provider

ITU International Telecommunications Union

IVR interactive voice response

JPEG Joint Photographic Experts' Group

kb/s kilobits per second

LAN local-area network

LAPB link-access procedure balanced

LATA local access and transport area

LCD liquid crystal diode

LDAP Lightweight Directory Access Protocol

LEC local exchange carrier

LPC linear predictive coding

LS link state

Mb/s megabits per second

MB megabytes

MC multipoint controller

MCU multipoint control unit
MFJ modified final judgment
MGCP Multimedia Gateway Control Protocol
MLQ maximum likelihood quantization
MOS mean opinion score
MP Multilink PPP
MTU maximum transmissible unit
NEBS network equipment building systems
NEC Nippon Electric Corporation
NT new technology
NTT Nippon Telephone and Telegraph
OAM operations, administration, and management
OC optical carrier
OCR optical character recognition
OS operating system
PAM pulse amplitude modulation
PBX Private Branch Exchange
PC personal computer
PCI peripheral component interconnect
PCM pulse code modulation
PCS personal communications service
PDH plesiochronous digital hierarchy
PDN public data network
PDU protocol data unit
PHB per-hop behavior
PIN personal identification number
PLP Packet Layer Protocol
POP point of presence
POTS plain old telephone service
PRI primary rate interface
PSPDN packet-switched public data network
PSTN public-switched telephone network
PVC permanent virtual circuit

QoS quality of service
RAM random-access memory
RAS remote-access server
RBOC Regional Bell Operating Company
RFC request for comments
RJ recommended jack
RLE run length encoding
RM resource management
ROM read-only memory
RR receive ready
RS recommended standard
RSVP Reservation Protocol
RTCP Real-Time Control Protocol
RTP Real-Time Protocol
SCP service control point
SDH synchronous digital hierarchy
SGML standard generalized markup language
SIP Session Initialization Protocol
SLA service level agreement
SNA systems network architecture
SONET synchronous optical network
SPC stored program control
SS7 signaling system seven
SSP service switching point
STM synchronous transfer mode
STS synchronous transport signal
SVC switched virtual circuit
TAPI telephony application program interface
TCAP transaction capabilities application part
TCP Transmission Control Protocol
TD transmitted data
TDM time division multiplexing
TEHO tail-end hop-off

TH transmission header

TIFF tagged image file format

TN transit node

ToS type of service

TTL time to live

TX transmit

UADSL universal ADSL

UDP User Datagram Protocol

UPS uninterruptible power supply

URL universal resource locator

VAD voice activation detection

VAR value-added reseller

VC virtual circuit

VLSI very large scale integration

VPN virtual public network

WAN wide-area network

WECO Western Electric Company

WFQ weighted fair queueing

WRED weighted random early discard

WWW World Wide Web

XML eXtensible markup language

INDEX

A

B

D

E

F

I

J

K

L

M

Q

R

S

T

W

X

Y

About the Authors

Walter J. Goralski has more than 25 years in the data communications field, including 14 years with AT&T. He is currently a Senior Member of Technical Staff with Hill Associates, a technical training and consulting firm in Colchester, Vermont, and an adjunct professor of computer science at Pace University Graduate School in New York. He is also the author of several books on ATM, the Internet, TCP/IP, and APPN as well as articles on data communications and other technology issues.

Matthew C. Kolon has more than 14 years of experience in MIS and as a consultant in computers and data communications. He is a Senior Member of Technical Staff at Hill Associates, where he teaches classes and develops course materials on routing and routing protocols, TCP/IP, the Internet, and LANs. He has also authored a number of articles on data communications.